Roofing

Level One

Trainee Guide
Second Edition

Pearson

NCCER

President and Chief Executive Officer: Boyd Worsham
Senior Director of Innovation and Advancement: Jennifer Wilkerson
Chief Learning Officer: Lisa Strite
Roofing Project Manager: John Esbenshade
Senior Manager of Projects: Chris Wilson
Senior Production Manager: Erin O'Nora
Testing/Assessment Project Manager: Elizabeth Schlaupitz
Technical Writers: Karyn Payne, Natalie Hasty
Managing Editor: Graham Hack
Desktop Publishing Manager: James McKay

Art Manager: Carrie Pelzer
Permissions Specialists: Kay Lewis, Amanda Smith
Project Assistant: Lauren Corley
Digital Content Coordinator: Rachael Downs
Production Specialists: Gene Page, Eric Caraballoso, Daphney Milian
Production Assistance: Edward Fortman, Joanne Hart, Olga Trofymenko
Editors: Jordan Hutchinson, Karina Kuchta, Hannah Murray, Zi Meng

Pearson

Director of Alliance/Partnership Management: Kelly Trakalo
Content Producer: Alexandrina B. Wolf
Assistant Content Producer: Alma Dabral
Digital Content Producer: Jose Carchi
Composition: NCCER
Printer/Binder: LSC Communications
Cover Printer: LSC Communications
Text Fonts: Palatino and Univers
Content Technologies: Gnostyx

Cover Image
Cover image courtesy of F.J.A. Christiansen Roofing Co., Inc., A Tecta America Company

Credits and acknowledgments for content borrowed from other sources and reproduced, with permission, in this textbook appear at the end of each module.

ISBN-13:   978-0-13-749099-8

To the Trainee

Roofs are the first line of defense against the elements. A roof means more than shelter; it means security. There is a wealth of career potential and opportunity in roofing for people with various skill sets and levels of education. The modules contained in this level address foundational knowledge essential for all roofing professionals.

New with *Roofing Level One*

NCCER and Pearson are pleased to present *Roofing Level One*. In partnership with the National Roofing Contractors Association (NRCA), this book has been modernized, enhanced, and expanded to best serve the industry. NRCA is the voice of roofing professionals and the leading authority in the roofing industry for information, education, technology, and advocacy.

This program offers two *interim credentials* in the following occupational areas of roofing: steep-slope roofing and low-slope roofing. Interim credentials provide employers specific training pathways for new personnel to achieve a full NCCER credential.

For more details on requirements for interim credentials and other information about this curriculum, visit the Roofing craft details page in the Program Resources section of NCCER's website.

We wish you success as you progress through this training program. If you have any comments on how NCCER might improve upon this textbook, please complete the User Update form located at the back of each module and send it to us. We will always consider and respond to input from our customers.

Our website, **www.nccer.org**, has information on the latest product releases and training.

Your feedback is welcome. You may email your comments to **curriculum@nccer.org** or send general comments and inquiries to **info@nccer.org**.

NCCER Standardized Curricula

NCCER is a not-for-profit 501(c)(3) education foundation established in 1996 by the world's largest and most progressive construction companies and national construction associations. It was founded to address the severe workforce shortage facing the industry and to develop a standardized training process and curricula. Today, NCCER is supported by hundreds of leading construction and maintenance companies, manufacturers, and national associations. The NCCER Standardized Curricula was developed by NCCER in partnership with Pearson, the world's largest educational publisher.

Some features of the NCCER Standardized Curricula are as follows:

- An industry-proven record of success
- Curricula developed by the industry, for the industry
- National standardization providing portability of learned job skills and educational credits
- Compliance with the Office of Apprenticeship requirements for related classroom training (*CFR 29:29*)
- Well-illustrated, up-to-date, and practical information

NCCER also maintains the NCCER Registry, which provides transcripts, certificates, and wallet cards to individuals who have successfully completed a level of training within a craft in NCCER's Curricula. *Training programs must be delivered by an NCCER Accredited Training Sponsor in order to receive these credentials.*

For information on NCCER's credentials and the NCCER Registry, contact NCCER Customer Service at 1-888-622-3720 or visit **www.nccer.org**.

Special Features

In an effort to provide a comprehensive and user-friendly training resource, this curriculum showcases several informative features. Whether you are a visual or hands-on learner, these features are intended to enhance your knowledge of the construction industry as you progress in your training. Some of the features you may find in the curriculum are explained below.

Introduction

This introductory page, found at the beginning of each module, lists the module Objectives, Performance Tasks, and Trade Terms. The Objectives list the knowledge you will acquire after successfully completing the module. The Performance Tasks give you an opportunity to apply your knowledge to real-world tasks. The Trade Terms are industry-specific vocabulary that you will learn as you study the module.

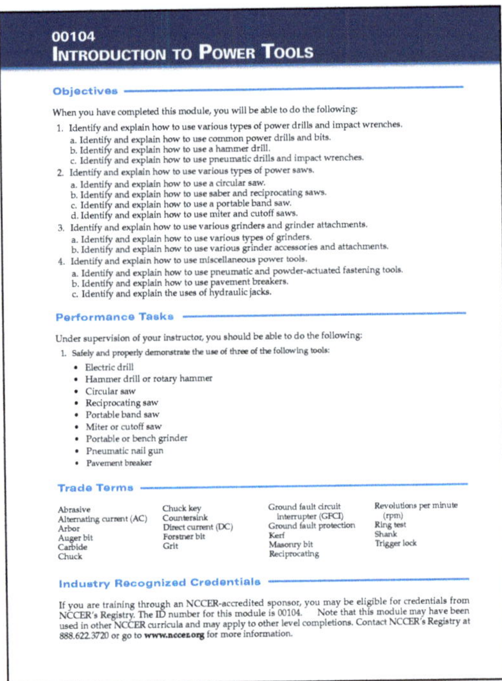

Trade Features

Trade features present technical tips and professional practices based on real-life scenarios similar to those you might encounter on the jobsite.

Bowline Trivia

Some people use this saying to help them remember how to tie a bowline: "The rabbit comes out of his hole, around a tree, and back into the hole."

Figures and Tables

Photographs, drawings, diagrams, and tables are used throughout each module to illustrate important concepts and provide clarity for complex instructions. Text references to figures and tables are emphasized with *italic* type.

Figure 31 Marking a cutting line.

Notes, Cautions, and Warnings

Safety features are set off from the main text in highlighted boxes and categorized according to the potential danger involved. Notes simply provide additional information. Cautions flag a hazardous issue that could cause damage to materials or equipment. Warnings stress a potentially dangerous situation that could result in injury or death to workers.

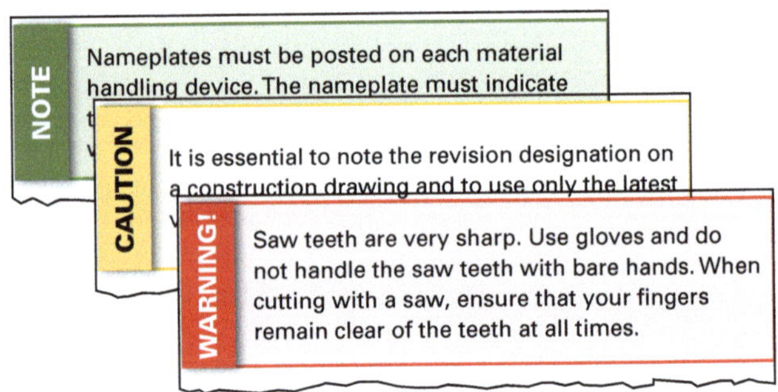

Going Green

Going Green looks at ways to preserve the environment, save energy, and make good choices regarding the health of the planet. Through the introduction of new construction practices and products, you will see how the greening of America has already taken root.

> ### ⌀ Going Green
>
> ## Big Horn Wind Farm
>
> The Big Horn wind farm uses 133 GE turbines, each rising 119 meters (389 feet) from the ground to the top of the blade tips. The wind farm uses only a

Did You Know?

The *Did You Know?* features offer hints, tips, and other helpful bits of information.

> **Did You Know?**
>
> ## How Blueprints Started
>
> The process for making blueprints was developed in 1842 by an English astronomer named Sir John F. Herschel. The method involved coating a paper with a special chemical. After the coating dried, an original hand drawing was placed on top of the paper. Both papers were then covered with a piece of glass and set in the sunlight for about an hour. The coated paper was developed much like a photograph. After a cold-water wash, the coated paper turned blue, and the lines of the drawing remained white.

Step-by-Step Instructions

Step-by-step instructions are used throughout to guide you through technical procedures and tasks from start to finish. These steps show you not only how to perform a task but also how to do it safely and efficiently.

> Perform the following steps to erect this system area scaffold:
>
> **Step 1** Gather and inspect all scaffold equipment for the scaffold arrangement.
>
> **Step 2** Place appropriate mudsills in their approximate locations.
>
> **Step 3** Attach the screw jacks to the mudsills.
>
> **Step 4** Adjust the screw jacks to near their lowest position.
>
> **Step 5** Determine the location of the highest base

Trade Terms

Each module presents a list of Trade Terms that are discussed within the text and defined in the Glossary at the end of the module. These terms are denoted in the text with **bold, blue type** upon their first occurence. To make searches for key information easier, a comprehensive Glossary of Trade Terms from all modules is located at the back of this book.

> During a rigging operation, the load being lifted or moved must be connected to the apparatus, such as a crane, that will provide the power for movement. The connector—the link between the load and the apparatus—is often a sling made of synthetic, chain, or wire rope materials. This section focuses on three types of slings:
>
> - Synthetic slings
> - Alloy steel chain slings
> - Wire rope slings

Review Questions

Review Questions are provided to reinforce the knowledge you have gained. This makes them a useful tool for measuring what you have learned.

> ### Review Questions
>
> 1. Identification tags for slings must include the _____.
>   a. type of protective pads to use
>   b. type of damage sustained during use
>   c. color of the tattle-tail
>   d. manufacturer's name or trademark
>
> 2. The type of wire rope core that is susceptible to heat damage at relatively low temperatures is the _____.
>   a. fiber core
>   b. strand core
>   c. independent wire rope core
>   d. metallic link supporting core
>
> 3. Synthetic slings must be inspected _____.
>   a. once every month
>   b. visually at the start of each work week
>   c. before every use
>   d. once wear or damage becomes apparent
>
> 4. An alloy steel chain sling must be removed from service if there is evidence that _____.
>   a. the sling has been used in different hitch configurations
>   b. replacement links have been used to repair the chain
>   c. the sling has been used for more than one year
>   d. strands in the supporting core have weakened
>
> 5. A piece of rigging hardware used to couple the end of a wire rope to eye fittings, hooks, or other connections is a(n) _____.
>   a. eyebolt
>   b. hitch
>   c. shackle
>   d. U-bolt
>
> 6. A lifting clamp is most likely to be used to move loads such as _____.
>   a. steel plates
>   b. piping b
>   c. conc
>
> 7. Chain hoists are able to lift heavy loads by utilizing a _____.
>   a. rope and pulley system
>   b. rigger's strength
>   c. stationary counterweight
>   d. gear system
>
> 8. Before attempting to lift a load with a chain hoist, make sure that the _____.
>   a. hoist is secured to a come-along
>   b. load is properly balanced
>   c. tag lines are properly anchored
>   d. tackle is connected to its power source
>
> 9. A hitch configuration that allows slings to be connected to the same load without using a spreader beam is a _____.
>   a. double-wrap hitch
>   b. choker hitch
>   c. bridle hitch
>   d. basket hitch
>
> 10. To make the emergency stop signal that is used by riggers, extend both arms _____.
>   a. horizontally with palms down and quickly move both arms back and forth
>   b. directly in front and then move both arms up and down repeatedly
>   c. vertically above the head and wave both arms back and forth
>   d. horizontally with clenched fists and move both arms up and down

NCCER Standardized Curricula

NCCER's training programs comprise more than 80 construction, maintenance, pipeline, and utility areas and include skills assessments, safety training, and management education.

Boilermaking
Cabinetmaking
Carpentry
Concrete Construction
Construction Craft Laborer
Construction Technology
Core: Introduction to Basic Construction Skills
Drywall
Electrical
Electronic Systems Technician
Heating, Ventilating, and Air Conditioning
Heavy Equipment Operations
Heavy Highway Construction
Hydroblasting
Industrial Coating and Lining Application Specialist
Industrial Maintenance Electrical and Instrumentation Technician
Industrial Maintenance Mechanic
Instrumentation
Ironworking
Manufactured Construction Technology
Masonry
Mechanical Insulating
Millwright
Mobile Crane Operations
Painting
Painting, Industrial
Pipefitting
Pipelayer
Plumbing
Reinforcing Ironwork
Rigging
Roofing
Scaffolding
Sheet Metal
Signal Person
Site Layout
Sprinkler Fitting
Tower Crane Operator
Welding

Maritime

Maritime Industry Fundamentals
Maritime Electrical
Maritime Pipefitting
Maritime Structural Fitter
Maritime Welding
Maritime Aluminum Welding

Green/Sustainable Construction

Building Auditor
Fundamentals of Weatherization
Introduction to Weatherization
Sustainable Construction Supervisor
Weatherization Crew Chief
Weatherization Technician
Your Role in the Green Environment

Energy

Alternative Energy
Introduction to the Power Industry
Introduction to Solar Photovoltaics
Power Generation Maintenance Electrician
Power Generation I&C Maintenance Technician
Power Generation Maintenance Mechanic
Power Line Worker
Power Line Worker: Distribution
Power Line Worker: Substation
Power Line Worker: Transmission
Solar Photovoltaic Systems Installer
Wind Energy
Wind Turbine Maintenance Technician

Pipeline

Abnormal Operating Conditions, Control Center
Abnormal Operating Conditions, Field and Gas
Corrosion Control
Electrical and Instrumentation
Field and Control Center Operations
Introduction to the Pipeline Industry
Maintenance
Mechanical

Safety

Fall Protection Orientation
Field Safety
Safety Orientation
Safety Technology

Supplemental Titles

Applied Construction Math
Tools for Success

Management

Construction Workforce Development Professional
Fundamentals of Crew Leadership
Mentoring for Craft Professionals
Project Management
Project Supervision

Spanish Titles

Acabado de concreto: nivel uno (*Concrete Finishing Level One*)
Aislamiento: nivel uno (*Insulating Level One*)
Albañilería: nivel uno (*Masonry Level One*)
Andamios (*Scaffolding*)
Carpintería: Formas para carpintería, nivel tres (*Carpentry: Carpentry Forms, Level Three*)
Currículo básico: habilidades introductorias del oficio (*Core Curriculum: Introductory Craft Skills*)
Electricidad: nivel uno (*Electrical Level One*)
Herrería: nivel uno (*Ironworking Level One*)
Herrería de refuerzo: nivel uno (*Reinforcing Ironwork Level One*)
Instalación de rociadores: nivel uno (*Sprinkler Fitting Level One*)
Instalación de tuberías: nivel uno (*Pipefitting Level One*)
Instrumentación: nivel uno, nivel dos, nivel tres, nivel cuatro (*Instrumentation Levels One through Four*)
Orientación de seguridad (*Safety Orientation*)
Paneles de yeso: nivel uno (*Drywall Level One*)
Seguridad de campo (*Field Safety*)

Acknowledgments

This curriculum was revised as a result of the vision and leadership of the following sponsors:

Advanced Roofing
Deer Park Roofing
Eagle Roofing
Florida Roofing and Sheet Metal Contractors
   Association

GAF
National Roofing Contractors Association
PDM Architects
The Arizona Roofer
University of Florida

This curriculum would not exist were it not for the dedication and unselfish energy of those volunteers who served on the Authoring Team. A sincere thanks is extended to the following:

Alison LaValley
Amy Staska
Becky Lemons
Brian Davis
Bryan Karel
Charles Goodman
Clay Thomas
Doug Duncan

Dylan Edwards
Glenn Watson
Grant Brock
Henry Staggs
Jack Moore
Jean-Paul Grivas
John Schehl
Jon Goodman

Josey Parks
Mike Silvers
Nick Sabino
Ray Coykendall
Rich Trewyn
Tom Shanahan
Tyler Allwood
Wade Shepherd

NCCER Partners

NCCER partnering organizations are national associations and organizations that share a common interest in the goals and objectives of NCCER. To learn more about NCCER business partners, go to **www.nccer.org/about-us/partners**.

You can also scan this code using the camera on your phone or mobile device to view these partnering organizations.

Contents

Module Eight

Sheet Metal in Roofing

Describes the properties of different types of metal that impact the way sheet metal is used in construction. Overviews the tools used to measure, layout, and install sheet-metal components in roof systems. Outlines commonly used sheet-metal roofing components and basic procedures used to prepare and install them. (Module ID 16107; 25 Hours)

Module Nine

Rigging Practices

Presents basic rigging, which refers to the preparation of a load for movement, as well as the preparation of hardware and other components used to connect the load to the crane. Rigging must be completed safely and effectively, resulting in a reliable connection to the load. An understanding of rigging fundamentals is essential to safely operate cranes and move/position heavy equipment, components, and structures. (Module ID 38102; 15 Hours)

Glossary

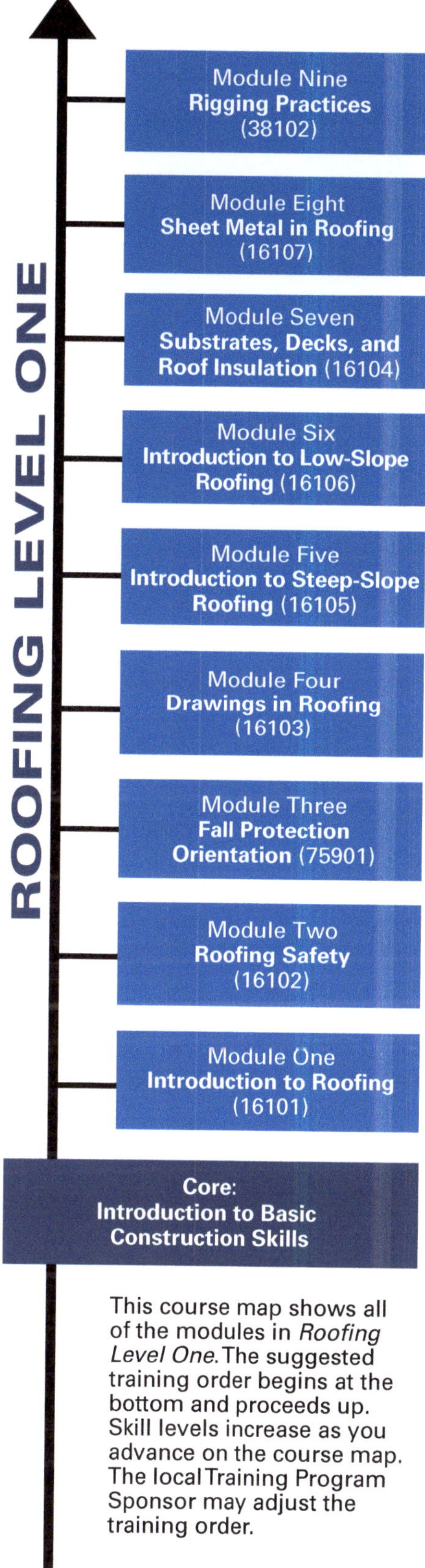

This course map shows all of the modules in *Roofing Level One*. The suggested training order begins at the bottom and proceeds up. Skill levels increase as you advance on the course map. The local Training Program Sponsor may adjust the training order.

Roofing

The Roofing curriculum has two possible career paths: Steep-Slope Roofer and Low-Slope Roofer. The tables below indicate the modules and registry ID numbers required for each credential. The tables also indicate the level of the craft where each module can be found.

Steep-Slope Roofer	Module ID	Level
Introduction to Roofing	16101	Level One
Roofing Safety	16102	Level One
Fall Protection Orientation	75901	Level One
Drawings in Roofing	16103	Level One
Introduction to Steep-Slope Roofing	16105	Level One
Introduction to Low-Slope Roofing	16106	Level One
Substrates, Decks, and Roof Insulation	16104	Level One
Sheet Metal in Roofing	16107	Level One
Rigging Practices	38102	Level One
Asphalt Shingle Roof Systems	16201	Level Two
Clay and Concrete Tile Roof Systems	16202	Level Two
Wood Roof Systems	16203	Level Two
Slate Roof Systems	16204	Level Two
Metal Roof Systems	16205	Level Two

Low-Slope Roofer	Module ID	Level
Introduction to Roofing	16101	Level One
Roofing Safety	16102	Level One
Fall Protection Orientation	75901	Level One
Drawings in Roofing	16103	Level One
Introduction to Steep-Slope Roofing	16105	Level One
Introduction to Low-Slope Roofing	16106	Level One
Substrates, Decks, and Roof Insulation	16104	Level One
Sheet Metal in Roofing	16107	Level One
Rigging Practices	38102	Level One
Metal Roof Systems	16205	Level Two
Thermoplastic Roof Systems	16206	Level Two
EPDM Roof Systems	16207	Level Two
Built-Up Roof Systems	16208	Level Two
Modified Bitumen Roof Systems	16209	Level Two
Liquid-Applied Roofing	16210	Level Two

Introduction to Roofing

OVERVIEW

Roofing professionals have a critical role in our society. A wide variety of roof systems are installed all over the world, keeping buildings comfortable and dry, and providing many other essential functions. Roofing is an exciting industry with many opportunities and specialty areas. Roofing professionals must be knowledgeable, qualified, and responsible to ensure safe and quality installation of roof systems.

Module 16101

Trainees with successful module completions may be eligible for credentialing through the NCCER Registry. To learn more, go to **www.nccer.org** or contact us at 1.888.622.3720. Our website, **www.nccer.org**, has information on the latest product releases and training.

Your feedback is welcome. You may email your comments to **curriculum@nccer.org**, send general comments and inquiries to **info@nccer.org**, or fill in the User Update form at the back of this module.

This information is general in nature and intended for training purposes only. Actual performance of activities described in this manual requires compliance with all applicable operating, service, maintenance, and safety procedures under the direction of qualified personnel. References in this manual to patented or proprietary devices do not constitute a recommendation of their use.

16101 V1.0

INTRODUCTION TO ROOFING

Objectives

Successful completion of this module prepares you to do the following:

1. Describe types of roofs and roof systems.
   a. Explain the difference between steep-slope roofs and low-slope roofs.
   b. List and describe common steep-slope roof systems.
   c. List and describe common low-slope roof systems.
   d. Identify specialty roof systems.
2. Explain the importance of safety in the roofing industry, and describe the obligations of the contractor, subcontractors, and yourself to ensure a safe work environment.
   a. Describe the OSHA 10 program.
   b. Explain hazard recognition and your role in it.
   c. Identify hazards that roofers encounter.
3. Understand the apprenticeship/training process for roofers.
   a. Describe modern apprenticeship training.
   b. Describe career paths, opportunities, and specialty areas available to roofers.
4. Understand the responsibilities of the employee and employer.
   a. Identify employee responsibilities.
   b. Identify employer responsibilities.

Performance Tasks

This is a knowledge-based module. There are no Performance Tasks.

Trade Terms

Abate
Bitumen
Built-up roof (BUR) systems
Deck
Flashings
Low-slope
Membrane
Membrane roof systems
Occupational Safety and Health Administration (OSHA)
On-the-job learning (OJL)

Personal fall arrest system (PFAS)
Personal protective equipment (PPE)
Ply
Safety culture
Single-ply roof systems
Site-specific safety program
Slope
Steep-slope
Underlayment
Water-shedding

Industry Recognized Credentials

If you are training through an NCCER-accredited sponsor, you may be eligible for credentials from NCCER's Registry. The ID number for this module is 16101. Note that this module may have been used in other NCCER curricula and may apply to other level completions. Contact NCCER's Registry at 1.888.622.3720 or go to **www.nccer.org** for more information.

Contents

This page is intentionally left blank.

1.0.0 ROOF SYSTEMS

Objective

Describe types of roofs and roof systems.

a. Explain the difference between steep-slope roofs and low-slope roofs.
b. List and describe common steep-slope roof systems.
c. List and describe common low-slope roof systems.
d. Identify specialty roof systems.

Trade Terms

Bitumen: A dark, cement-like substance found in asphalts, tars, and pitches. May also refer to any material composed mainly of bitumen, such as asphalt or coal tar.

Built-up roof (BUR) systems: Roof systems in which multiple layers of bitumen, reinforcement, and surfacing material are applied in place on the deck.

Deck: A structural component of the roof of a building, capable of safely supporting the weight of the roof or waterproofing system, as well as the additional live loads required by the governing building codes. The deck provides the substrate to which the roof or waterproofing system is applied.

Flashings: Components used to weatherproof or seal edges of a roof system at perimeters, penetrations, walls, expansion joints, valleys, drains, and other places where the roof covering is interrupted or terminated.

Low-slope: A category of roofs that generally includes weatherproof membrane types of roof systems installed on slopes of 3:12 or less.

Membrane: A flexible or semiflexible roof covering or waterproofing whose primary function is to exclude water.

Membrane roof systems: Roof systems containing a membrane material (weatherproof covering). Membrane roof systems are usually installed as low-slope systems.

Occupational Safety and Health Administration (OSHA): An agency of the US Department of Labor whose mission is to set occupational safety and health standards for all places of employment, enforce these standards, ensure that employers provide and maintain a safe workplace for all employees, and provide research and educational programs to support safe working practices.

Ply: A layer of felt in a built-up membrane roof or waterproofing system.

Single-ply roof systems: Roof systems in which the primary roof covering is a single layer of flexible membrane material (i.e., EPDM or thermoplastic).

Slope: The angle of a roof surface, usually expressed as a ratio of vertical rise to horizontal length (sometimes referred to as *run*). When dimensions are given in inches, slope may be expressed as a ratio of rise over a distance of 12" (for example, 4:12) or as an angle in degrees.

Steep-slope: A category of roofing that generally includes water-shedding types of roof coverings installed on slopes greater than 3:12.

Underlayment: An asphalt-saturated felt or other composite or synthetic sheet material (sometimes self-adhering) installed between a roof deck and roof covering, usually used in a steep-slope roof construction. Underlayment is primarily used to separate the roof covering from the roof deck, shed water, and provide fire protection and secondary weather protection.

Water-shedding: Able to depend on gravity for drainage to prevent water from getting through the roof system. Steep-slope roof systems are generally water-shedding.

Roofing professionals are vital to our society—every building needs a roof. Whether you are in a school, store, business, or the comfort of your home, the roof over your head helps keep you protected from the elements. Roofs not only protect the structure beneath them; they are also essential to the functionality of a building. They can direct light and rainwater, capture solar energy, insulate a building, and much more.

Roofing professionals install roof systems on many types of properties. They must have a keen eye for details and the precision of a fine craftsman. In addition to roof system installation, much roofing work involves repair and reroofing. This keeps the demand for roofers relatively stable, even during economic and construction downturns.

Registered apprenticeship programs, like the NCCER Roofing program, provide an option to advance your career in the roofing industry without a university degree (*Figure 1*). There are many specialty areas available to roofers because of the wide variety of roof systems. For example, asphalt shingle systems (the most common in residential

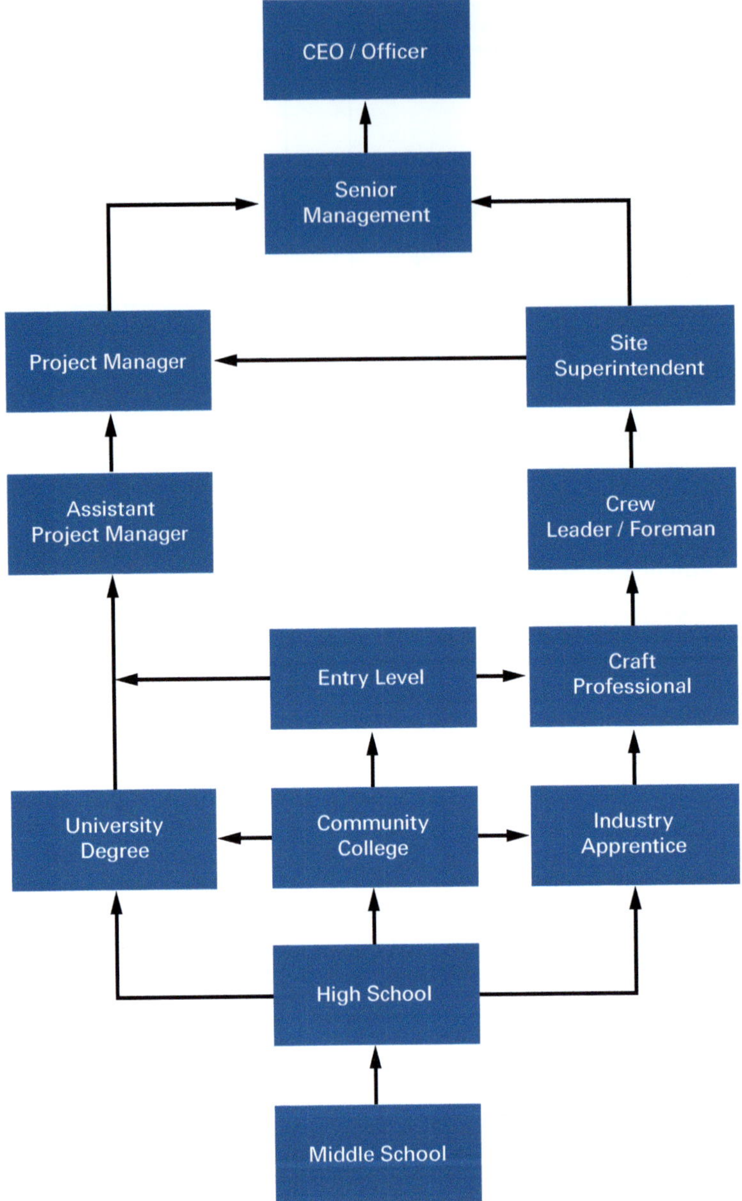

Figure 1 Career path for roofers.

settings) require a specific skill set. Alternatively, membrane roof systems seen on flat roofs in commercial settings involve larger roofing crews and different steps, tools, and skills to install.

1.1.0 Steep-Slope and Low-Slope Roofs

A roof assembly is comprised of many interacting components, including the roof deck, air or vapor retarder (if applicable), insulation, and primary covering designed to weatherproof the structure (such as a membrane or shingles). In general, a roof assembly consists of the structural deck and roof system.

A roof system—the part installed by roofing professionals—includes every component above the roof deck. Roof systems are categorized as low-slope or steep-slope based on the roof's degree of incline, known as its slope. Low-slope roofs (sometimes informally called *flat roofs*) are often seen on larger buildings (*Figure 2*). Steep-slope roofs are common on residential homes (*Figure 3*).

The slope of a roof is one of the primary differences among roof assemblies. Roof slope allows gravity to move water off a roof, promoting drainage. Without slope, a roof could hold water, causing significant water damage or even building collapse.

NCCER – *Roofing*

The Fifth Facade

The word *facade* comes from the French word for *face*. In architecture, the roof is often called *the fifth facade* because it is just as vital to the artistic design of the building as the four walls holding it up.

Figure Credit: iStock@africanpix

The most common method used to express the slope of a roof is a ratio between the vertical rise of the roof and the horizontal run, as illustrated in *Figure 4*. This is usually expressed in inches of rise per foot (12") of run, or in feet of rise per 12' of run. In this example, the rise is 8' and the run is 12', so the slope of the roof is 8:12. Additional examples of different roof slopes are shown in *Figure 5*.

Figure 2 Low-slope roof.

The NCCER Roofing program defines low- and steep-slope roofs as follows:

- *Low-slope* — 3:12 slope or less
- *Steep-slope* — Above 3:12 slope

Different safety precautions are required depending on the slope of the roof. It is vital that you always follow applicable safety procedures.

> **NOTE**
>
> Throughout NCCER's Roofing program, the term *steep-slope* refers to a roof slope of above 3:12. This differs from the Occupational Safety and Health Administration (OSHA) definition of steep-slope roofs (slope above 4:12). When installing or repairing roof systems, always check company, local, state, and federal codes. If any of the applicable codes contradict each other, follow the strictest one.

1.2.0 Steep-Slope Roof Systems

Steep-slope roof systems are generally water-shedding, whereas low-slope roofs are weatherproof. Steep-slope roofs use slope and gravity as a primary method of keeping water out of the building (*Figure 6*). Steep-slope roof systems are typically composed of individual

Figure 3 Steep-slope roof.

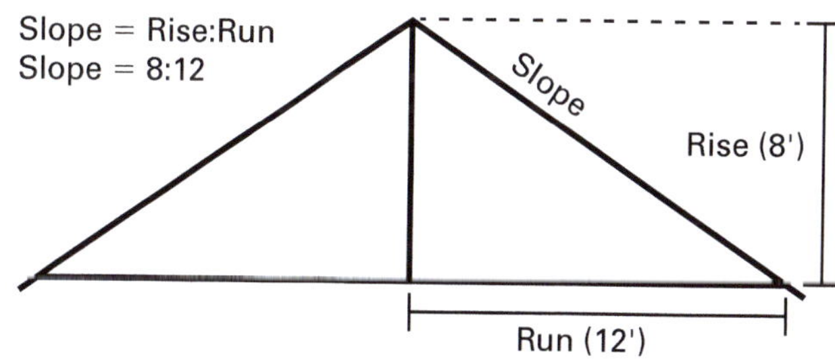

Figure 4 Roof slope.

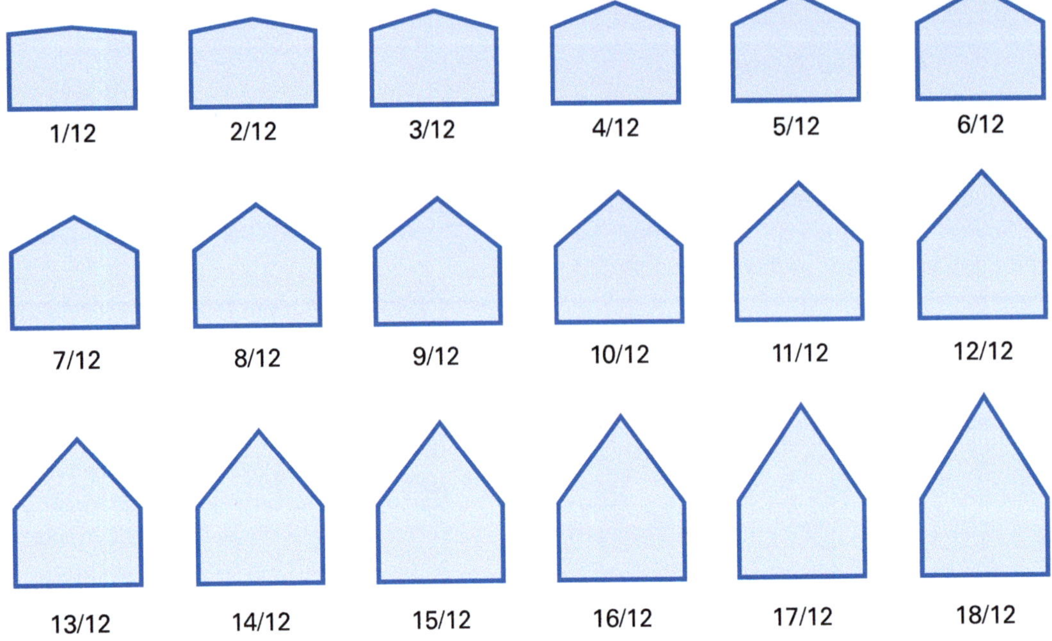

Figure 5 Examples of different roof slopes.

pieces or shingles installed in an overlapping pattern over a water-shedding underlayment.

The following are common types of steep-slope roof systems:

- Asphalt shingle systems
- Metal roof systems
- Clay and concrete tile
- Wood systems
- Slate systems
- Synthetic systems

1.2.1 Asphalt Shingle Systems

Asphalt shingles (*Figure 7*) are the most common residential roof system, accounting for nearly 70 percent of domestic roofing installations. Asphalt shingles are made of asphalt over a reinforcing mat covered with a surfacing of colored ceramic or mineral granules. They became popular in the early twentieth century because they are more fire-resistant and less expensive than wood. They also offer more colors, shapes, and patterns.

Asphalt shingles are installed in overlapping rows, resulting in a multilayered, water-shedding roof covering. Asphalt shingles are installed over an underlayment, which is laid continuously over a roof deck. An asphalt-shingle roof system is composed of asphalt shingles, an underlayment, and related components such as fasteners and flashings.

Asphalt shingles are generally nailed to wood roof decks, typically consisting of wood panels (plywood or oriented strand board), wood planks, or wood boards fastened to the underlying structural framing.

1.2.2 Metal Roof Systems

Types of steep-slope metal roof systems include metal shingles or architectural metal panels. Metal shingles (*Figure 8*) are designed to imitate the appearance of other shingles or tiles at a fraction of the weight. Architectural metal panel roof systems (*Figure 9*) are made with overlapping or interlocking metal sheets. Metal roof systems are known for being long-lasting and durable.

Figure 6 Steep-slope roof shedding water mechanically.

Figure 7 Asphalt shingle roof system.

Figure 8 Metal shingle roof system.

Figure 9 Architectural metal panel roof system.

1.2.3 Concrete and Clay Tile Systems

Clay tile (*Figure 10*) is made by molding and firing clay at high temperatures. These tiles are available in many different shapes, sizes, and colors. Concrete tiles are made from a mixture of cement, sand, and water. These tiles sometimes have lugs on their bottom surface so that they can be anchored to a batten strip during installation. Both clay and concrete tile systems may include accessory pieces to help seal different areas of the roof.

1.2.4 Wood Systems

Wood shake and wood shingle roof systems are less common than other systems, accounting for only a small percentage of roof systems (*Figure 11*). They are often used in regions of the United States for aesthetic reasons to match the time period of old houses.

The terms *shake* and *shingle* refer to the look of the wood roofing unit or piece. A shake typically has a rough and textured surface, while a shingle's surface is sawn flat and smooth. Wood shakes are generally thicker in their dimensions and appearance than wood shingles. Cedar is a commonly used wood for roof systems. Other types include cypress, redwood, and oak.

Ancient Tile Roofs

Tile has been a common steep-slope roofing material for centuries. Roof tiles from ancient Egypt and Babylonia were made of sunbaked slabs of beaten clay. Hand-formed and fired clay tile also was common in Asia. After the formed clay tiles had set, they were cured in kilns or furnaces. Firing the clay hardened it and made it durable.

Figure Credit: iStock@Natalia Lipatova

Figure 10 Clay tile roof system.

Figure 11 Wood shake roof system on a nineteenth-century house.

1.2.5 Slate Systems

Slate is a naturally occurring roofing material made from rocks (*Figure 12*). Slate systems come in a variety of textures and colors depending on their source materials. Roofs may be constructed with rough or smooth slate, designed with varying lengths and thicknesses.

Figure 12 Slate shingle roof system.

1.2.6 Synthetic Systems

Synthetic roof systems entered the market in the early 1990s. They are manufactured with recycled plastic or rubber made to look like asphalt, slate, or wood shingles. Synthetic systems are much lighter than most roof systems, and they have fire and algae resistant characteristics. *Figure 13* shows an example of synthetic wood shingles.

1.3.0 Low-Slope Roof Systems

Low-slope roofs (*Figure 14*) are weatherproof instead of water-shedding. They still have some slope to move water off the roof, but they are not steep enough for water-shedding roof systems such as shingles. They generally rely on a weatherproof membrane covering, or roof membrane, to keep water from entering the structure. Systems that employ a roof membrane are called *membrane roof systems*.

Common low-slope roof systems include built-up roof (BUR) systems, polymer-modified bitumen systems, single-ply systems, and liquid-applied systems. Structural metal panels, which are sometimes used on steep-slope roofs, can also provide a weatherproof covering in low-slope applications. Structural metal roof systems are shown in *Figure 15* and *Figure 16*.

Figure 13 Synthetic wood shingles.

Figure 14 Low-slope roofs.

Figure 15 Structural metal panel roof system.

Figure 16 Roofing professional coating a structural metal roof system.

1.3.1 BUR and Modified-Bitumen Systems

There are two roof system categories that use bitumen as a primary element: built-up systems and polymer-modified bitumen systems. Bitumen is either asphalt-based or coal tar-based. Asphalt comes from petroleum, and coal tar comes from coal. Bitumen is used for waterproofing and is a good adhesive.

Built-up systems (*Figure 17*) are the oldest type of weatherproof membrane. They are often used on commercial and industrial buildings. Built-up systems consist of three general parts:

- *Bitumen* — Asphalt or coal tar pitch that provides waterproofing.
- *Reinforcement* — Ply sheets, often made of felt, that are layered with bitumen to reinforce it and create the roof membrane. These sheets are often referred to as *felts*.
- *Surfacing material* — Protects the membrane from ultraviolet (UV) degradation and weather erosion. A variety of materials are used for surfacing, such as a mineral aggregate, a liquid-applied coating, or a sheet material.

As the name implies, a built-up roof membrane is constructed at the jobsite by assembling the layers directly on the roof (*Figure 18*).

Figure 17 Built-up roof system.

A polymer-modified bitumen membrane is also made up of bitumen, reinforcement, and surfacing. However, unlike BUR systems, the components are assembled into sheets at a factory before being installed. The bitumen in these systems is modified polymer to make it stronger. Polymer-modified bitumen sheets can be installed by heating the bitumen on the back of the sheet, adhering it to the substrate. The asphalt can also be hot-mopped or cold-applied in front of the sheet, which is then rolled into place.

1.3.2 Single-Ply Systems

Single-ply roof systems use a weatherproof membrane made of flexible sheets of rubber or plastic compounds that often include a reinforcing mat or fabric. Common types of single-ply membrane materials include EPDM (*Figure 19*) and thermoplastic (*Figure 20*). EPDM membranes are stable and can be adhered for strong, reliable service. Thermoplastic membranes, such as PVC or TPO, are lighter in color and can be easily reshaped.

1.3.3 Liquid-Applied Systems

Liquid-applied roof membranes (*Figure 21*) are made of polymer-based resin or polymer-modified compounds. They are generally applied in two coats that include reinforcements, such as polyester fabric or fleece. The resin or compound cures to form a weatherproof membrane. Liquid-applied systems may include additional surfacing, such as aggregate or a coating. These roof systems have gained popularity, especially in reroofing projects.

Figure 18 Installing a built-up roof system.

Figure 19 EPDM roof system.

Mercedes-Benz Stadium

Atlanta's Mercedes-Benz Stadium has one of the most interesting roofs in the United States. A 360-degree scoreboard is integrated into the roof structure. The roof is comprised of eight 500-ton moving panels that take 12 motors to operate. The roof takes eight minutes to open and is designed to resemble a camera shutter.

The stadium is the first professional sports venue to achieve a Leadership in Energy and Environmental Design (LEED) Platinum rating from the US Green Building Council. One of its energy-saving features is an efficient stormwater-management system. Water runoff from the roof is diverted to a rainwater cistern and reused for the building's mechanical system and landscape irrigation.

Figure Credit: File: Mercedes Benz Stadium time lapse capture 2017-08-13.jpg is licensed under the Creative Commons Attribution 3.0 Unported license.

1.4.0 Specialty Systems

Some roof systems have specialized coverings or unique components, such as solar panels or even greenery and walkways. The following are some specialized roof systems:

- *Vegetative systems* — A vegetative roof system (*Figure 22*) consists of vegetative and growth medium installed on top of a waterproof membrane and other necessary components that provide insulation and other benefits. This is an increasingly popular system that is environmentally friendly. Vegetative roof systems are sometimes referred to as *green roof systems*.
- *Spray polyurethane foam (SPF) systems* — An SPF roof system consists of foam that is sprayed in place on a roof to create a membrane, which is typically covered with a coating (*Figure 23*).
- *Photovoltaic (PV) roof systems* — PV roof systems (*Figure 24*) incorporate solar panels or solar shingles to generate power from sunlight.

Figure 20 Thermoplastic roof system.

Figure 21 Liquid-applied roof system.

Figure 22 Vegetative roof system.

Figure 23 SPF roof system.

Figure 24 Solar shingles.

1.0.0 Section Review

1. Which of the following roof slopes is considered low-slope?

   a. 45°
   b. Half pitch
   c. Five in twelve
   d. 1:12

2. Which type of shingles are made of recycled plastic or rubber?

   a. Metal shingles
   b. Synthetic shingles
   c. Slate shingles
   d. SPF roofing

3. Which type of roof covering can be used on both low-slope and steep-slope roofs?

   a. Structural metal panels
   b. Slate shingles
   c. Wood shakes
   d. EPDM

4. SPF roof systems are _____.

   a. made of solar shingles
   b. sprayed in place on a roof
   c. installed with heat welding
   d. made of concrete

2.0.0 SAFETY

Objective

Explain the importance of safety in the roofing industry, and describe the obligations of the contractor, subcontractors, and yourself to ensure a safe work environment.

a. Describe the OSHA 10 program.
b. Explain hazard recognition and your role in it.
c. Identify hazards that roofers encounter.

Trade Terms

Abate: To reduce or minimize.

Personal fall arrest system (PFAS): A system that activates during a fall to catch a worker and prevent them from hitting the ground. PFASs include multiple components, including a body harness, anchorage device, lifeline, and connectors.

Personal protective equipment (PPE): Equipment or clothing designed to prevent or reduce injuries.

Safety culture: The culture created when the whole company sees the value of a safe work environment.

Site-specific safety program: A safety program developed for a jobsite that identifies and takes into account any specific potential hazards that may be encountered.

All roofing professionals are obligated to work safely and make sure anyone they supervise or work with is also working safely. Contractors are obligated to maintain a safe workplace for all employees.

Creating and maintaining a safety culture is an ongoing process that includes a sound safety structure and attitude and relates to both organizations and individuals. While you are ultimately responsible for yourself, safety is everyone's responsibility (*Figure 25*).

Some contractors have safety committees. If you work for such a contractor, you are obligated to follow the safety committee's rules for proper work procedures and practices.

If you notice unsafe equipment or conditions, report it directly to your supervisor. Ensure the equipment will not be used by anyone else by following lockout tagout procedures or removing the equipment.

Substance Abuse

Statistics show that employees experiencing problems with drugs and alcohol have a higher rate of accidents. Workers who are impaired by these substances on a jobsite are a danger to themselves and to fellow workers. In addition, substance abusers are absent more often and are less productive, and the quality of their work suffers. Drugs and alcohol have no place in the workplace—especially on the roof.

In 2018 alone, more than 67,300 people died in the United States from an overdose. According to the US Bureau of Labor Statistics, overdose deaths that occur on the job are on the rise. Addiction is an illness that can be treated. Resources to overcome it can be found at **resources.facingaddiction.org** or by calling the confidential national hotline at **1-800-622-HELP (4357)**.

If you see something that is not safe, always report it—do not ignore it. It will not correct itself. While you may not have the authority to correct unsafe conditions encountered on the jobsite, you are obligated to report the issue.

Contractors know that any lost time while making conditions safe is nothing compared to the effects that an accident has on morale and productivity. Or worse, ignoring an unsafe condition could result in serious consequences, including property damage, injury to yourself or others, loss of employment, or loss of life. OSHA encourages employees to report hazardous conditions and requires employers to prevent and address them. This applies to every part of the construction industry. Whether you work for a large contractor or a small subcontractor, it is always a good idea to report unsafe conditions.

Figure 25 Safety is everyone's responsibility.

The easiest way to report unsafe conditions is to tell your supervisor. If your supervisor ignores the unsafe condition, report it to the next highest supervisor. If it is the owner who is being unsafe, let them know your concerns. If nothing is done about it, report it to OSHA. If you are worried about your job being on the line, think about it in terms of your life, or someone else's life, being on the line.

The US Congress passed the Occupational Safety and Health Act in 1970. This act created OSHA, which is part of the US Department of Labor. The mission of OSHA is to set occupational safety and health standards for all places of employment, enforce these standards, ensure that employers provide and maintain a safe workplace for all employees, and provide research and educational programs to support safe working practices. OSHA was created with the stated purpose "to assure so far as possible every working man and woman in the Nation safe and healthful working conditions and to preserve our human resources."

2.1.0   OSHA 10

The OSHA Outreach Training Program for the Construction Industry, often called *OSHA 10*, is a recommended volunteer program that teaches workers about their rights, contractor responsibilities, and complaint procedures, as well as how to identify, abate, avoid, and prevent job-related hazards. Even though OSHA considers this to be a voluntary program, some jurisdictions, contractors, and unions require this outreach training program before allowing anyone to work on construction sites. This can also fulfill their safety training goals.

The OSHA 10 program is a 10-hour construction industry program intended for entry-level workers. It covers a variety of construction safety and health hazards encountered on the jobsite, with emphasis on hazard identification, avoidance, control, and prevention, and not on the OSHA standards themselves. Mandatory topics included in the OSHA 10 program include:

- Introduction to OSHA
- Focus Four Hazards (leading causes of death in construction)
  - Falls
  - Electrocution
  - Struck-by
  - Caught-in/between
- Personal protective equipment (PPE) and life-saving equipment
- Health Hazards in Construction

In addition, some of the following topics may be addressed, depending on the contractor and/or type of construction work the contractor is performing:

- Confined Spaces for Construction
- Excavations
- Scaffolding
- Materials Handling, Storage, Use, and Disposal
- Cranes, Derricks, Hoists, Elevators, and Conveyors
- Stairways and Ladders
- Hand and Power Tools
- Bloodborne Pathogens
- Fire Protection and Fire Prevention

2.2.0   Site Safety and Hazard Recognition

The process of hazard recognition, evaluation, and control is the foundation of an effective safety program. When hazards are identified and assessed, they can be addressed quickly, reducing the hazard potential. Simply put, the more aware you are of your surroundings and the dangers in them, the less likely you are to be involved in an accident. A site-specific safety program should be developed.

There are many ways to recognize hazards and potential hazards on a jobsite. Some techniques are more complicated than others. In order to be effective, they all must answer this question: what could go wrong with this situation or operation? No matter what hazard recognition technique you use, answering that question in advance will save lives and prevent equipment damage.

2.2.1   Job Safety Analysis (JSA)

Performing a job safety analysis (JSA), sometimes called a *job hazard analysis (JHA)*, is one approach to hazard recognition. In a JSA, the task to be performed is broken down into its individual parts or steps and then each step is analyzed for its potential hazards. Once a hazard is identified, certain actions or procedures are recommended that will correct that hazard before an accident occurs.

For example, during a JSA, it is determined that workers will need to install flashing along the roof edge. To alert workers of the roof edge, warning lines (ropes and high-visibility flags) will be erected at a minimum of six feet from the roof edge. If workers are required to go outside of the warning line, they must be tied off to an approved anchor and use a harness, lifeline, and a rope grab. Otherwise, a safety monitor is required to observe them while they work. The rope grab should be positioned in such a way as to prevent a fall from the roof. *Figure 26* shows an example of a form used to conduct a job safety analysis.

2.3.0 Roofing Hazards

Plans for all roofing work, including new construction, replacement, repairs, and maintenance, should always consider and ensure the safety of roofing workers, building occupants, the public, the environment, and the building itself.

Building occupants and passers-by must be protected from injury or significant disturbance during a roofing project. Potential dangers include loud noises, fumes, falling objects inside or outside a building, and dangerous debris on the ground from roof removal. Roofing contractors must identify any hazards that may concern building occupants at the beginning of the project and warn the building owner before starting work.

Roofing workers should go home safe and healthy every day. Worker safety also makes the roofing contractor's business more profitable by reducing insurance expenses and increasing productivity.

Provisions must be made to safeguard the building from damage, which could be caused by fire, equipment, or environmental loads like snow, ice, or rain.

Protecting the environment is an issue for roofing contractors in regards to materials that are used to install a new roof system, as well as disposal of damaged or removed roofing material. For example, installing some roof systems requires materials that are flammable or combustible or may hold risks of toxicity if used improperly. Roofing contractors must be aware of the hazards posed by products and comply with regulations for their use, storage, and disposal.

OSHA is the government agency regulating safety in the roofing industry. These regulations protect workers and keep them safe. Building codes primarily work to protect the general public by providing criteria regulating design,

<table>
<tr><td colspan="4" align="center">JOB SAFETY ANALYSIS FORM   Page __ of __</td></tr>
<tr><td colspan="2">Job Title:
Job Location:
PPE:
Tools, Materials & Equipment:</td><td colspan="2">Date of Analysis:
Conducted By:
Staffing:
Duration:</td></tr>
<tr><td align="center">Step</td><td align="center">Hazard</td><td align="center">Quality Concern</td><td align="center">New Procedure or Protection</td></tr>
<tr><td></td><td></td><td></td><td></td></tr>
</table>

Figure 26 Job safety analysis (JSA) form.

Natural Disasters

Roofing professionals are key after natural disasters like hurricanes, tornados, hailstorms, and strong winds. Most wind damage begins on the edges of roof systems. Wind can loosen and raise roof materials, allowing water to get into the building. Debris such as branches can also puncture roof systems. After a severe weather, roofing contractors help homeowners determine what repairs need to be made.

Figure Credit: National Aeronautics and Space Administration (NASA)

construction, quality of materials, maintenance, and use and occupancy of buildings. Building codes generally focus on preventing fires, wind damage, water damage, and building collapse from internal or external forces.

When arriving at a new jobsite, the first thing roofing workers do is visually inspect the deck to make sure it is clean, dry, and in good condition. During this inspection, the worker ensures that all necessary structural components are present for flashing and obtains the dimensions to start installing the underlayment. It is important to note that roofers are not responsible for the structural integrity of the deck. Workers who install or repair roof systems are not generally responsible for the structural integrity of the deck. If you find something that does not look structurally sound, such as fasteners not being fully driven, stop work immediately and notify your supervisor.

Falling is one of the greatest hazards to roofing workers. All roofers must know how to properly don and doff (put on and take off) a **personal fall arrest system (PFAS)**.

> **NOTE**
>
> Fall protection is covered in more detail in NCCER Module 75901, *Fall Protection Orientation.*

2.0.0 Section Review

1. OSHA's ultimate role is to _____.
   a. inspect jobsites for safety violations
   b. fine companies that violate safety regulations
   c. distribute safety equipment to workers
   d. ensure that employers maintain a safe workplace

2. What is the foundation of an effective safety program?
   a. Inspections
   b. Installation techniques
   c. Material handling
   d. Hazard recognition, evaluation, and control

3. What government agency regulates safety in the roofing industry?
   a. CDC
   b. OSHA
   c. JSA
   d. PFAS

3.0.0 TRAINING AND APPRENTICESHIP

Objective

Understand the apprenticeship/training process for roofers.

a. Describe modern apprenticeship training.
b. Describe career paths, opportunities, and specialty areas available to roofers.

Trade Term

On-the-job learning (OJL): Job-related learning an apprentice acquires while working under the supervision of journey-level workers. Also called *on-the-job training (OJT)*.

Apprentice training goes back thousands of years, and its basic principles have not changed. It is a means for a person entering the craft to learn from those who have mastered the craft. In addition, it focuses on learning by doing, giving trainees a chance to apply their knowledge in practical situations.

3.1.0 Modern Apprenticeship Training

Many craftworkers learn their skills through a registered apprenticeship program, which this training material supports. Properly executed apprenticeship programs produce knowledgeable, skilled, and well-rounded craftworkers.

3.1.1 The Office of Apprenticeship

The US Department of Labor (DOL) Office of Apprenticeship sets the minimum standards for training programs across the country. These programs rely on mandatory classroom instruction and on-the-job learning (OJL). They require at least 144 hours of classroom instruction per year and 2,000 hours of OJL per year.

NCCER uses the minimum DOL standards as a foundation for comprehensive curricula, providing trainees with in-depth classroom and OJL experience.

3.1.2 On-the-Job Learning (OJL)

All apprenticeship standards require work-related OJL. This training may begin after graduation from high school or before graduation as a part of a youth apprenticeship program.

The OJL is broken down into specific tasks in which the apprentice receives hands-on training. In addition, a specified number of hours is required in each task. In a competency-based program, it may be possible to shorten this time by testing out of specific tasks through a series of performance exams. The required OJL is traditionally acquired in increments of 2,000 hours per year.

The apprentice must log all work time and turn it in to the apprenticeship committee so that accurate time control can be maintained. After each 1,000 hours of related work, the apprentice typically will receive a pay increase as prescribed by the apprenticeship standards.

For those entering an apprenticeship program, a high school or technical school education is desirable, but not required. Courses in shop, mechanical drawing, and general mathematics are helpful. Manual dexterity, good physical conditioning, and quick reflexes are important. The ability to solve problems quickly and accurately and to work closely with others is essential. You must also maintain a high awareness of safety concerns.

A prospective apprentice may be required to submit certain information to the apprenticeship committee such as the following:

- Aptitude test (General Aptitude Test Battery or GATB Form Test) results (usually administered by the local Employment Security Commission)
- Proof of educational background (candidate should have school transcripts sent to the committee)
- Letters of reference from past employers and friends
- Proof of age
- If the candidate is a veteran, a copy of Form DD214
- A record of technical training received that relates to the construction industry and/or a record of any pre-apprenticeship training

Note that informal OJL provided by employers is usually less thorough than OJL provided through a formal apprenticeship program. The degree of training and supervision in this type of program often depends on the size of the employer. A small contractor may provide training in only one area, whereas a large company may be able to provide training in several areas.

3.1.3 The NCCER Roofing Program

The NCCER Roofing curriculum is a two-year program that provides trainees with industry-driven training and education. Like every craft training program offered by NCCER, this program was developed by and for the construction industry. It adopts a purely competency-based teaching approach. This means that trainees must show the instructor that they possess the knowledge and skills needed to safely perform the hands-on tasks that are covered in each module.

Trainees who successfully complete a module receive credit in NCCER's Registry. NCCER's Registry system can then verify training and skills for workers as they move from state to state, from company to company, or even within a company.

Modules are typically bundled together to form a level that represents the total number of instructional hours for one year of the program (for example, *Roofing Level One*). Each level contains the DOL's minimum of 144 hours of instruction. Successfully completing all the modules in a level results in a level completion credential in the NCCER Registry. NCCER also offers interim credentials for certain combinations of modules. See the *Appendix* for examples of the credentials issued by NCCER.

Whether you enroll in an NCCER program or another apprenticeship program, make sure you work for an employer or sponsor who supports a nationally standardized training program that includes credentials to confirm your skill development.

3.2.0 Roofing Career Paths

Roof systems all over the world must be installed and maintained by someone qualified to do the work. This creates a great opportunity for work that is rewarding both personally and financially. Roofing contractors often specialize in specific roof systems, employing roofing professionals who gain experience and expertise in these systems.

Roofing is a complex industry made up of many types of professionals who all have a common goal to create long-lasting, trouble-free roof systems. The highest quality roof systems are designed, manufactured, installed, and maintained when everyone involved in a project understands the roles and responsibilities of all parties. Some key groups in the roofing industry include the following:

- *Professional roofing contractors* — Professional roofing contractors specialize in how to install the products and materials that make up high-performance roof systems.
- *Associations* — Associations are organizations founded by contractors and other members of the industry to act as a worldwide voice of the industry.
- *Designers* — Roof system designers include licensed architects, engineers, and other consultants who specialize in determining which materials work best together and what products will enhance the life of a roof system.
- *Product performance and testing organizations* — Various organizations conduct testing to determine whether a product or material meets established performance criteria. Examples include ASTM International, Underwriters Laboratories (UL), and FM Global.
- *Manufacturers* — Manufacturers design, test, and create the products and materials used in roof systems.

The roofing industry is filled with opportunities. The NCCER Roofing Curriculum provides the opportunity to earn interim credentials, which offer two training pathways in the specific occupational areas of roofing (steep-slope roofing and low-slope roofing). These training pathways ensure that you are prepared to safely and successfully install various types of roof systems.

National Roofing Contractors Association

The National Roofing Contractors Association (NRCA) is one of the construction industry's most respected trade associations. It is the voice of roofing professionals and the leading authority in the roofing industry for information, education, technology, and advocacy.

Through NRCA's ProCertifications®, experienced and skilled roofing workers can advance their careers by becoming certified in specific roof system installations. ProCertifications® currently offered by NRCA include the following:

- Asphalt Shingles Installer
- Thermoplastic Systems Installer
- EPDM Systems Installer
- Roofing Foreman
- Qualified Assessor

For more information, visit NRCA's website at **www.nrca.net**.

Figure Credit: Courtesy of the National Roofing Contractors Assoc.

3.0.0 Section Review

1. Entry into an apprenticeship program is likely to require a(n) ______.

   a. GATB Form Test
   b. license
   c. Scholastic Aptitude Test
   d. OSHA 40-hour course

2. The NCCER roofing training program applies DOL standards and is a ______.

   a. one-year program
   b. two-year program
   c. three-year program
   d. four-year program

4.0.0 EMPLOYEE AND EMPLOYER RESPONSIBILITIES

Objective

Understand the responsibilities of the employee and employer.

a. Identify employee responsibilities.
b. Identify employer responsibilities.

The safe and cost-effective installation and repair of roofing requires close collaboration between the employer and the employees. All workers must understand the responsibilities of providing a safe and productive workplace.

4.1.0 Employee Responsibilities

To be successful, you must be able to use current trade materials, tools, and equipment to finish the task quickly and efficiently. You must stay up-to-date on technical advancements and continually gain the skills to use them. A professional never takes chances when it comes to personal safety or the safety of others.

4.1.1 Professionalism

The term *professionalism* broadly describes the desired overall behavior and attitude expected in the workplace. Professionalism is too often absent from the construction site and various trades. Many would argue that professionalism must start at the top in order for an organization to be successful. It is true that management's support of professionalism is important to its success in the workplace, but it is just as important that individuals recognize personal responsibilities for professionalism.

Professionalism includes honesty, productivity, safety, civility, cooperation, teamwork, clear and concise communication, being on time, and coming prepared to work. It can be demonstrated in a variety of ways every minute you are on the job. For example, leaving nails and debris on a building owner's property after a roofing project reflects poorly on you and your company. Leaving the property clean and tidy shows professionalism.

Professionalism is a personal responsibility that benefits both the employer and the employee.

The construction industry is what each individual chooses to make of it. Choose professionalism, and the industry image will follow.

4.1.2 Honesty

Honesty and personal integrity are important traits of successful professionals. Professionals pride themselves on performing a job well while being punctual and dependable. Each job is completed in a professional way, never by cutting corners or reducing materials. A valued professional maintains work attitudes and ethics that protect property belonging to employers and customers, reducing the chance of damage or theft.

Honesty and success go hand-in-hand for both the employer and the professional roofer. It is not simply a choice between good and bad but a choice between success and failure. Dishonesty will always catch up with you. Whether you steal materials, tools, or equipment from the jobsite or simply lie about your work, it will not take long for your employer to find out.

If you plan to be successful and enjoy continuous employment, consistency of earnings, and being sought after as opposed to seeking employment, then start out with the basic understanding of honesty in the workplace. You will reap the benefits.

4.1.3 Loyalty

Employees expect employers to look out for their interests, to provide them with steady employment, and to promote them to better jobs as openings occur. Employers feel that they also have a right to expect loyalty from their employees—whether that be keeping their interests in mind, speaking well of the employer to others, or keeping any minor troubles or business matters confidential. Both employers and employees should keep in mind that loyalty is not something to be demanded; it is something to be earned.

4.1.4 Willingness to Learn

Every company and jobsite has its own way of doing things. Employers expect their workers to be willing to learn these ways. You must be willing to adapt to change and learn new methods and procedures as quickly as possible. Sometimes a change in safety regulations or the purchase of new equipment makes it necessary for even experienced employees to learn new methods and operations. Successful people take every opportunity to learn more about their trade.

4.1.5 Taking Responsibility

Most employers expect their employees to see
what needs to be done and do it. After an assignment is received and the procedure and safety
guidelines are fully understood, you should assume the responsibility for that task without further reminders.

Ethical Principles for Members of the Construction Trades

Honesty — Be honest and truthful in all dealings.
Conduct business according to the highest
professional standards. Faithfully fulfill all contracts
and commitments. Do not deliberately mislead or
deceive others.

Integrity — Demonstrate personal integrity and the
courage of your convictions by doing what is right,
even if there is pressure to do otherwise. Do not
sacrifice your principles because it seems easier.

Loyalty — Be worthy of trust. Demonstrate
fidelity and loyalty to companies, employers and
sponsors, coworkers, trade institutions, and other
organizations.

Fairness — Be fair and just in all dealings. Do
not take undue advantage of another's mistakes
or difficulties. Fair people are open-minded and
committed to justice, equal treatment of individuals,
and tolerance for and acceptance of diversity.

Respect for others — Be courteous and treat all
people with equal respect and dignity.

Obedience — Abide by laws, rules, and regulations
relating to all personal and business activities.

Commitment to excellence — Pursue excellence
in performing your duties, be well-informed and
prepared, and constantly try to increase your
proficiency by gaining new skills and knowledge.

Leadership — By your own conduct, seek to be a
positive role model for others.

4.1.6 Cooperation

To cooperate means to work together. In our modern business world, cooperation is the key to getting things done. Learn to work as a member of a
team with your employer, supervisor, and fellow
workers in a common effort to get the work done
efficiently, safely, and on time.

4.1.7 Rules and Regulations

Employees can work well together only if there is
some understanding about the nature of work to
be done, when and how it will be done, and who
will do it. Rules and regulations are a necessity in
any work situation and must be followed by all
employees.

4.1.8 Tardiness and Absenteeism

Tardiness means being late for work, and absenteeism means being off the job for one reason or
another. While occasional absences are unavoidable, consistent tardiness and frequent absences
are an indication of poor work habits, unprofes-
sional conduct, and a lack of commitment.

Although workers may not be paid when they
are absent or tardy, there is still a cost to the employer. For example, the worker's health insurance must still be paid, even though the worker
is not on site. In addition, jobs are budgeted and
scheduled based on a certain workforce size. If
you are not there, work is not being done and
deadlines are not being met. It is important for
you to be at work, on time, every day. If you must
be absent, call in as soon as possible so that your
employer can find a replacement.

4.1.9 Safety

In exchange for the benefits of your employment
and your own well-being, you are obligated to
work safely. You are also obligated to make sure
anyone you supervise or work with is working
safely. Your employer is obligated to maintain a
safe workplace for all employees. Safety is everyone's responsibility.

You have a responsibility to maintain a safe
working environment. This means two things:

- Follow your company's rules for proper working procedures and practices.
- Report any unsafe equipment and conditions
  directly to your supervisor.
- Maintain your equipment in a safe, usable condition.

If you see something unsafe while on the job,
report it! Do not ignore it. It will not correct itself.
In the end, even if you do not think an unsafe
condition affects you, it does. Always report unsafe conditions. Do not think your employer will
be angry because your productivity suffers while
the condition is being reported. On the contrary,
your employer will be more likely to criticize you
for not reporting a problem.

OSHA regulations require you to report hazardous conditions. This applies to every part of the construction industry. Whether you work for a large contractor or a small contractor, you are obligated to report unsafe conditions.

Tips for a Positive Attitude

Here is a short checklist of things that you should keep in mind to help you develop and maintain a positive attitude:

- Remember that your attitude follows you wherever you go.
- Helpful suggestions and compliments are much more effective than negative ones.
- Look for the positive characteristics of your co-workers and supervisors.

4.2.0 Employer Responsibilities

Just as the employee has responsibilities on the job, the employer also has responsibilities. OSHA requires each employer to provide a safe and hazard-free work environment. OSHA also requires that employees comply with OSHA rules and regulations that relate to their conduct on the job. OSHA can perform spot inspections of jobsites, impose fines for violations, and even stop work from proceeding until the jobsite is safe.

According to OSHA standards, you are entitled to on-the-job safety training. As a new worker, you must be:

- Oriented to the company policies and procedures
- Trained how to do your job safely
- Oriented to the specific job that you are working on
- Trained on the required PPE (some PPE is employer-provided)
- Warned about specific hazards pertaining to the project
- Supervised for safety while performing the work

Federal and state safety inspectors have the legal authority to impose fines for safety violations. States may also have their own safety regulations and agencies to enforce them. In states that do not develop such regulations and agencies, federal OSHA standards are mandatory.

OSHA standards are listed in *29 CFR 1926, OSHA Safety and Health Standards for the Construction Industry* (sometimes called *OSHA Standard 1926*). Other safety standards that apply to construction are published in *29 CFR 1900–1910*.

Some of the important requirements that OSHA places on employers in the construction industry include the following:

- The employer must perform frequent and regular inspections of equipment and conditions on the jobsite.
- The employer must train all employees to recognize and avoid unsafe conditions and to know the regulations that pertain to the job so they may control or eliminate any hazards.
- No one may use any tools, equipment, machines, or materials that do not comply with *29 CFR 1926*.
- The employer must ensure that only qualified individuals operate tools, equipment, and machines. For some tools and equipment, such as powder-actuated tools (PATs) and construction forklifts, specialized training is required. Only workers that have successfully completed the training can use such tools or equipment.
- The employer must post, in an easily seen area, signs informing employees of their rights and responsibilities.
- The employer must ensure there are no serious hazards on the jobsite and make sure the workplace complies with OSHA rules and regulations.
- Warning signs, posters, and labels must be posted in all required areas.
- The employer must instruct all employees to recognize and avoid unsafe conditions; also, to know the regulations that pertain to the job so that employees may control or eliminate any hazards.
- The employer must provide medical training and examinations when required by OSHA.
- Employers with more than 10 employees must keep records of work-related injuries and illnesses. These records must be available to employees.
- Employers must not discriminate against employees who are exercising their rights under OSHA regulations.

Additional employer responsibilities are described in the Americans with Disabilities Act of 1990 (ADA). If a worker has a disability but is qualified to do the job, that worker has equal rights to employment. The US Equal Employment Opportunity Commission, along with state and local civil rights agencies, enforce ADA regulations.

4.0.0 Section Review

1. Which of the following terms broadly describes the desired overall behavior and attitude expected in the workplace?

    a. Professionalism
    b. Collaboration
    c. Honesty
    d. On-the-job behavior

2. OSHA standards for the construction industry are covered in _____.

    a. *29 CFR 1900–1910*
    b. *29 CFR 1970*
    c. *29 CFR 1926*
    d. *29 CFR 1965–1983*

1. Which of the following is the *most* common residential roof system?

    a. Asphalt shingles
    b. Slate
    c. PVC
    d. BUR

2. Which of the following is the *oldest* type of weatherproof membrane?

    a. EPDM
    b. TPO
    c. PVC
    d. BUR

3. Who is responsible for safety?

    a. Clients
    b. Only employers
    c. Only employees
    d. Everyone

4. All roofers *must* know how to properly ______.

    a. install a deck
    b. inspect trusses
    c. don and doff a PFAS
    d. create a safety procedure

5. Which of the following is something a prospective apprentice may be required to do?

    a. Submit proof of age
    b. Be a veteran
    c. Take a prescribed amount of college credits
    d. Graduate high school

6. Which of the following specializes in installing roof systems?

    a. Contractors
    b. Manufacturers
    c. Owners
    d. Associations

7. Minimum standards for apprentice training programs are established by ______.

    a. OSHA
    b. the DOL
    c. NCCER
    d. the employer

8. Which of the following is *true* about tardiness and absenteeism?

    a. It is never acceptable to be absent, even when you are contagious.
    b. It is okay to be a little late for work as long as you make up the time.
    c. If you must be absent, call in early so that your employer can find a replacement.
    d. If you are not being paid, your absence does not cost the company anything.

9. If you see a safety violation at your jobsite, you should ______.

    a. ignore it unless it affects you directly
    b. make a mental note to avoid the area in future
    c. report it to your supervisor
    d. assume it is okay as long as no one has been injured

10. According to OSHA standards, all employees are entitled to ______.

    a. show up late for work
    b. a JSA
    c. on-the-job safety training
    d. a safety committee

Fill in the blank with the correct term learned from your study of this module.

1. ________________ is the federal government agency established to ensure a safe and healthy workplace.

2. Job-related learning acquired while working is known as ________________.

3. A category of roofs that generally includes weatherproof membrane systems is ________________.

4. A flexible roof covering whose primary function is to exclude water is called a(n) ________________.

5. A system that activates during a fall to catch a worker and prevent them from hitting the ground is called a(n) ________________.

6. To reduce or minimize is to ________________.

7. Roof systems in which multiple layers of bitumen, reinforcement, and surfacing material are applied in place on the deck are called ________________.

8. A roof that is able to depend on gravity for quick drainage of water is considered ________________.

9. Roof systems containing a weatherproof covering are called ________________.

10. The structural component of a roof is the ________________.

11. A dark, cement-like substance found in asphalts is called ________________.

12. Clothing designed to prevent or reduce injuries is called ________________.

13. The angle of a roof surface is called its ________________.

14. ________________ is when the whole company sees the value of a safe work environment.

15. An asphalt-saturated felt or other composite or synthetic sheet material installed between the roof deck and covering is called a(n) ________________.

16. Components used to weatherproof or seal edges of a roof system are called ________________.

17. A category of roofing that generally includes water-shedding roof coverings is called ________________.

18. Roof systems in which the principal roof covering is one layer of flexible membrane material are called ________________.

19. A program developed for a jobsite that identifies any potential hazards that may be encountered is called a(n) ________________.

20. A layer of felt or thin sheet of wood in a built-up membrane roof or waterproofing system is a(n) ________________.

Trade Terms

Abate
Bitumen
Built-up roof (BUR) systems
Deck
Flashings
Low-slope
Membrane
Membrane roof systems
Occupational Safety and Health Administration (OSHA)
On-the-job learning (OJL)
Personal fall arrest system (PFAS)
Personal protective equipment (PPE)
Ply
Safety culture
Single-ply roof systems
Site-specific safety program
Slope
Steep-slope
Underlayment
Water-shedding

Cornerstone of Craftsmanship

Jon Goodman
Subject Matter Expert
NRCA

How did you choose a career in the industry?

I transitioned into a roofing career from a business background in the electronics industry. I started out at Hewlett Packard. They were paying for my education, and I worked my way up from entry level to supervision. While there in the 1980s, I was promoted to production management for an emerging technology, but the economy dropped in the electronics business and I hit a dead end. I decided to transition to the roofing industry because there were more opportunities for advancement.

Who inspired you to enter the industry?

I had a college friend who owned a roofing company in Wisconsin who kept asking me to come and join him. Just as I was at a turning point and wondering what direction to go with my life, he made the offer again.

What do you enjoy most about your job?

I love having the opportunity to use all my skills. At the electronics giant, I was a supervisor in a line of about 24 people manufacturing sophisticate analytical devices that cost $50K a piece. But I had no idea what they did and had no input in their design or function. There were specialists to do everything. When I went to work for the roofing company, it was a family-owned business. They taught me what they knew, and I taught them what I knew. I was able to have my hands in everything, and I was using all my skills. I was involved in a wide variety of things, like sales, training, production management, public speaking, traveling, etc., and every day was different and exciting. I fell in love with it.

What types of training have you been through?

I earned a bachelor's degree while a married father of two working full time. Within the roofing industry, I've collected certificates from OSHA 10 and 30, Asbestos, NRCA Qualified Assessor ProCertification®, NRCA Qualified Trainer, TRTile certification, and ESL.

How important is education and training in construction?

Education and training are at the core of the construction trades. They are the catalyst that creates quality of products, confidence of consumers, safety of workers, and loyalty of employees.

What kinds of work have you done in your career?

I chopped wood and dug graves in high school. I sorted mail at the United Parcel Service, was a Production Supervisor at Hewlett Packard's Signal Analysis division, and was in the roofing trades for 31 years, culminating as an Owner and President of a roofing company in Appleton, WI by the time I retired for the first time.

Tell us about your present job.

Presently, I am the staff writer for NRCA's TRAC (Training for Roof Application Careers) program, creator and illustrator of HOTPs, and co-instructor of the Qualified Trainer Conference, Roof Repair and Maintenance Conference, and Roof Design and Consideration program. I also hold NRCA ProCertification® certificates for Qualified Assessor, Asphalt Shingles, Thermoplastic, and EPDM.

What do you enjoy most about your job?

Encouraging people to consider the roofing trade as an incredible place to build a career and a life. I love to be involved in people's lives, whether it's interacting with customers or mentoring employees. I'm invited to step into their world and make a difference. I like to make a dent where I'm sent.

What factors have contributed most to your success?

The hand of God providing my passion, drive, and curiosity. I'm curious about everything, which can be a blessing and a fault. I like to try my hand at most things, and I'm not afraid to fail or tackle things I don't know how to do.

Would you suggest construction as a career to others? Why?

Absolutely. We are the builders and caretakers of the nation's infrastructure. Roofers provide shelter to our families, our businesses, our churches, and our schools. Opportunities abound. Prosperity is abundant. We take pride in the work to which we apply our hands, heads, and hearts.

What advice would you give to those new to the field?

Find the company that cares about you. You'll recognize them when they offer a vision for your future and a plan to get you there. Ask them about that vision and plan in your employment interview.

Interesting career-related fact or accomplishment:

My company won the contractor of the year award for Habitat for Humanity, held numerous roof company marketing events in Lambeau Field, met Sean Penn at a fish taco truck in Hawaii while on an NRCA teaching assignment, and also got lost in Yokosuka, Japan with no language skills while on another NRCA teaching assignment.

How do you define craftsmanship?

I think this quote sums it up well:

> *"He who works with his hands is a laborer.*
> *He who works with his hands and his head is a craftsman.*
> *He who works with his hands and his head and his heart is an artist."*
> —Saint Francis of Assisi

Trade Terms Introduced in This Module

Abate: To reduce or minimize.

Bitumen: A dark, cement-like substance found in asphalts, tars, pitches, and asphaltites. May also refer to any material composed mainly of bitumen, such as asphalt or coal tar.

Built-up roof (BUR) systems: Roof systems in which multiple layers of bitumen, reinforcement, and surfacing material are applied in place on the deck.

Deck: A structural component of the roof of a building, capable of safely supporting the weight of the roof or waterproofing system, as well as the additional live loads required by the governing building codes. The deck provides the substrate to which the roof or waterproofing system is applied.

Flashings: Components used to weatherproof or seal edges of a roof system at perimeters, penetrations, walls, expansion joints, valleys, drains, and other places where the roof covering is interrupted or terminated.

Low-slope: A category of roofs that generally includes weatherproof membrane types of roof systems installed on slopes of 3:12 or less.

Membrane: A flexible or semiflexible roof covering or waterproofing whose primary function is to exclude water.

Membrane roof systems: Roof systems containing a membrane material (weatherproof covering). Membrane roof systems are usually installed as low-slope systems.

Occupational Safety and Health Administration (OSHA): An agency of the US Department of Labor whose mission is to set occupational safety and health standards for all places of employment, enforce these standards, ensure that employers provide and maintain a safe workplace for all employees, and provide research and educational programs to support safe working practices.

On-the-job learning (OJL): Job-related learning an apprentice acquires while working under the supervision of journey-level workers. Also called *on-the-job training (OJT)*.

Personal fall arrest system (PFAS): A system that activates during a fall to catch a worker and prevent them from hitting the ground. PFASs include multiple components, including a body harness, anchorage device, lifeline, and connectors.

Personal protective equipment (PPE): Equipment or clothing designed to prevent or reduce injuries.

Ply: A layer of felt or thin sheet of wood in a built-up membrane roof or waterproofing system.

Safety culture: The culture created when the whole company sees the value of a safe work environment.

Single-ply roof systems: Roof systems in which the primary roof covering is a single layer of flexible membrane material (i.e., EPDM or thermoplastic).

Site-specific safety program: A safety program developed for a jobsite that identifies and takes into account any specific potential hazards that may be encountered.

Slope: The angle of a roof surface, usually expressed as a ratio of vertical rise to horizontal length (sometimes referred to as *run*). When dimensions are given in inches, slope may be expressed as a ratio of rise over a distance of 12" (for example, 4:12) or as an angle in degrees.

Steep-slope: A category of roofing that generally includes water-shedding types of roof coverings installed on slopes greater than 3:12.

Underlayment: An asphalt-saturated felt or other composite or synthetic sheet material (sometimes self-adhering) installed between a roof deck and roof covering, usually used in a steep-slope roof construction. Underlayment is primarily used to separate the roof covering from the roof deck, shed water, and provide fire protection and secondary weather protection.

Water-shedding: Able to depend on gravity for quick drainage to prevent water from entering into the roof system. Steep-slope roof systems are generally water-shedding.

SAMPLES OF NCCER TRAINING CREDENTIALS

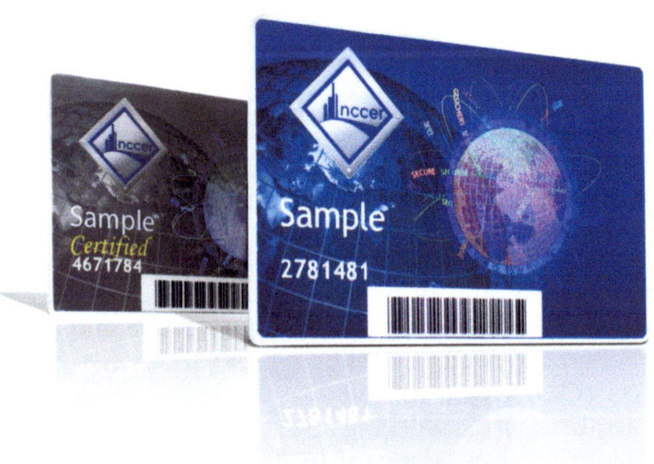

Additional Resources

This module presents thorough resources for task training. The following reference material is suggested for further study.

29 *CFR* 1900–1910, **www.ecfr.gov**.

29 *CFR* 1926, **www.ecrf.gov**.

Build Your Future, **www.byf.org**.

Careers in Roofing, **careersinroofing.com**.

Occupational Outlook Handbook, Roofers, Bureau of Labor Statistics, US Department of Labor, **www.bls.gov/ooh/construction-and-extraction/roofers.htm**.

National Roofing Contractors Association (NRCA), **www.nrca.net**.

Figure Credits

iStock@quangpraha, Module Opener

Courtesy of the National Roofing Contractors Assoc., Figures 2, 17, 18, 20, and 22

iStock@Ivan Elizondo, Figure 6

iStock@Ratchat, Figure 7

iStock@Volodymyr Shtun, Figure 8

iStock@titine974, Figure 9

Patrick D Murphy Co., Inc., Architects, Figures 10, 15, and 19

iStock@nobtis, Figure 12

Courtesy of Dayus Roofing Inc., Figure 13

iStock@Svetlana123, Figure 14

Courtesy of GAF, Figures 16, 21, and 23

iStock@Petmal, Figure 24

Section Review Answer Key

SECTION 1.0.0

Answer	Section Reference	Objective
1. d	1.1.0	1a
2. b	1.2.6	1b
3. a	1.3.0	1c
4. b	1.4.0	1d

SECTION 2.0.0

Answer	Section Reference	Objective
1. d	2.0.0	2
2. d	2.2.0	2b
3. b	2.3.0	2c

SECTION 3.0.0

Answer	Section Reference	Objective
1. a	3.1.2	3a
2. b	3.1.3	3a

SECTION 4.0.0

Answer	Section Reference	Objective
1. a	4.1.1	4a
2. c	4.2.0	4b

NCCER CURRICULA — USER UPDATE

NCCER makes every effort to keep its textbooks up-to-date and free of technical errors. We appreciate your help in this process. If you find an error, a typographical mistake, or an inaccuracy in NCCER's curricula, please fill out this form (or a photocopy), or complete the online form at **www.nccer.org/olf**. Be sure to include the exact module ID number, page number, a detailed description, and your recommended correction. Your input will be brought to the attention of the Authoring Team. Thank you for your assistance.

Instructors – If you have an idea for improving this textbook, or have found that additional materials were necessary to teach this module effectively, please let us know so that we may present your suggestions to the Authoring Team.

NCCER Product Development and Revision

13614 Progress Blvd., Alachua, FL 32615

Email: curriculum@nccer.org
Online: www.nccer.org/olf

❏ Trainee Guide   ❏ Lesson Plans   ❏ Exam   ❏ PowerPoints   Other _______________

Craft / Level: ____ Copyright Date: __________

Module ID Number / Title: ____

Section Number(s): ____

Description: ____

____

____

____

____

Recommended Correction: ____

____

____

____

____

Your Name: ____

Address: ____

____

Email: ____ Phone: _______________________

This page is intentionally left blank.

Roofing Safety

OVERVIEW

Roofing work involves handling hazardous materials and lifting heavy objects, as well as working at elevated locations, near power lines, and around fire hazards. Carefully considering and accounting for these hazards before and during a roofing project ensures roofing professionals' safety. Most construction injuries and deaths are caused by falls. It is crucial to be familiar with and use a variety of fall protection systems when on the job.

Module 16102

Trainees with successful module completions may be eligible for credentialing through the NCCER Registry. To learn more, go to **www.nccer.org** or contact us at 1.888.622.3720. Our website, **www.nccer.org**, has information on the latest product releases and training.

Your feedback is welcome. You may email your comments to **curriculum@nccer.org**, send general comments and inquiries to **info@nccer.org**, or fill in the User Update form at the back of this module.

This information is general in nature and intended for training purposes only. Actual performance of activities described in this manual requires compliance with all applicable operating, service, maintenance, and safety procedures under the direction of qualified personnel. References in this manual to patented or proprietary devices do not constitute a recommendation of their use.

16102 V1.0

16102
ROOFING SAFETY

Objectives

Successful completion of this module prepares you to do the following:

1. Identify roofing hazards in addition to related safety considerations and practices.
   a. Explain the purpose of a job safety analysis (JSA).
   b. Classify types of personal protective equipment (PPE) and explain when they should be used.
   c. Describe proper housekeeping practices on a jobsite to ensure public safety.
   d. Designate proper lifting and material-handling techniques.
2. Express how to handle, transport, store, and work safely around hazardous materials.
   a. Illustrate hazard communication requirements.
   b. Describe fire protection and prevention methods.
   c. Identify safety concerns and precautions associated with adhesives and solvents.
3. Illustrate fall protection methods.
   a. Classify systems used for fall protection.
   b. Explain fall protection in low- and steep-slope roofing.
   c. Describe safety procedures for ladder inspection and setup.

Performance Tasks

Under supervision, you should be able to do the following:

1. Find personal protective equipment (PPE) requirements on a safety data sheet (SDS).
2. Select the correct type of gloves to use on a given task.
3. Properly don a personal fall arrest system (PFAS).

Trade Terms

Guardrails
Lifeline
Occupational Safety and Health Administration (OSHA)
Penetration
Personal fall arrest system (PFAS)
Personal protective equipment (PPE)

Safety data sheet (SDS)
Safety monitoring
Safety nets
Slope
Suspension trauma

Industry Recognized Credentials

If you are training through an NCCER-accredited sponsor, you may be eligible for credentials from NCCER's Registry. The ID number for this module is 16102. Note that this module may have been used in other NCCER curricula and may apply to other level completions. Contact NCCER's Registry at 1.888.622.3720 or go to **www.nccer.org** for more information.

Contents

This page is intentionally left blank.

1.0.0 ROOFING HAZARDS AND PRECAUTIONS

Objective

Identify roofing hazards in addition to related safety considerations and practices.

a. Explain the purpose of a job safety analysis (JSA).
b. Classify types of personal protective equipment (PPE) and explain when they should be used.
c. Describe proper housekeeping practices on a jobsite to ensure public safety.
d. Designate proper lifting and material-handling techniques.

Performance Tasks

1. Find personal protective equipment (PPE) requirements on a safety data sheet (SDS).
2. Select the correct type of gloves to use on a given task.

Trade Terms

Occupational Safety and Health Administration (OSHA): An agency of the US Department of Labor whose mission is to set occupational safety and health standards for all places of employment, enforce these standards, ensure that employers provide and maintain a safe workplace for all employees, and provide research and educational programs to support safe working practices.

Personal protective equipment (PPE): Equipment or clothing designed to prevent or reduce injuries.

To work safely, roofing professionals must stay alert to potential hazards, take proper precautions, and practice the basic rules of safety. You must always be safety-conscious and report any unsafe conditions to your supervisor and co-workers. Safety should become a habit. Keeping a safe attitude on the job will directly reduce the number and severity of accidents. Remember that your own safety is ultimately up to you.

The **Occupational Safety and Health Administration (OSHA)** is a government agency that requires employers to maintain a safe work environment for their employees. OSHA saves statistics related to construction site incidents, injuries, and fatalities to educate the public about the most hazardous areas of construction work. The top four most life-threatening hazards on a construction site, often referred to as the *Fatal Four* or *Focus Four*, are as follows (*Figure 1*):

- Falls
- Struck-By Objects
- Electrocutions
- Caught-In/Between

> **NOTE**
>
> Visit OSHA's website at **www.osha.gov** for the most recent Fatal Four statistics.

According to OSHA, eliminating the Fatal Four would save approximately 591 workers' lives in the US each year. OSHA has developed safety regulations for the construction industry in *29 CFR 1926*. Employers are required to follow these regulations to provide a safe work environment. Employees are also responsible for protecting their own safety by following safe policies and procedures, such as wearing the appropriate **personal protective equipment (PPE)**.

Examples of common hazards encountered in roofing work include the following:

- *Fall hazards* — Falls are the leading cause of death in the construction industry. Roofing workers are at even higher risk, as they work at elevated locations daily. At-risk workers must follow safety guidelines for using fall protection.
- *Falling objects* — Tools and materials must be properly secured when working at elevation to protect workers and others below. In addition, hazards caused by tools and materials must be considered in the pre-job planning efforts.
- *Crane hazards* — Heavy objects must sometimes be lifted to or lowered from the roof using a crane. This process is dangerous and heavily regulated. OSHA has comprehensive guidelines for crane operations, communication, signaling, and rigging.
- *Overhead power lines* — Overhead power lines must be carefully identified when working on roofing projects, not only for the workers, but also for equipment operators moving materials to various parts of the roof. OSHA requires that a safe distance of at least 10' (3 m) must be maintained between a power line and a piece of equipment for line voltages of 50kV or less. For voltages above 50kV, the distance must be increased by 0.4' (12 cm) for each 1kV of voltage.

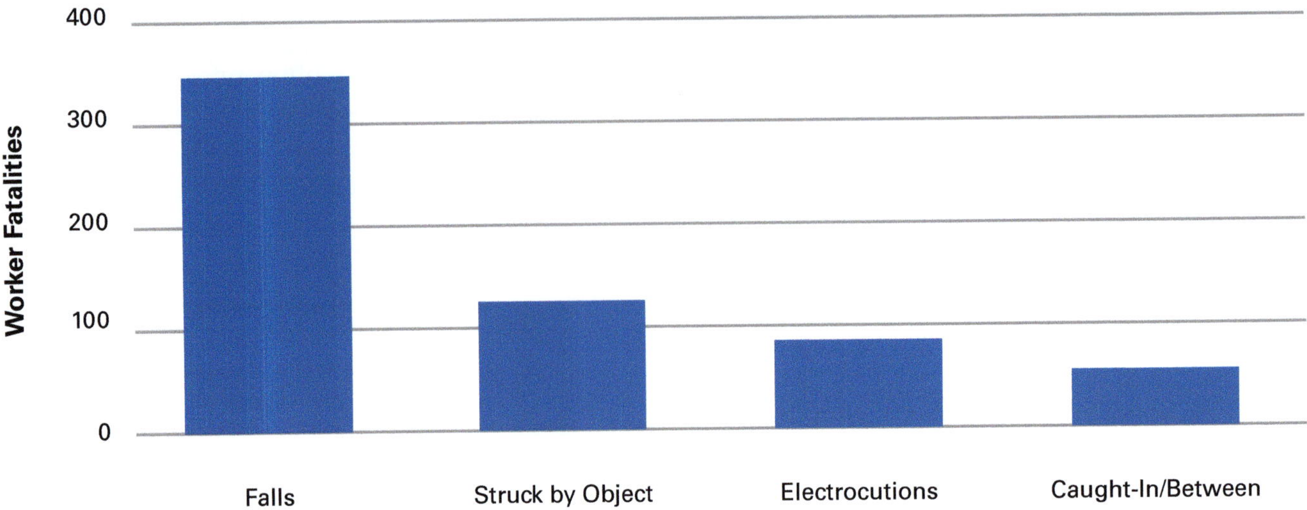

Figure 1 OSHA Fatal Four statistics.

- *Bad decking* — Before beginning a reroofing project, it is important to check the underside of the roof deck for dangerous areas, electrical wires, and obvious defects such as holes. You must be aware of your surroundings and able to recognize clear and present hazards, such as faulty decking, open holes, or cracked concrete. Decks may also be slippery due to frost, snow, or rain.
- *Structural integrity* — Every roof is designed with a maximum load capacity. If this capacity is exceeded, the roof could collapse. The weight of roofing materials must not exceed the anticipated load. When possible, loads should be placed directly over rafters or trusses.
- *Hot bitumen* — Hot bitumen, or hot asphalt and coal tar, must be handled with extreme care and proper PPE. Workers must follow safety guidelines when dealing with kettles and tankers. For example, warning lines must be set up around the kettle or tanker area to keep people away.
- *Torches* — Torches must be handled with great care, and protective safety glasses and gloves must be worn while torching. When attaching hoses or appliances to a torch, be sure the torch head is off and hold the cylinder upright. National Roofing Contractors Association (NRCA) offers a Certified Roofing Torch Applicator (CERTA) Training Program that teaches participants how to properly use PPE and first aid associated with torch safety.
- *Hazardous materials* — Roofers deal with a variety of hazardous substances, such as adhesives and fuels. The handling, storage, and transportation of these materials are regulated by OSHA. Containers of hazardous materials must be labeled.

Consider a job safety analysis (JSA) before beginning a roofing project to identify the potential hazards and recommend actions that can be taken to eliminate or minimize each hazard.

While employers are required to provide a safe environment for employees, it is equally important for you to take responsibility for your own safety by following guidelines, wearing PPE, and maintaining a tidy work environment. Most construction-related injuries and deaths result from a failure to follow safe work practices or to recognize a hazard.

Substance Abuse

Statistics show that employees with drug and/or alcohol addictions have a higher rate of accidents. Workers who are impaired by these substances on a jobsite are a danger to themselves and those around them. In 2018 alone, more than 67,300 people died in the US from an overdose.

According to the US Bureau of Labor Statistics, overdose deaths that occur while working are on the rise. Addiction is an illness that can be treated. Resources to overcome addiction can be found at **resources.facingaddiction.org** or by calling the confidential national hotline at **1-800-622-HELP (4357)**.

1.1.0 Job Safety Analysis (JSA)

A job safety analysis, also known as *job hazard analysis (JHA)*, is an approach to hazard recognition. In a JSA, the task is broken down into its individual parts or steps, and then each step is analyzed for potential hazards. Once a hazard is identified, certain actions or procedures are recommended for correction. *Figure 2* shows one example of a JSA form.

Figure 3 shows a portion of another JSA form that uses three columns: Sequence of Tasks, Potential Hazards, and Preventive Measures (sometimes referred to as *mitigation*). The column headings mirror the three basic steps in the job safety analysis process:

Step 1 Identify the individual tasks, from first to last, that must be performed to do the job.

Step 2 Identify potential hazards for each task.

Step 3 Identify the most effective preventive measures for each of the potential hazards.

It is important to write legibly when filling out a JSA form. Increasingly, workers with smartphones or tablets use applications and online forms to complete JSA forms electronically (*Figure 4*).

The final step in conducting a JSA is to develop practical methods of making the job safer. The goal is to prevent the occurrence of potential incidents and minimize the risk and severity of injury should an incident occur. This is best achieved by applying a hierarchy of risk-control methods (*Figure 5*). With this approach, the greater the severity of the hazard/risk, the more effort that should be applied to eliminate it. The hierarchy of risk control methods is as follows, in order from most effective to least effective:

1. *Elimination* — Remove the hazard. For example, redesigning a process to eliminate the use of a hazardous material.

JOB SAFETY ANALYSIS FORM

Job Title:
Job Location:
PPE:
Tools, Materials & Equipment:

Date of Analysis:
Conducted By:
Staffing:
Duration:

Page __ of __

Step	Hazard	Quality Concern	New Procedure or Protection

© PSA, Inc. / NCCER / JSA Proc./Forms

Figure 2 Job safety analysis (JSA) form.

JOB SAFETY ANALYSIS WORKSHEET

Job:

Analyzed by: Date:
Reviewed by: Date:
Approved by: Date:

Sequence of Tasks	Potential Hazards (Energy type and contacts)	Preventive Measures (Barriers)

Figure 3 Alternate JSA form.

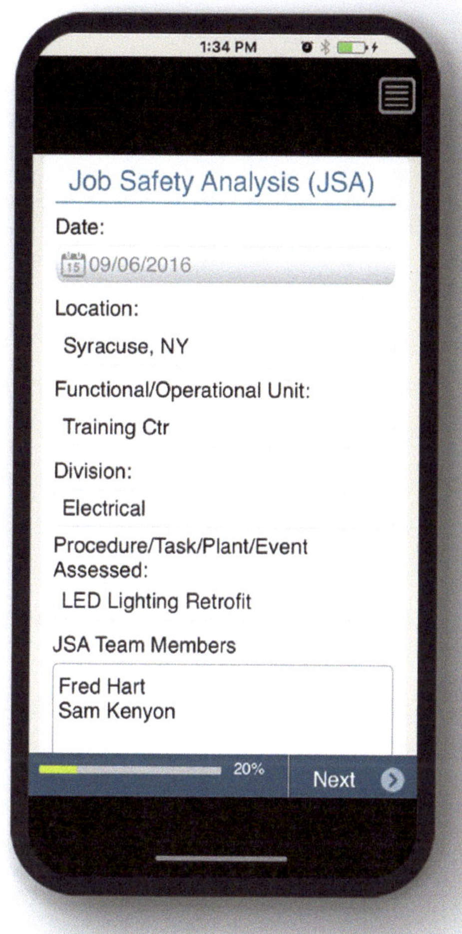

Figure 4 Electronic JSA form.

2. *Substitution* — Replace the hazard. For example, substituting a solvent-based paint with a water-based paint to prevent exposure to harmful fumes.

3. *Engineering controls* — Isolate the hazard. For example, preventing unauthorized access to a high-voltage area by fencing around electrical gear.

4. *Awareness* — Warn workers of the hazard. For example, alerting a crew if a walking surface is unstable or slippery.

5. *Administrative controls* — Change the way people work. For example, displaying warning signs and labels around a hazard.

6. *Personal protective equipment* — Provide appropriate personal protective equipment. For example, ensuring that workers have appropriate gloves, hard hats, and safety shoes.

Elimination, substitution, and engineering controls are the most effective methods because they are usually applied at the source of the hazard. Awareness, administrative controls, and PPE are less effective because they are more likely to be affected by human error (for example, a worker forgetting to put on PPE).

1.2.0 Hazard Control and PPE

Hazard control and personal protective equipment (PPE) are safety approaches used to protect workers from danger on the job. It is important to educate yourself on relevant PPE and hazard-control methods before beginning a job. Consult your safety data sheet (SDS), local building codes, and manufacturer instructions for appropriate safety information.

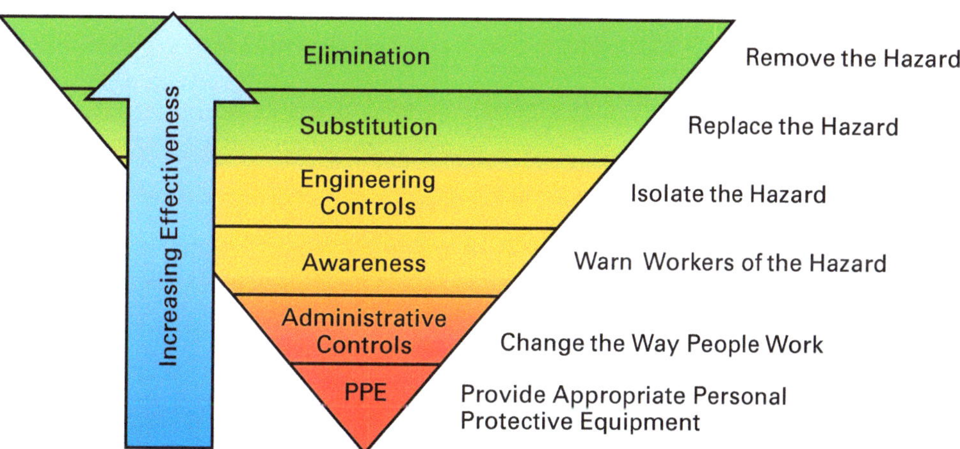

Figure 5 Hierarchy of risk-control methods.

1.2.1 Hazard Control

Hazard control reduces safety risk through education on potential dangers and prevention of unsafe situations.

Hazard-control methods that apply to all weather conditions include the following:

- Check and comply with any federal, local, and state code requirements when working on roofs.
- Always use an OSHA-rated ladder sized appropriately for the job and inspect it for defects and damage before use.
- Brush or sweep the roof periodically to remove any accumulated dirt or debris.
- Remove any unused tools, cords, and other loose items from the roof. They can be a serious hazard.
- Be alert to any other potential hazards, such as live power lines.

Because roofing takes place outdoors, you may work in especially hot or cold temperatures for extended periods of time. Take precautions to avoid heat exhaustion and exposure to the sun's ultraviolet (UV) rays. In cold weather, beware of frostbite, clumsiness caused by stiff joints and muscles, and slippery surfaces.

Hazard-control methods concerning weather include the following:

- Rain, frost, and snow make roofs slippery. If possible, wait until the roof is dry. If waiting for it to dry is not possible, wear special roofing footwear with skid-resistant cleats in addition to fall protection equipment.
- Take time to adapt to your weather conditions before you begin working.
- Wear light clothing that is made of natural fibers in hot conditions. Be sure to wear layers in cold conditions.
- Use sunblock with a sun protection factor (SPF) of 30 or higher on exposed skin.
- Drink adequate amounts of water to prevent dehydration, especially in dry parts of the country.
- Electrolyte-filled drinks help regulate muscles and neurons in your body. Drinking these prevents exhaustion and promotes muscle function and hydration.

1.2.2 PPE

Basic safety practices can reduce or remove hazards before work begins. PPE protects workers from hazards when they cannot be removed.

Observe the following guidelines to ensure your safety and the safety of others:

- Wear fall protection devices.
- Wear a hard hat.
- If possible, wear tinted glasses or goggles.
- Wear boots or shoes that are in good condition and appropriate for the job.
- Wear eye protection appropriate for the task at hand. Examples of such equipment include safety glasses with side shields, goggles, or a face shield to avoid eye injuries from hot asphalt or coal tar, chemicals, power tools, flying nails, or tear-off debris.
- Wear gloves to protect against cuts, burns, and exposures.

Respiratory Protection

Regulations for respiratory protection are found in *29 CFR 1910.134*. Contractors must develop and implement a written respiratory protection program when the permissible exposure limit (PEL) of airborne contaminants could be exceeded or when they require workers to use respirators.

It is crucial to use the correct type of respirator for a given job—one job may require an air-purifying respirator, and another job could require a supplied-air respirator. While air-purifying respirators clean the air that a wearer is breathing, supplied-air respirators generate breathable air within the respirators' system and require training before use. For roof installation, air-purifying respirators are much more commonly used than supplied-air respirators.

An example of an air-purifying respirator sometimes used in roofing, known as an N95 mask, is shown in *Figure 6*.

Eye and Face Protection

The eyes and face are extremely sensitive, and any blunt trauma, chemical exposure, or radiation can cause irreversible damage, especially to eyesight. *Figure 7* shows an example of eye protection.

The types of eye and face protection available include:

- Spectacles (with metal or plastic frames and side shields)
- Safety glasses (flexible or cushioned fitting)
- Face shield (plastic or mesh window)

Any of these eye and face protectors can be made to fit individual prescriptions, which alleviates the issue of finding protective wear to fit snugly over glasses.

Figure 6 N95 respirator mask.

Figure 7 Eye protection.

Hand Protection

Employees must use hand protection when they are exposed to hazards related to their work process or environment, chemical or radiological hazards, or mechanical irritants. Hand protection must be selected to prevent injury from absorption of chemicals or other physical contact that could cause cuts, abrasions, punctures, burns, or other injuries.

Always be sure you are wearing the appropriate gloves for a given job. For general roofing work—which includes handling all types of roofing materials, carrying equipment, and dealing with hot asphalt—leather gloves are usually appropriate. Cut-resistant gloves, such as those made with Kevlar fiber, are helpful to wear when using sharp tools (*Figure 8*). Some kinds of heavy cotton gloves, preferably with leather palms, are also acceptable.

Neoprene, butyl, PVC, rubber, or nitrile gloves should be used for the application of chemical solvents and adhesives to prevent absorption. Gloves must be selected carefully for chemical resistance because not all gloves are resistant to every solvent, cleaner, or adhesive used in roofing.

Protective Clothing

Maximum safety in the roofing workplace requires full-body covering: a long-sleeved, button-down shirt that is tucked into pants and buttoned at the cuffs. It should only have one button undone at the neck at most Long sleeves are essential in protecting a worker's arms, and the button-down style allows for easy removal in case of an emergency to prevent extended contact with hot materials.

Workers must wear long pants to protect their legs. Because cuffs are ideal pooling locations for substances such as hot bitumen, pants should not have cuffs. Kneepads are useful when employees are required to spend extended periods on their knees.

1.3.0 Housekeeping and Public Safety

One of the easiest ways to prevent accidents is to maintain a neat and orderly jobsite. A construction site littered with materials and debris is unsafe and presents unnecessary risks to workers. It is important for workers and vehicles to have pathways that are clear of obstructions, such as debris and unattended tools.

Figure 8 Cut-resistant gloves.

Many local regulations prohibit or regulate the burning of waste to prevent uncontrolled fire.

Lean Construction

The term *lean construction* describes construction methods that are designed to eliminate waste throughout a project's lifecycle. Methods of lean construction include proper storage of equipment, staging only equipment that will be needed that day, and ensuring that nothing hits the floor or is wasted. Lean construction also encourages all parties to share best practices and establishes a partnership between suppliers and end-users to reduce scheduling delays and the overordering or underordering of supplies and materials. Additionally, keeping your jobsite orderly creates a safer workplace.

Source: www.leanconstruction.org

A littered jobsite is not only unsafe—it also reflects poorly on the company or contractor performing the job. A professional roofing company will maintain the building owner's property in a clean and orderly manner throughout the course of the project. Extra care must be exercised to keep surrounding properties clean during tear-off work.

Roofing operations must ensure the safety of the public as well as workers. For some projects, an entry-level worker may be assigned as a ground monitor. A ground monitor watches for pedestrian traffic near the worksite to make sure no one enters an unsafe area. If ladders are left up overnight, they must be barricaded so children and pedestrians do not climb them.

Materials and equipment should be assembled in a ground staging area before use. This ensures things are kept organized and out of the way when not in use.

OSHA requires that all scraps, debris, and materials be disposed of regularly and stored properly. These standards, found in *29 CFR 1926.25*, discuss the cleanliness of the general work area to prevent employees from stepping on nails, twisting their ankles, or otherwise being injured, as well as to prevent trash fires.

All debris, especially scrap lumber with protruding nails, must be removed from work areas, passageways, and stairs. Combustible scrap must be safely moved from work areas on a regular basis. Oily rags, flammable liquids, and other hazardous waste must be separated and thrown into designated containers. Containers used for hazardous or flammable materials must have lids, and they must be emptied regularly.

Disposal from elevated surfaces is addressed in *29 CFR 1926.252*. These regulations ensure work areas are maintained properly and no one is struck by falling debris. Any time materials must be dropped more than 20' (9 m) on the outside of a building, they must be sent down an enclosed chute (*Figure 9*). An enclosed chute is defined as a slide, closed in on all sides, through which materials are moved from a roof or other elevated area to a lower area.

If the roof height exceeds 20' (9 m) and a chute is not feasible, alternative methods of lowering the material must be implemented so it is not dropped to the ground. Debris can be lowered by hoist or crane in debris boxes, slings, or pallets. If materials are being dropped inside a building through a hole and without a chute, the area below must be completely barricaded.

Figure 9 Enclosed chute for lowering construction debris.

All scraps, waste, and trash must be removed from the immediate work area as work progresses. However, waste on lower levels should not be removed until work is stopped at the higher level. This prevents potentially dangerous situations that require workers to communicate on different levels of a building.

1.4.0 Proper Lifting and Material Handling

Back injuries are the most common and one of the most debilitating injuries roofing workers experience. Many back injuries are not caused by a single incident but by the cumulative stress of using poor lifting techniques. Over time, these bad habits cause back pain to develop. Most back injuries can be prevented by learning proper lifting techniques and following safe material-handling guidelines.

The best method of preventing back injuries is to learn to lift properly and get help with heavy loads. The following tips can help you avoid back strain:

- If possible, use equipment such as hoists, dollies, handcarts, and forklifts to lift objects.
- Never carry anything up a ladder except what is on your tool belt. Carrying heavy material up a ladder is bad for your back and the ladder, dangerous to others, and could cause you to fall.
- Frequently stretch to loosen and prepare your muscles for work.
- Before you move an object, check the intended path of travel and ensure it is free of obstructions.

An example of proper lifting is shown in *Figure 10*. Keep the following precautions in mind when lifting.

- Make the lift smoothly and with control.
- Look straight ahead while lifting.
- Move your feet to pivot; do not twist or you may injure yourself.

- Constantly scan the path ahead for obstructions. If you cannot see your path over or around the object being carried, get help to transport the object.
- Avoid lifting objects over your head.
- Never lift over the side or tailgate of a pickup truck.
- Do not twist your body when lifting an object or setting it down.
- Never reach over an obstacle to lift a load.
- Do not step over objects in your way.

Remember that you do not have to do all the heavy lifting yourself. Getting assistance from other employees will help you do your job more efficiently and prevent injuries. Use the following guidelines when lifting large or heavy objects:

- Always check the load to find out how heavy it is. Do not try to pick it up. Instead, lift a corner to get a sense of its weight. If it is too heavy or bulky, get help.
- Try to rotate jobs with other employees. Switching jobs avoids repetition and saves on long-term aches and pains, as well as injury.
- Communicate with the person assisting you, so you lift and set the load down at the same time.
- If possible, both persons should try to carry the load while facing forward.
- If the load is heavy, take rest breaks.

While some objects may seem lightweight, they can still be dangerous to carry without help. For example, large, lightweight insulation boards can be difficult to carry in the wind. If insulation materials catch the wind, they may be damaged or cause workers to lose their footing.

Many roofing materials are very heavy in their shipment packaging. Material-handling operations should ensure personnel are not overburdened. In many cases, materials should be removed from their original packaging and divided into lighter loads.

NCCER – *Roofing*

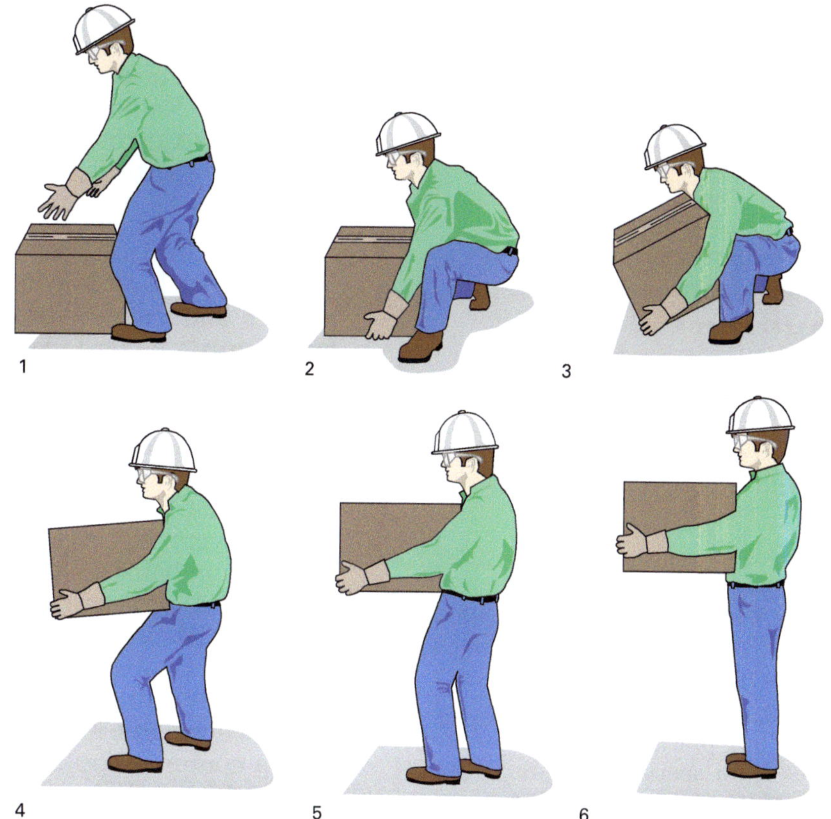

Figure 10 Proper lifting.

1.0.0 Section Review

1. Which of the following causes the *most* fatalities in the construction industry?

   a. Falling objects
   b. Hazardous materials
   c. Falls
   d. Electrocution

2. Which of the following is an example of PPE that may be needed during a roofing project?

   a. Gloves
   b. Safety data sheets
   c. An OSHA-rated ladder
   d. An underlayment

3. An enclosed chute *must* be used any time debris must be lowered more than ______.

   a. 10' (3 m)
   b. 20' (6 m)
   c. 30' (9 m)
   d. 40' (12 m)

4. When preparing to lift a large object, you should ______.

   a. contact your supervisor
   b. avoid asking coworkers for help
   c. carry it a few feet to see if you need help
   d. make sure the intended path of travel is free of obstructions

2.0.0 HAZARDOUS MATERIALS

Objective

Express how to handle, transport, store, and work safely around hazardous materials.

  a. Illustrate hazard communication requirements.

  b. Describe fire protection and prevention methods.

  c. Identify safety concerns and precautions associated with adhesives and solvents.

Trade Terms

Penetration: Any construction, such as pipes, conduits, or HVAC supports, passing through a roof or waterproofing system.

Safety data sheet (SDS): A form that lists the hazards, safe handling practices, and emergency control measures for a specific substance. Also called a *material safety data sheet (MSDS)*.

Roofing workers encounter a variety of materials during a roofing project. Hazardous materials pose such a risk to health and safety that their transport, packaging, and identification are regulated by the US Department of Transportation (DOT). Roofing contractors must be aware of the nature of the materials that are being transported by their employees. They must follow rules that govern the vehicles and drivers involved in the process.

The mishandling of hazardous materials could cause injuries or cost lives. Penalties for noncompliance with hazardous materials regulations could include civil fines of up to hundreds of thousands of dollars and imprisonment for up to five years.

2.1.0 Hazard Communication

Hazardous materials are substances that have been established by federal regulators as posing unreasonable risk to health, safety, and property when transported. There are nine classes of hazardous materials:

- *Class 1* — Explosives
- *Class 2* — Gases
- *Class 3* — Flammable and Combustible Liquids
- *Class 4* — Flammable Solids
- *Class 5* — Oxidizing Substances and Organic Peroxides
- *Class 6* — Poisons
- *Class 7* — Radioactive Materials
- *Class 8* — Corrosives
- *Class 9* — Miscellaneous Hazardous Materials

Many of the substances roofing contractors encounter are in Class 9 (e.g., high-temperature materials such as hot bitumen) or Class 3 (e.g., mastics and cutback products). Substances that fall within a hazard class may be further specified by a division within the class.

Hazardous materials are assigned identification numbers along with proper shipping names, which are assigned in the Hazardous Materials Table in *49 CFR 172.101*.

Employers are required to establish a written, comprehensive hazardous communications (HAZCOM) program that includes provisions for container labeling and employee safety training activities. Hazard communication is aligned with the Globally Harmonized System of Classification and Labeling of Chemicals (GHS) that helps provide a common approach to chemical classification and hazard communication. OSHA also requires that a **safety data sheet (SDS)** be available for everyone's use (*Figure 11*).

Caulks, solvents, adhesives, coatings, and mastics are just some of the roofing products that may potentially be classified as hazardous materials. It is important to be aware of the following potentially hazardous materials and familiar with procedures for safely handling them:

- *Fuels* — Since fuels are flammable, they must be handled carefully to avoid catching fire. For example, electrical tools can ignite a fire if gasoline is spilled.

Figure 11 HAZCOM.

- *Adhesives and solvents* — These are used by roofing workers on a routine basis and can produce dangerous vapors or cause chemical burns.
- *Silica dust* — Silica can be found in a variety of clay and concrete products used in roofing. Silica dust is produced when these products are cut or ground during installation. During these processes, workers must wear respiratory protection to prevent serious health issues such as silicosis.
- *Asbestos* — Asbestos is known to cause serious health problems. It may be present in roof cement, coatings, or mastics in older buildings. Roofing contractors must determine that there is no asbestos on a jobsite before work begins.
- *Hot asphalt* — Hot asphalt can be extremely dangerous if it is not handled safely. In addition to the risk of burns, hot asphalt also contains fumes that can cause headaches, skin rashes, fatigue, and other dangerous side effects. Roofing professionals must use the proper PPE when transporting and applying hot asphalt, as designated by local code and SDS.

2.2.0 Fire Prevention and Flammable Liquids

Employers are required by OSHA to implement fire protection and prevention programs in the workplace. Fire is a chemical reaction that requires the following three elements to be present:

- *Heat (or an ignition source)* — Ignition sources can include any material, equipment, or operation that emits a spark or flame—including obvious items, such as torches, as well as less obvious items, such as static electricity and grinding operations. Equipment or components that radiate heat, such as kettles, torches, heat welders, catalytic converters, and mufflers, can also be ignition sources.
- *Fuel* — Fuel sources include combustible materials, such as wood, paper, trash, and clothing; flammable liquids, such as gasoline or solvents; and flammable gases, such as propane or natural gas.
- *Oxygen* — Oxygen comes from the air in the atmosphere. Air contains approximately 21 percent oxygen.

These three elements typically are referred to as the *fire triangle* (*Figure 12*). Fire is the result of the reaction between the fuel and oxygen in the air. Scientists developed the concept of a fire triangle to aid in understanding of the cause of fires and how they can be prevented and extinguished. Heat, fuel, and oxygen must combine in a precise way for a fire to start and continue to burn. If one element of the fire triangle is not present, fire will not start.

To prevent fires, the three elements of the fire triangle must be kept separate. Posting and enforcing "no smoking" signs around flammable liquids and gases is one method of achieving this goal. Another method is to have fire watches on all work involving torch-applied materials for a minimum of two hours after the last torch is turned off.

2.2.1 Storage of Flammables

Proper storage and handling of flammable and combustible liquids helps prevent fires from occurring. Only approved, closed containers for storage of flammable or combustible liquids may be used under OSHA rules.

Use the following guidelines for storing flammable liquids:

- Always keep flammables away from any ignition sources, such as heat, flames, or smoking materials.

Figure 12 Fire triangle.

Asbestos Use

Asbestos was identified as a health risk many years ago. However, countries have reacted to the hazard in different ways. While some countries banned its use in products and construction decades ago, other countries, such as India, continue to use asbestos without regulation of any form. In fact, asbestos use in India has risen in the last decade.

The World Health Organization (WHO) estimates that 125 million people worldwide are exposed to asbestos each year in the workplace. Even in the United States, asbestos may still be encountered in old buildings, flooring, and insulation components. Be aware of the regulations concerning the use of asbestos in your region and stay alert to sources that might cause harm if not avoided.

- Make sure there is good ventilation in the storage area.
- Always keep fire doors and walkways clear.
- Do not store flammable and combustible liquids near stairways or exits.
- No more than 25 gallons of flammable or combustible liquids may be stored indoors outside of an approved storage cabinet.
- No more than 60 gallons of flammable and no more than 120 gallons of combustible liquids may be stored in a single cabinet, and no more than three cabinets may be located in a single storage area.
- Outdoor portable storage tanks containing flammable or combustible liquids must be stored at least 20' (9 m) from any building.
- Always keep flammable liquids in approved safety containers when not in use. This does not apply to highly viscid (thick) materials.
- Always bond and ground containers when transferring materials to prevent a buildup of static electricity, which potentially could create a spark. The bond wire connecting the two containers should be set up prior to the transfer. The ground wire should be connected to one of the containers and to a proper ground. The ground will eliminate any static electrical charges created.

2.2.2 Liquefied Petroleum Gas

Liquefied petroleum (LP) gas is used widely in the roofing industry to heat kettles and torches. Because LP gas is a compressed gas, fairly large quantities can be stored in relatively small containers. LP gas expands at a ratio of 270 to 1. This means that one liquid drop of LP gas would expand to a gas state of 270 times greater volume.

Because it is denser and heavier than air, LP gas collects in low-lying areas. If you suspect a leak in a cylinder, do not use fire to attempt to find the hole. Instead, use soapy water and look for bubbles.

If an LP gas fire breaks out, employees should evacuate the area immediately and call the fire department. Fighting an LP gas fire requires specialized training that only the fire department can provide. If a fire starts on a jobsite, call the fire department as soon as possible.

Use the following guidelines when handling and storing LP cylinders:

- Cylinders must be marked "flammable gas."
- Cylinders should not be dropped or allowed to strike each other.
- When storing, using, or transporting cylinders, keep them securely fastened in an upright position and ensure the container valve is closed and covered with a safety cap or collar.
- Do not store LP cylinders indoors.
- Never hoist cylinders by attaching lines to valves or collars; a hoisting cage or cradle is the safest way to hoist a cylinder.
- Cylinders should be moved by means of a hand truck. If it is absolutely necessary to move one by hand, roll it on its edge—never drag it.
- Make sure the pressure regulator is properly adjusted and is not damaged.
- Check the hose prior to use for cuts, cracks, or worn places.
- Use a heat shield to protect the containers when they are mounted on a kettle.
- When in doubt, always consider cylinders full and handle them accordingly.
- When not in use, turn off the fuel supply at the tank.
- When in use, keep all LP gas tanks at least 10' (3 m) away from the kettle.
- Keep bulk propane and storage tanks at least 25' (7.6 m) away from the kettle, tanker, or building.
- If a propane cylinder frosts, it means the cylinder is too small. Never turn it on its side or use a torch to defrost the cylinder.

2.2.3 Torch-Applied Roofing Materials

Torch-applied roofing materials (*Figure 13*) pose a serious fire hazard to roofing contractors and building owners. Sometimes the hazards are obvious, such as torching to a combustible deck or near flammable liquids. Other concerns may be less obvious, such as torching around a drain or penetration where flames can be drawn into a building.

Torches are extremely dangerous tools. Torch heads reach a temperature of 2,000°F (1,093°C). One moment of carelessness can result in severe injury or property damage. Make sure you completely understand how to operate a torch before using one. Installation of torch-applied products creates the risk of fire, including smoldering fires. Torch-applied products must be applied only by professional roofing applicators trained in proper torch application and safety procedures. Use the following precautions when applying torch-applied roof materials:

- Always inspect the torch head, valves, hoses, gauges, connections, and fittings for any defects. Never use any equipment, including the LP gas cylinders, unless it is in good working order.
- Wear leather gloves and eye protection when working with torches.
- Observe all applicable manufacturer guidelines.

Figure 13 Torch-applied roofing materials.

- Always keep propane cylinders at least 10' (3 m) away from any torch.
- Only use a spark igniter to light the torch; never use a match or lighter.
- Never torch a combustible deck or material.
- Never leave a lit torch unattended.
- Never point a torch at anyone.
- Never torch to areas that cannot be seen fully.
- Do not use torches near vents or air intakes.
- Never use a torch to heat a propane tank that begins to frost on the outside.
- Always have appropriate fire extinguishers within easy reach.
- When you are finished with the torch, close the propane cylinder valve. Squeeze the trigger on the torch to burn off any remaining propane in the hose.
- A fire watch should be assigned during lunch breaks and for a minimum of two hours after torching to make sure nothing catches fire.

Contractors can obtain certification through the Certified Roofing Torch Application program administered by the NRCA.

2.3.0 Adhesives and Solvents

Adhesives and solvents are some of the most hazardous substances roofing workers encounter. Because these materials are used so frequently, workers do not always realize the dangers they present. If proper precautions are not used, a worker can easily be overcome by their vapors or suffer chemical burns.

Use the following guidelines when handling adhesives and solvents:

- Always wear the proper PPE, including rubber gloves.
- Wear respiratory protection when required.
- Make sure the work area is well-ventilated. If the area is not well-ventilated, use fans to help circulate the air.
- Read the label and follow the manufacturer's recommendations for applications and handling.
- Never smoke around adhesives or solvents; the vapors can ignite.
- Always be aware of other workers and contractors in the work area. Acetylene torches, electric welders, and other flame-producing equipment may ignite solvent vapors.
- Check to see that fire extinguishers are easily accessible.

- When cleaning tools, use solvent-free cleaners if possible. If you use solvents, remember to check the SDS and wear gloves, goggles and any other PPE required. Work in a well-ventilated area. Wear respiratory protection when required. Properly dispose of the rags used for cleaning.
- To prevent combustion, static charges should be avoided whenever adhesives are being used.

Use the following guidelines for the storage of adhesives and solvents:

- Read the manufacturer's label or SDS for the required storage specifications.
- Fire extinguishers must be accessible at all times.
- Make sure that the containers are securely and separately stored where the labels are easy to read. Post "no smoking" signs. If several containers of the same material are stored in a single area, one large and easy-to-read label may be placed in an obvious location.
- Always make sure the lids and tops of all solvents and adhesives are tightly closed before storing.
- Store the different solvents and adhesives separately and away from other chemicals.
- Immediately and properly dispose of all empty pails and containers.
- Flammable materials should not be stored near building stairways or exits.

2.0.0 Section Review

1. An SDS is a(n) _____.
    a. system used to prevent falls from roofs
    b. form that lists hazards and safety precautions for a substance
    c. net used to catch fallen tools
    d. adhesive

2. Which of the following is *not* likely to be classified as hazardous?
    a. Asbestos
    b. Fuels
    c. Wood boards
    d. Silica dust

3. What three components comprise the fire triangle?
    a. Heat, oxygen, and nitrogen
    b. Smoke, heat, and oxygen
    c. Fuel, wood, and heat
    d. Heat, fuel, and oxygen

3.0.0 FALL SAFETY

Objective

Illustrate fall protection methods.
   a. Identify systems used for fall protection.
   b. Explain fall protection in low- and steep-slope roofing.
   c. Describe safety procedures for ladder inspection and setup.

Performance Task

   3. Properly don a personal fall arrest system (PFAS).

Trade Terms

Guardrails: Railings installed around the perimeter of a worksite to prevent falls from a roof's edge.

Lifeline: A component consisting of a flexible line that is either connected vertically to an anchorage at one end (vertical lifeline) or connected horizontally to an anchorage at both ends (horizontal lifeline). A lifeline serves as a means for connecting other components of a personal fall arrest system to the anchorage.

Personal fall arrest system (PFAS): A system that activates during a fall to catch a worker and prevent them from hitting the ground. PFAS include multiple components, including a body harness, anchorage device, lifeline, and connectors.

Safety monitoring: A safety system that assigns monitoring duties to a competent person who has the knowledge and authority to make safety decisions.

Safety nets: Nets installed around and underneath a working area that can catch falling objects or workers.

Slope: The angle of a roof surface, usually expressed as a ratio of vertical rise to horizontal length (sometimes referred to as *run*). When dimensions are given in inches, slope may be expressed as a ratio of rise over a distance of 12" (for example, 4:12) or as an angle in degrees.

Suspension trauma: A serious medical condition that can occur when a worker is suspended for a short period of time in a body harness.

Fall protection is an essential component of roofing safety. In 2018, OSHA reported that deaths caused by falls accounted for 33.5 percent (roughly one-third) of construction-related fatalities. Nearly 90 percent of these fatal falls occurred when there was no fall protection system in place. Since most of these falls occur from or through roofs (*Figure 14*), it is critical that each stage of the roofing process takes fall-related safety precautions into consideration.

> **NOTE**
>
> Fall safety and fall protection systems are covered in more detail in NCCER Module 75901, *Fall Protection Orientation.*

Roof systems are categorized as low-slope or steep-slope based on slope of the roof. The slope is a ratio expressing the vertical rise of the roof over a horizontal distance. The height of the roof is referred to as the *rise* of the slope, and the horizontal distance is referred to as the *run*. For example, a roof with 7" (10 cm) of rise for every 12" (30 cm) of run has a slope of 7:12.

OSHA defines a roof with a slope less than or equal to 4:12 as *low-slope*. Roofs with a slope greater than 4:12 are defined as *steep-slope*. OSHA uses this distinction when regulating fall protection measures, such as mandated warning-line systems used for low-slope roofing.

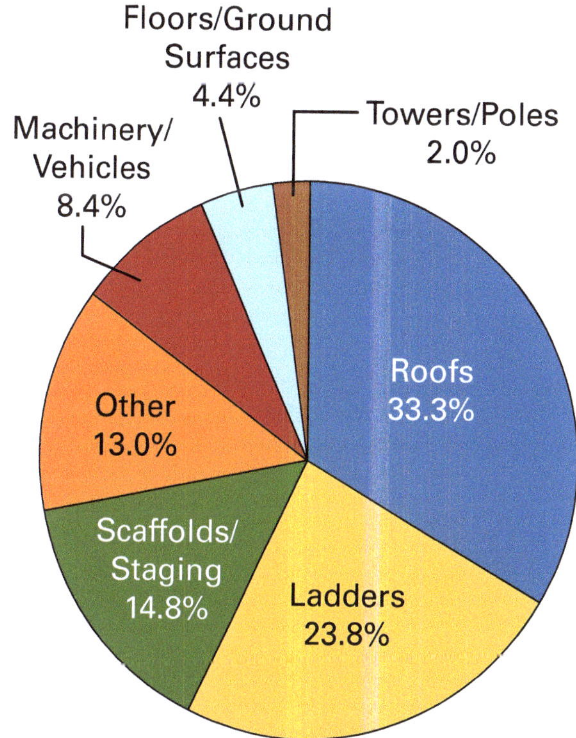

Figure 14 Causes of death from falls.

3.1.0 Fall Protection Systems

Employers are required to implement fall protection systems any time an employee is exposed to a walking or working surface at elevations of 6' (1.8 m) or greater. These systems must be prepared by a qualified safety coordinator who, through a degree or extensive experience, has demonstrated a successful ability to plan for and resolve issues concerning the safety requirements of a new roofing project. Some common fall protection systems include the following:

- **Personal fall arrest system (PFAS)**
- **Guardrails**
- Guarding screens
- Warning lines
- **Safety monitoring**
- **Safety nets**

3.1.1 Fall Arrest and Travel Restraint

Personal fall arrest systems (*Figure 15*) are a broad category of systems used to catch an employee in the case of a fall. These systems usually include a body harness and a secure anchorage point, and a connecting device between them (such as a lanyard). Fall arrest systems also include a deceleration device to stop the fall. Fall arrest systems must be inspected before each use to ensure they are functioning properly.

A travel restraint system uses a lanyard, a lifeline, and a safety harness or belt to prevent an employee from accidentally wandering to an unguarded edge and risking a fall. This system hooks a worker's safety harness to a lifeline either directly or through a rope grab. Travel-restraint systems are adjusted to allow employees range over the areas that require their attention while preventing them from passing into a hazardous zone.

Figure 15 PFAS.

While travel restraint systems prevent falls by restricting movement, fall arrest systems protect an employee after a fall occurs by slowing down and stopping the fall before the employee reaches the ground (*Figure 16*).

All personal fall arrest systems must have the ABCs of Fall Arrest Systems (*Figure 17*):

- *Anchorage/anchorage connector* — Anchorage connectors secure a worker's lanyard or lifeline to the anchorage point. While planning a fall arrest system, the anchorage connector must be carefully selected and placed. It must be easily accessible, safely located, and capable of supporting at least 5,000 lb (2,250 kg) per employee attached in case of a fall.
- *Body wear* — A full-body harness that evenly distributes potential fall force over the user's upper thighs, chest, and shoulders. This harness also provides a point of connection, such as a D-ring that connects the user to their linked device. A D-ring can always be found integrated into the back of a full-body harness (*Figure 18*); however, harnesses may also have additional D-rings located around the shoulder, hip, or sternum areas.

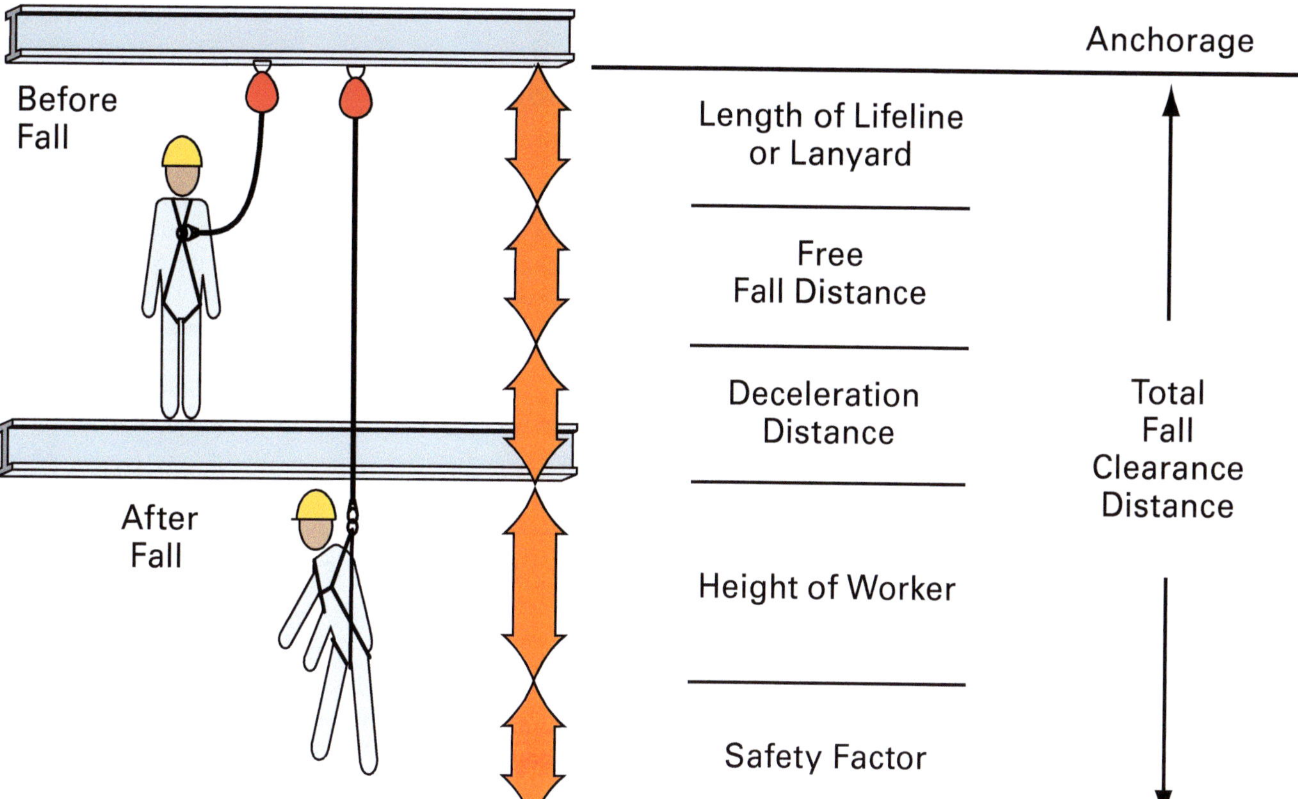

Figure 16 PFAS in action.

Figure 17 ABCs of personal fall arrest systems.

- *Connecting devices* — Connecting devices, such as a lanyard or lifeline, connect a user's body harness to a designated anchorage point (*Figure 19*). If a fall occurs, the connecting device contains a braking mechanism that typically activates within three feet of a drop to reduce movement impact and prevent swinging falls.

Even after a PFAS arrests a fall, the suspended employee's safety remains at risk until he or she is lowered to the ground. Suspension trauma is one of the most significant risks to workers awaiting or attempting rescue. Therefore, every elevated jobsite should have an established rescue and retrieval plan prepared by a qualified individual.

Some basic options for these final steps are self-rescue, assisted self-rescue, and mechanically aided rescue. Self-rescue is when workers pull themselves to safety. An example of assisted self-rescue is a coworker fetching the suspended employee a ladder to safely descend. Mechanically aided rescue may involve using a lift to retrieve a worker after their fall. Those at the worksite must assess each rescue situation to select the safest next steps.

Suspension Trauma

Suspension trauma can occur after a fall has been prevented by a PFAS and the worker is suspended in a hanging position while awaiting rescue. Since a full-body harness works by supporting body weight through its leg straps, the worker may experience the slowing or stopping of blood circulation to the legs as the straps grip them tightly. This reduction of circulation forces the heart to work harder to provide blood to the body's vital organs. Suspension trauma can lead to circulatory shock, potentially resulting in a loss of consciousness, nausea, and a drop in blood pressure and heart rate.

OSHA reports that suspension trauma could be fatal within a short time after a fall. A fall protection plan that accounts for this dangerous window of time is an essential safety component that must be considered when preparing for the rescue of suspended workers.

3.1.2 Guardrails

Guardrails (*Figure 20*) are placed around the perimeter of a construction area to provide a safety barrier that prevents falls from the roof's edge. A rooftop worksite should have a railing system running along its open perimeters.

Always consider the following requirements when installing guardrail systems:

- Guardrails must rise between 39" and 45" (99 and 115 cm) above the working surface.
- Midrails should be installed halfway between the working surface and top rail.
- Any openings in the guardrail perimeter should be blocked off when they are not in use.
- All guardrails must be smooth to prevent injury or clothes from snagging.
- Toprails must be able to withstand at least 200 lb (90 kg) of force.

3.1.3 Guarding Screens

Guarding screens are used to prevent falls through openings on the roof's surface (*Figure 21*). Falls through roofs are a major roofing safety hazard. Guarding screens add barriers to open areas of the roof, such as holes created for skylight or chimney installations (*Figure 22*).

Figure 19 Body harness connected to anchorage point.

Figure 18 D-ring on a body harness.

Figure 20 Guardrails.

Figure 21 Skylight guarding screen.

- If mechanical equipment is not in use, the warning line must be at least 6' (1.8 m) from every edge.

3.1.5 Safety Monitoring

In the instance that a worker needs to perform tasks outside a warning line, a safety monitor should be designated. This safety monitor is an individual who has demonstrated competency in recognizing fall hazards and can communicate effectively with their coworkers when a potential danger is noted. Make sure to check with local regulations for limits on how many workers can be observed by a monitor at one time. It is crucial that the safety monitor observes from the same working level as those he or she is monitoring. No additional responsibilities should be assigned to a safety monitor that may distract them from the monitoring task at hand. Once all workers have safely returned inside a secured perimeter, the safety monitor may resume work on other tasks.

3.1.6 Safety Nets

Safety nets (*Figure 24*) should be installed as closely as possible to the working surface and placed no more than 30' (9 m) below a potential fall point. Net openings should not exceed 6" (15 cm) on any side, and there must be a border rope surrounding the perimeter of net webbing that can withstand up to 5,000 lb (2,230 kg).

Safety nets must be inspected weekly and after each use to ensure there is no damage that would prevent them from working correctly. They must also have sufficient space underneath the area where they are installed to prevent a falling worker or object from contacting any surface other than the net. While safety nets are not commonly used in the roofing industry, they may be used on a jobsite by other trades.

Figure 22 Guarding screen.

3.1.4 Warning Lines

Warning-line systems (*Figure 23*), commonly used on low-slope roofs, place a barrier around all sides of a roof work area to caution employees when they are nearing an unprotected edge. Warning lines must be flagged every 6' (1.8 m) with high-visibility material and stand no less than 34" (86 cm), with their highest point reaching no more than 39" (99 cm) from the working space.

OSHA mandates that warning lines be constructed with ropes, wires, or chains and have a tensile strength of 500 lb (230 kg). In some circumstances, warning lines may be used in place of or in combination with guardrails and PFAS.

Use the following guidelines when dealing with mechanical equipment in a warning-line system:

- No mechanical equipment should be kept outside the warning-line perimeter.
- If mechanical equipment is being used, OSHA requires the warning line that is positioned perpendicularly to the mechanical equipment's direction of travel to be at least 10' (3 m) from the roof's edge.

NCCER – *Roofing*

Figure 23 Warning-line system.

Figure 24 Safety-net system.

3.2.0 Fall Protection on Low- and Steep-Slope Roofs

Various kinds of fall protection are required depending on the slope of the roof. OSHA defines low- and steep-slope roofs as follows:

- *Low-slope* — 4:12 slope or less
- *Steep-slope* — Above 4:12 slope

3.2.1 Low-Slope Roofs

Depending on the project, low-slope roof systems may require a range of fall protection methods. Some projects benefit from a combination of available safety options; for example, a low-slope roofing safety plan may combine guardrails with a warning-line system.

Warning-line systems are usually required to be placed around the entire low-slope work zone, whether that area encompasses the whole roof or just a section of it. If the roof is 50' (15.2 m) wide or less, a safety-monitoring system may be used alone in place of warning lines.

If a low-slope roof requires a worker's attention outside of the warning-line zone, it is crucial that an appointed safety monitor supervise that worker until he or she returns to the area secured by warning lines.

3.2.2 Steep-Slope Roofs

Since steep slopes add an extra dimension, there are fewer safety options available for steep-slope construction than for low-slope. Steep-slope roofing reduces available fall protection systems to just include PFAS, safety-net systems, and guardrail systems with toeboards. It may also be difficult or impossible to stand on steep-slope roofs without additional support and fall protection. Chicken ladders and roof brackets provide roofers with a platform where they can step.

If a guardrail system is used for fall protection in steep-slope roofing, OSHA mandates that additional toe boards (*Figure 25*) must be erected along the perimeters of a working surface to prevent tools or debris from sliding down the roof and creating a fall hazard to those walking below the working area.

Figure 25 Guardrails with toe boards.

3.3.0 Ladder Safety

Ladders are used daily to perform work in elevated locations. You can reduce safety risks by carefully inspecting ladders before you use them and by using them properly (*Figure 26*).

Basic components of a portable straight ladder are shown in (*Figure 27*). Portable ladders must be capable of withstanding four times their maximum intended loads (unless they are of extra heavy-duty type IA or IAA metal or plastic, in which case they must support 3.3 times their maximum intended loads).

The NIOSH Ladder Safety app (*Figure 28*) is available to download for free and designed to improve extension and ladder safety. This app, designed for smart devices, provides guides and tools to prevent common causes of safety issues with ladders.

DOs

- Be sure your ladder has been properly set up and is used in accordance with safety instructions and warnings.
- Wear shoes with non-slip soles.

- Keep your body centered on the ladder. Hold the ladder with one hand while working with the other. Never let your belt buckle pass beyond either ladder rail.

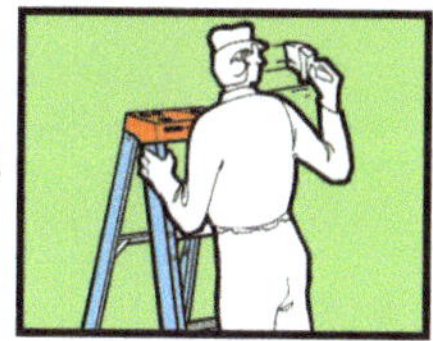

- Move materials with extreme caution. Be careful pushing or pulling anything while on a ladder. You may lose your balance or tip the ladder.

- Get help with a ladder that is too heavy to handle alone. If possible, have another person hold the ladder when you are working on it.

- Climb facing the ladder. Center your body between the rails. Maintain a firm grip.
- Always move one step at a time, firmly setting one foot before moving the other.

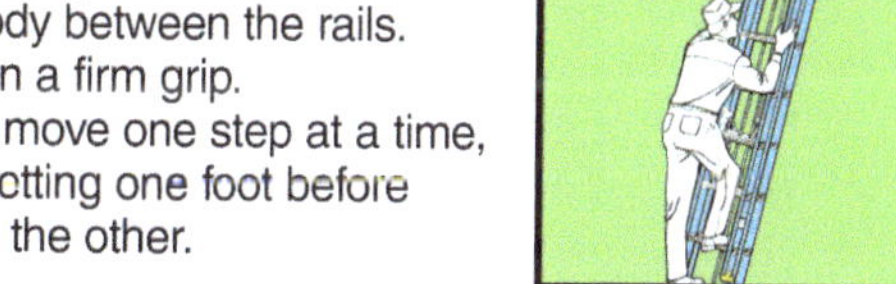

- Haul materials up on a line rather than carry them up an extension ladder.
- Use extra caution when carrying anything on a ladder.

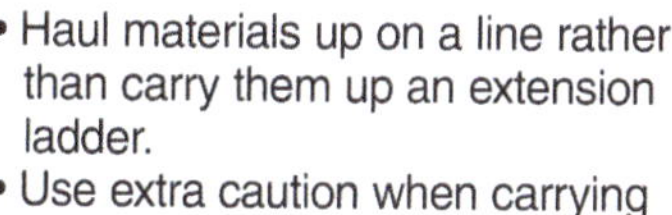

Read ladder labels for additional information.

DON'Ts

- DON'T stand above the highest safe standing level.
- DON'T stand above the second step from the top of a stepladder and the 4th rung from the top of an extension ladder. A person standing higher may lose their balance and fall.

- DON'T climb a closed stepladder. It may slip out from under you.
- DON'T climb on the back of a stepladder. It is not designed to hold a person.

- DON'T stand or sit on a stepladder top or pail shelf. They are not designed to carry your weight.
- DON'T climb a ladder if you are not physically and mentally up to the task.

- DON'T exceed the Duty Rating, which is the maximum load capacity of the ladder. Do not permit more than one person on a single-sided stepladder or on any extension ladder.

- DON'T place the base of an extension ladder _too close_ to the building as it may tip over backward.
- DON'T place the base of an extension ladder _too far away_ from the building, as it may slip out at the bottom.

Please refer to the 4 to 1 Ratio Box.
- DON'T over-reach, lean to one side, or try to move a ladder while on it. You could lose your balance or tip the ladder.

Climb down and then reposition the ladder closer to your work!

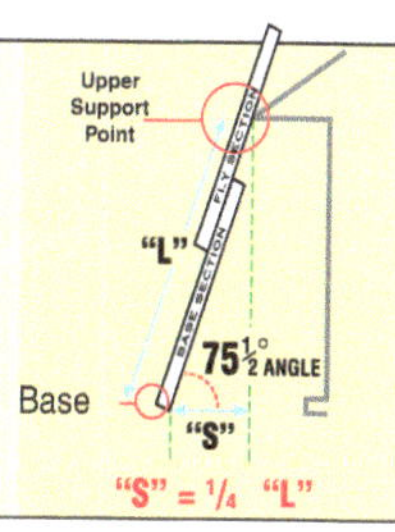

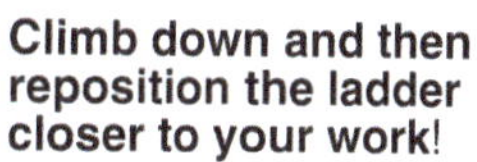

Figure 26 Ladder safety Dos and Don'ts.

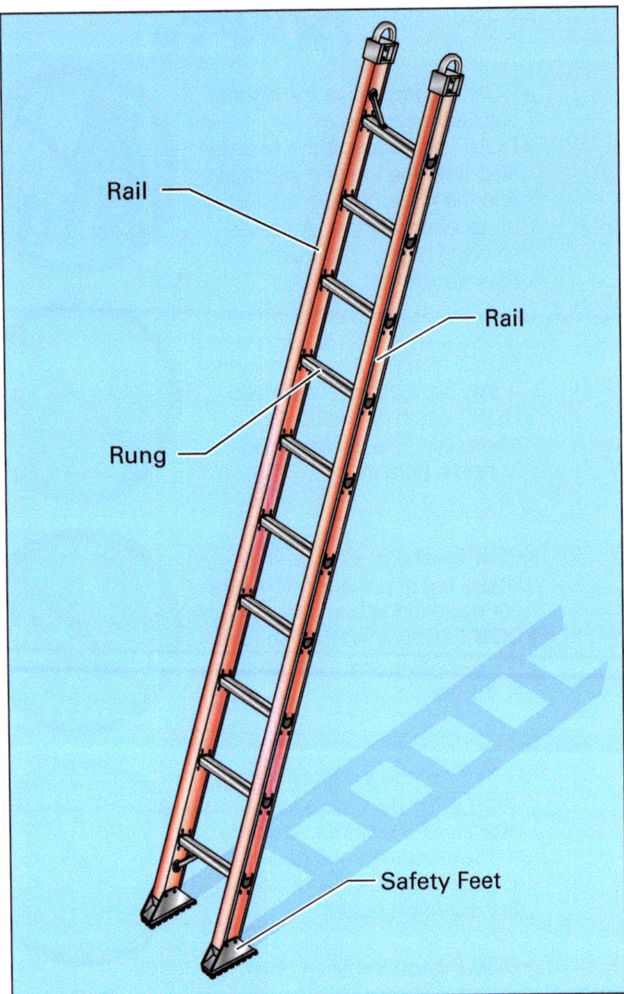

Figure 27 Components of a ladder.

Use the following general guidelines when inspecting, setting up, and using ladders:

- Rungs and Cleats
  - The rungs on portable ladders should be 10" to 14" apart.
  - The rungs on step ladders should be 8" to 12" apart.
  - The rungs on extension ladders should be 8" to 18" apart on the base sections and 6" to 12" on the extension sections.
  - The rungs should be shaped or coated so they are slip-resistant.

- Siderails
  - The siderails on all ladders should be at least $11\frac{1}{2}$" apart.

- Maintenance
  - Ladders should be free of all slipping hazards, such as oil and grease.
  - The ladder should be inspected by a competent person on a regular basis.
  - None of the ladders in service should have any defects.
  - All repaired ladders should be restored to their original condition.

- Setup
  - The ground should be stable and level.
  - If the surface is not stable, the feet should be secured.
  - Barricades should be placed where pedestrians, machines, or equipment could displace the base of the ladder.
  - The ladder's base should be 1' away from the building for every 4' of the building's height (or for every 8' if the ladder is job-made).
  - The top of the ladder should extend 3' higher than the roof and be secured and equipped with a secure grab rail.
  - The areas at the top and bottom of the ladder should be clear of debris or other objects.
  - The rails should be balanced equally against the upper level.

- Use
  - There should be no electrical wires in areas where portable metal ladders are being used.
  - Employees must face the ladder, maintain one hand on the ladder at all times, and not carry difficult loads when they climb up or down the ladder. Use backpacks or methods such as laddervators to leave your hands free for climbing.
  - Employees must not use the top of a stepladder as a step or climb the cross bracing.
  - Ladders must be removed at the end of the day to prevent access to the roof by others.

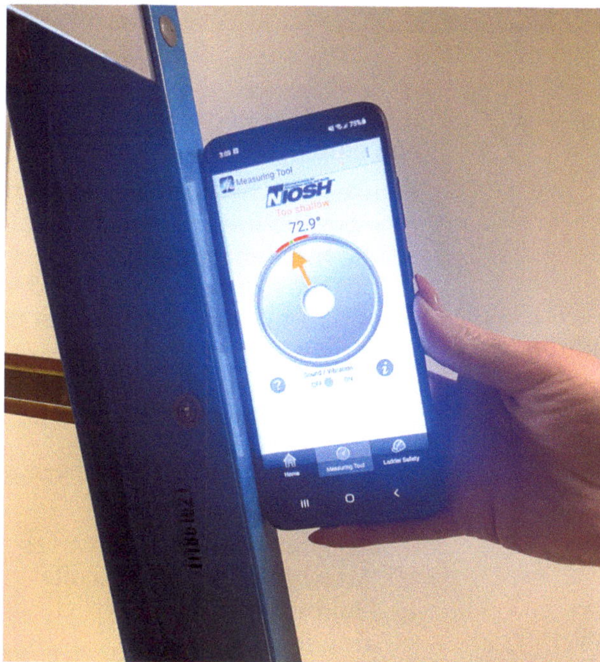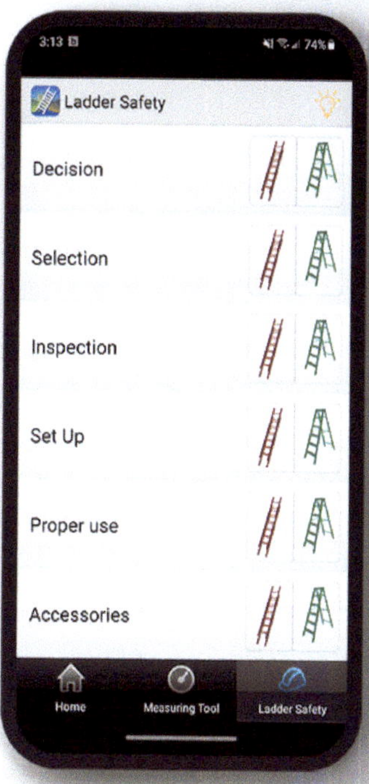

Figure 28 NIOSH Ladder Safety App.

3.0.0 Section Review

1. Which fall safety system carries the risk of suspension trauma?

   a. Guardrail systems
   b. Personal fall arrest systems
   c. Safety-monitoring systems
   d. Safety-net systems

2. If a low-slope roofing system requires a worker's attention outside of the warning-line zone, ______.

   a. the workers must extend the warning-line zone to encompass the area that requires attention
   b. a worker may travel beyond the warning-line zone under the supervision of a safety monitor
   c. the workers may travel beyond the warning-line zone if they promptly address the issue
   d. no actions should be taken until a safety coordinator creates a new plan

3. Which of the following guidelines applies when using ladders?

   a. Always maintain at least one hand on the ladder when climbing.
   b. Ladders must be used any time you ascend or descend more than 6'.
   c. Ladders must always be placed at a 45-degree angle.
   d. The top of the ladder should not extend more than 2' above the edge of the roof.

1. Equipment worn to prevent or reduce injuries is called _____.

    a. DOL
    b. OSHA
    c. PPE
    d. JSA

2. Safety regulations for the construction industry can be found in _____.

    a. *29 CFR 1926*
    b. *29 CFR 1820*
    c. *42 CFR 600*
    d. *49 CFR 105*

3. For power line voltages of 50kV or less, OSHA requires which minimum distance to be maintained between equipment and power lines?

    a. 5' (1.5 m)
    b. 10' (3 m)
    c. 20' (6 m)
    d. 30' (9.1 m)

4. An approach to hazard recognition that breaks down a job into tasks, identifies hazards associated with each task, and identifies preventive measures for each hazard is called a(n) _____.

    a. JSA
    b. SDS
    c. Hierarchy of controls
    d. HAZCOM

5. A slide through which materials are safely dropped from a roof is called a(n) _____.

    a. D-ring
    b. debris box
    c. safety net
    d. enclosed chute

6. Suspected asbestos-containing materials discovered during work *must* be _____.

    a. immediately reported to a supervisor
    b. carefully disposed of by the roofing workers
    c. handled by the fire department
    d. removed as soon as they are discovered

7. If a liquefied petroleum (LP) gas fire breaks out, employees should _____.

    a. use a fire hydrant to put out the flames
    b. evacuate the area immediately and call the fire department
    c. call the fire department and work together to control the flames until help arrives
    d. wait for instructions from a safety coordinator on how to proceed

8. Which of the following is the safest way of transporting LP cylinders?

    a. Attaching lines to the valves
    b. Attaching lines to the collars
    c. Holding the top of the cylinder and dragging it
    d. Using a hoisting cage or cradle

9. Which of the following guidelines should be followed when installing torch-applied roofing materials?

    a. Always keep propane cylinders close by.
    b. Never torch near vents or air intakes.
    c. Always light the torch with a match.
    d. When you are finished with a torch, use a lighter to burn off remaining propane in the hose.

10. Employers are required to implement a fall protection system in the construction industry any time an employee is exposed to a walking or working surface at elevations of _____.

    a. 4' or greater (1.2 m or greater)
    b. 5' or greater (1.5 m or greater)
    c. 6' or greater (1.8 m or greater)
    d. 7' or greater (2.1 m or greater)

11. The D-ring can *always* be found at which part of a body harness?

    a. Shoulders
    b. Back
    c. Hips
    d. Sternum

12. Which of the following fall protection systems provides a safety barrier that must rise
between 39" and 45" (99 and 115 cm) above
the working surface?

a. Personal fall arrest systems (PFAS)
b. Guardrail systems
c. Safety-monitoring systems
d. Safety-net systems

13. Which of the following fall protection systems requires that there is sufficient space
underneath the installation area to prevent
falling workers or items from hitting an additional surface?

a. Warning lines
b. Guardrail systems
c. Safety-monitoring systems
d. Safety-net systems

14. How often should safety nets be inspected
for damage?

a. Daily and after each use
b. Weekly and after each use
c. Monthly and after each use
d. Annually and after each use

15. Which of the following are the safety systems available for steep-slope roofing
projects?

a. Safety monitoring, PFAS, and guardrails
with toeboards
b. Safety monitoring, guardrails with toeboards, and warning lines
c. Warning lines, guardrails with toeboards,
and PFAS
d. PFAS, safety nets, and guardrails with
toeboards

Fill in the blank with the correct term learned from your study of this module.

1. The agency whose mission is to set and enforce safety standards for all places of employment is called the ________________.

2. Clothing used to prevent or reduce injuries is called ________________.

3. A form that lists hazards, safe handling practices, and emergency control measures for a specific substance is called a(n) ________________.

4. A serious medical condition that can occur when a worker is suspended for a short time in a body harness is called ________________.

5. A(n) ________________ is a safety device that activates during a fall to catch workers and prevent them from hitting the ground.

6. ________________ are installed around the perimeter of a work site to prevent falls from a roof's edge.

7. A safety system that assigns monitoring duties to a competent person is called ________________.

8. The angle of a roof's surface, expressed as a ratio of the units of vertical rise to the units of horizontal run, is called ________________.

9. A component consisting of a flexible line that serves as a means of connecting other components of a personal fall arrest system to the anchorage is called a(n) ________________.

10. ________________ are installed around and underneath a working area to catch falling objects or workers.

11. Any construction passing through a roof or waterproofing system (such as a pipe, conduit, or HVAC support) is called a(n) ________________.

Trade Terms

Guardrails
Lifeline
Occupational Safety and Health
   Administration (OSHA)

Penetration
Personal fall arrest system (PFAS)
Personal protective equipment (PPE)
Safety data sheet (SDS)

Safety monitoring
Safety nets
Slope
Suspension trauma

Cornerstone of Craftsmanship

Rich Trewyn
Subject Matter Expert
NRCA

How did you choose a career in the industry?

After graduating with a degree in Occupational Safety and Health, I went on to work for a consulting firm. I was asked to train a group of roofing professionals at one point, and the rest is history. Shortly after that, I was asked to work for the company where I had previously done training.

Who inspired you to enter the industry?

My father. He was in safety before me, and I aspired to be more like him. I was very fortunate to have worked with him before he passed away.

What types of training have you been through?

I have been through every type of training you can imagine, from roles in leadership to chemical-hazard recognition. Training in the safety industry has a wide variety of topics, and I have gotten to be a part of most of those.

How important is education and training in construction?

Very important, period. A well-trained workforce is the key to a successful industry.

What kinds of work have you done in your career?

My first work began shortly after college. I worked as a consultant and performed a variety of training programs, including programs for roofing workers. After that, I began my career in the roofing industry, starting with a local company with offices in Milwaukee, WI and Chicago, IL. That company grew into one of the nation's largest roofing companies, and I took on a new role. After a change of pace and a couple of years in the manufacturing industry, I took the job at NRCA.

Tell us about your present job.

My position as the Director of Enterprise Risk Management at the National Roofing Contractors Association (NRCA) supports the development of NRCA's enterprise risk management initiative. This includes support for the Enterprise Risk Management Task Force and related project work. My position supports the department's safety programs, products and services, as well as development and execution where necessary. The position also provides internal safety advisement around office functions, procedures, and policies.

This position is an evolving one where the manager will research, design, and when appropriate, direct activities that aid in the understanding of enterprise risk management while supporting committee activity that advances the association's ability to assist members' efforts to implement effective ERM programs. For example, developing and delivering ERM efforts that educate members via written materials (such as articles) and/or educational vehicles (such as videos or on-site classes).

What do you enjoy most about your job?

The interaction with our members is my favorite part of the job. When you find a solution for something that a member thought was impossible to solve, it is an extremely rewarding experience. My start in the industry began as a consultant which had me teaching a variety of subjects to several different groups. That experience molded me with the confidence to take on any challenge in front of me.

Would you suggest construction as a career to others? Why?

Yes, I have recommended it as a career to many people, including my own kids. The variety of challenges and day-to-day changes make for non-stop action on the job.

What advice would you give to those new to the field?

Take advantage of any training offered. Soak up and absorb anything you can. Construction is so diverse that it is good to have knowledge about a variety of subject matters if possible.

Interesting career-related fact or accomplishment:

One career-related accomplishment that I am very proud of is that before the age of 40, I became the National Safety Director for the nation's largest roofing company and managed over 30 safety professionals.

How do you define craftsmanship?

In roofing, craftsmanship is not easily definable. There are so many skills and such a diverse subject field that to be a true craftsman is to have a wide variety of knowledge and skills in all areas of the roofing trade. That is not to say that one cannot be a craftsman in one specific designation.

Trade Terms Introduced in This Module

Guardrails: Railings installed around the perimeter of a worksite to prevent falls from a roof's edge.

Lifeline: A component consisting of a flexible line that is either connected vertically to an anchorage at one end (vertical lifeline) or connected horizontally to an anchorage at both ends (horizontal lifeline). A lifeline serves as a means for connecting other components of a personal fall arrest system to the anchorage.

Occupational Safety and Health Administration (OSHA): An agency of the US Department of Labor whose mission is to set occupational safety and health standards for all places of employment, enforce these standards, ensure that employers provide and maintain a safe workplace for all employees, and provide research and educational programs to support safe working practices.

Penetration: Any construction, such as pipes, conduits, or HVAC supports, passing through a roof or waterproofing system.

Personal fall arrest system (PFAS): A system that activates during a fall to catch a worker and prevent them from hitting the ground. PFAS include multiple components, including a body harness, anchorage device, lifeline, and connectors.

Personal protective equipment (PPE): Equipment or clothing designed to prevent or reduce injuries.

Safety data sheet (SDS): A form that lists the hazards, safe handling practices, and emergency control measures for a specific substance. Also called a *material safety data sheet (MSDS)*.

Safety monitoring: A safety system that assigns monitoring duties to a competent person who has the knowledge and authority to make safety decisions.

Safety nets: Nets installed around and underneath a working area that can catch falling objects or workers.

Slope: The angle of a roof surface, usually expressed as a ratio of vertical rise to horizontal length (sometimes referred to as *run*). When dimensions are given in inches, slope may be expressed as a ratio of rise over a distance of 12" (for example, 4:12) or as an angle in degrees.

Suspension trauma: A serious medical condition that can occur when a worker is suspended for a short time in a body harness.

Additional Resources

This module presents thorough resources for task training. The following reference material is suggested for further study.

29 CFR 1910, **www.ecfr.gov**.

29 CFR 1926, **www.ecfr.gov**.

CPWR (The Center for Construction Research and Training), Mental Health and Addiction Resources, **www.cpwr.com/research/research-to-practice-r2p/r2p-library/other-resources-for-stakeholders/mental-health-addiction**.

National Roofing Contractors Association (NRCA), **www.nrca.net**.

NCCER Module 75901, *Fall Protection Orientation*.

Occupational Outlook Handbook, Roofers, Bureau of Labor Statistics, US Department of Labor, **www.bls.gov/ooh/construction-and-extraction/roofers.htm**.

Occupational Safety and Health Administration (OSHA), US Department of Labor, **www.osha.gov**.

OSHA 3755-05, *Protecting Roofing Workers*. US Department of Labor. **www.osha.gov/Publications/OSHA3755.pdf**.

Figure Credits

Courtesy of the National Roofing Contractors Assoc., Module Opener, Figures 9, 13, 15, 17–19, 21–23, 25

GoCanvas.com, Figure 4

"N95醫療級防疫口罩" by Mark Chang is licensed under Creative Commons Attribution 2.0 Generic (CC BY 2.0)., Figure 6

iStock@sturti, Figure 7

iStock@Jens_Lambert_Photography, Figure 8

Courtesy of Guardian Fall Protection, Figure 20

elcosh.org, Figure 24

Section Review Answer Key

SECTION 1.0.0

Answer	Section Reference	Objective
1. c	1.0.0	1
2. a	1.2.2	1b
3. b	1.3.0	1c
4. d	1.4.0	1d

SECTION 2.0.0

Answer	Section Reference	Objective
1. b	2.1.0	2a
2. c	2.1.0	2a
3. d	2.2.0	2b

SECTION 3.0.0

Answer	Section Reference	Objective
1. b	3.1.1	3a
2. b	3.2.1	3b
3. a	3.3.0	3c

This page is intentionally left blank.

NCCER CURRICULA — USER UPDATE

NCCER makes every effort to keep its textbooks up-to-date and free of technical errors. We appreciate your help in this process. If you find an error, a typographical mistake, or an inaccuracy in NCCER's curricula, please fill out this form (or a photocopy), or complete the online form at **www.nccer.org/olf**. Be sure to include the exact module ID number, page number, a detailed description, and your recommended correction. Your input will be brought to the attention of the Authoring Team. Thank you for your assistance.

Instructors – If you have an idea for improving this textbook, or have found that additional materials were necessary to teach this module effectively, please let us know so that we may present your suggestions to the Authoring Team.

NCCER Product Development and Revision

13614 Progress Blvd., Alachua, FL 32615

Email: curriculum@nccer.org
Online: www.nccer.org/olf

❑ Trainee Guide ❑ Lesson Plans ❑ Exam ❑ PowerPoints Other ___________________

Craft / Level: ____ Copyright Date: __________

Module ID Number / Title: ____

Section Number(s): ____

Description: ____

____

____

____

____

Recommended Correction: ____

____

____

____

____

Your Name: ____

Address: ____

____

Email: ____ Phone: _____________________

This page is intentionally left blank.

Fall Protection Orientation

OVERVIEW

Fall protection is more than wearing a harness on the jobsite. Workers are exposed to many potentially life-threatening hazards on job sites, and death by falling is one of the greatest risks. Proper training of every craft professional, coupled with enforcement of safety regulations and standards, is the most effective way to reduce this risk and protect the lives and well-being of the construction workforce.

Module 75901

Trainees with successful module completions may be eligible for credentialing through NCCER's Registry. To learn more, go to **www.nccer.org** or contact us at 1.888.622.3720. Our website has information on the latest product releases and training, as well as online versions of our *Cornerstone* magazine and Pearson's product catalog.

Your feedback is welcome. You may email your comments to **curriculum@nccer.org**, send general comments and inquiries to **info@nccer.org**, or fill in the User Update form at the end of this module.

This information is general in nature and intended for training purposes only. Actual performance of activities described in this manual requires compliance with all applicable operating, service, maintenance, and safety procedures under the direction of qualified personnel. References in this manual to patented or proprietary devices do not constitute a recommendation of their use.

75901 V1

Fall Protection Orientation

Objectives

When you have completed this module, you will be able to do the following:

1. Describe the importance of safety and the process of hazard recognition and control.
   a. Explain the importance of safety awareness and how to develop a safety-conscious attitude.
   b. Identify methods of recognizing, reducing, and preventing hazards to prevent workplace incidents and accidents.
2. Identify the reasons for the OSHA regulations that govern the elevated work in construction.
   a. Understand the role of OSHA in regulating the construction workplace and know where to find regulations, training, and safety requirements for elevated work.
3. Identify incidents, accidents, causes, costs, and consequences related to falls in the workplace.
   a. Identify the categories of workplace incidents and accidents.
   b. Describe the costs of workplace incidents and accidents.
   c. Describe the causes of workplace incidents and accidents and identify hazards that lead to falls.
4. Identify the safety concerns and requirements related to elevated work.
   a. Identify the work areas most likely to present fall hazards.
   b. Identify safety precautions related to elevated masonry work.
   c. Identify safety precautions related to elevated steel erection work.
5. Identify the appropriate fall protection and lifesaving equipment and describe their proper use.
   a. Describe the procedures for using fall protection equipment.
   b. Describe the requirements for anchor points.
   c. Describe the requirements for full-body harnesses and identify the steps for donning a harness.
   d. Describe the requirements for lanyards, lifelines, and deceleration devices.
   e. Identify barriers, guardrails, safety nets, and climbing devices used to prevent falls.
6. Identify and describe the safe use of ladders, stairs, and scaffolds.
   a. Describe the hazards related to use of ladders and stairs in the workplace.
   b. Identify the safety requirements and proper use of various types of ladders and stairs.
   c. Identify types of scaffolds and describe how to use them safely.
7. State the guidelines for the safe operation of aerial lifts.
   a. Identify the types of aerial lifts and precautions for safe use.

Performance Tasks

This is a knowledge-based module. There are no Performance Tasks.

Trade Terms

Accident
Aerial lifts
Body harness
Capacity
Competent person
Controlled access zone (CAZ)
Controlled decking zone (CDZ)
Cut
Excavations
Free fall
Guarded
Hand line
Incident
Lanyard
Lifeline
Limited access zones
Management system

Maximum intended load
Midrail
Occupational Safety and Health Administration
  (OSHA)
On-the-job learning (OJL)
Personal fall arrest system (PFAS)
Personal protective equipment (PPE)
Planked
Putlogs
Qualified person
Safety culture
Scaffolding
Self-retracting lanyard (SRL)
Six-foot rule
Tensile strength
Top rail

Industry Recognized Credentials

If you are training through an NCCER-accredited sponsor, you may be eligible for credentials from NCCER's Registry. The ID number for this module is 75901. Note that this module may have been used in other NCCER curricula and may apply to other level completions. Contact NCCER's Registry at 888.622.3720 or go to **www.nccer.org** for more information.

Contents

1.0.0 WORKPLACE SAFETY AND HAZARD CONTROL

Objective

Describe the importance of safety, the causes of workplace incidents, and the process of hazard recognition and control.

a. Explain the importance of safety awareness and how to develop a safety-conscious attitude.
b. Identify methods of recognizing, reducing, and preventing hazards to prevent workplace incidents and accidents.

Trade Terms

Accident: As defined by OSHA, an unplanned event that results in personal injury or property damage.

Competent person: As defined by OSHA, one who is capable of identifying existing and predictable hazards in the surroundings or working conditions which are unsanitary, hazardous, or dangerous to employees, and who has authorization to take prompt corrective measures to eliminate them.

Incident: As defined by OSHA, an unplanned event that does not result in personal injury but may result in property damage or is worthy of recording.

Occupational Safety and Health Administration (OSHA): An agency of the US Department of Labor whose mission is to assure safe and healthful working conditions for employees by setting and enforcing standards and by providing training, outreach, education and assistance.

On-the-job learning (OJL): The learning an apprentice acquires while working on the job under the supervision of journey workers. Commonly called *on-the-job training* (*OJT*).

Planked: Having pieces of material 2 in. (5 cm) thick or greater and 6 in. (15 cm) wide or greater used as flooring, decking, or scaffold decks.

Qualified person: As defined by OSHA, one who, by possession of a recognized degree, certificate, or professional standing, or who by extensive knowledge, training, and experience, has successfully demonstrated his ability to solve or resolve problems relating to the subject matter, the work, or the project.

Safety culture: The culture created when the whole company sees the value of a safe work environment.

Scaffolding: A temporary built-up framework or suspended platform or work area designed to support workers, materials, and equipment at elevated or otherwise inaccessible job sites.

Work at construction and industrial job sites can be hazardous. A job-site incident or accident may be caused by at-risk behavior, poor planning, lack of training, or failure to recognize the hazards. To help prevent incidents, every company must have a proactive safety program. Safety must be incorporated into all phases of the job and involve employees at every level, including management.

When you take a job, you have an obligation to your employer, co-workers, family, and yourself to work safely. You also have an obligation to make sure anyone you work with is working safely. Employers are likewise obligated to maintain a safe workplace for all employees. The ultimate responsibility for on-the-job safety, however, rests with you. Safety is part of everyone's job. In this module, you will learn to ensure your safety, and that of the people you work with, by obeying the following rules:

- Follow safe work practices and procedures, both regulatory and corporate.
- Inspect safety equipment before use.
- Use safety equipment properly.

On a typical job site, there are often many workers from a variety of crafts in one place. These workers are all performing different tasks and operations. As a result, the job site is constantly changing, and hazards are continually emerging. These hazards can jeopardize your safety.

Safety training is provided to make you aware that hazards exist all around you every day. The time you spend learning and practicing safety procedures can save your life and the lives of others.

Safety is a learned behavior and attitude. It is a way of working that must be incorporated into the company as a culture. A safety culture is created when all the workers at a job site or in an organization see the value of a safe work environment and support it through their actions.

Creating and maintaining a safety culture is an ongoing process that includes a sound safety structure and attitude, and that relates to both organizations and individuals. Everyone in the company, from management to laborers, must be responsible for safety every day they come to work.

There are many benefits to having a safety culture. Companies with strong safety cultures usually have the following characteristics:

- Fewer at-risk behaviors
- Lower incident and accident rates
- Less turnover
- Less absenteeism
- Higher productivity

A strong safety culture can also improve a company's safety record, which leads to winning more bids and keeping workers employed. Contractors with poor safety records are sometimes excluded from bidding, so good safety performance is essential. Factors that contribute to a strong safety culture include the following:

- Embracing safety as a core value
- Strong leadership
- Establishing and enforcing high standards of performance
- The commitment and involvement of all employees
- Effective communication and commonly understood and agreed-upon goals
- Using the workplace as a learning environment
- Encouraging workers to have a questioning attitude and empowering them to stop work when faced with potential hazards
- Good organizational learning and responsiveness to change
- Providing timely response to safety issues and concerns
- Continually monitoring performance
- Positive reinforcement when proper safety practices are demonstrated by employees

1.1.0 Accident Prevention and Safety Awareness

According to the Bureau of Labor Statistics, there were 991 fatalities in the construction industry in 2016, and 384 were due to falling. A total of 93 workers were struck by an object, 82 were electrocuted, and 72 were caught or compressed in moving equipment. Most accidents, but not all, can be prevented by workers following safety procedures.

The proper use of fall protection equipment at all times helps prevent fatalities and injuries from falls. The most common circumstance of falling is from a roof. This is probably because many roofs are sloped and do not provide good anchorages for fall protection, especially in the case of residential roofs. Ladders, scaffolding, and staging locations present other common circumstances of falls. The relative instability of both supports makes them places where great caution should be employed. Tying off ladders and properly securing fall protection systems significantly reduces the risk of an accident due to falling.

The same rules apply to all forms of threats. Keep a watch on all directions, especially overhead. If a crane brings an object close, keep an eye on it. If a machine is operating nearby, be aware of the hazards that could occur. A spinning wheel or drum that catches a worker's sleeve or pants leg can pull the worker into it. An object being moved can strike a column or other obstruction and shift suddenly. Always be ready to stay out of the way of danger. Follow all safety procedures and remain alert at all times.

1.1.1 Developing Safety Awareness

As an employee, it is important to have an awareness of safety. Always understand the hazards that exist and the reasons for the safety rules. Practice safety procedures until they become second nature. The safe way is always the right way to perform any activity. The following is a partial list of hazards that construction workers may encounter:

- Falling from heights
- Being caught between a stationary object and a moving truck or railway car when unloading steel
- Being struck by falling materials, tools, and equipment or swinging hooks, chains, and cables
- Hearing damage from continuous exposure to high noise levels
- Burns and eye damage from welding, cutting, and chipping
- Sprains and strains from improper lifting and carrying methods
- Electrical shock from poorly grounded or defective tools and equipment
- Injuries due to defective tools and equipment
- Injuries due to improper tool or equipment use

Stay focused on the job being performed. If tired or distracted, immediately stop what you are doing. Just a moment's lapse in attentiveness when working above the ground can cause a serious accident. The same guidelines should be observed when approaching co-workers. Wait until they have finished a particular operation before distracting them.

NCCER – *Fall Protection Orientation*

1.1.2 Maintaining a Safety Conscious Attitude

Observe all warning signs. Use tools and equipment as prescribed and maintain good housekeeping. Safety conscious workers can greatly reduce risks by anticipating hazards and observing general safety precautions such as the following:

- Comply with Occupational Safety and Health Administration (OSHA) requirements for head and foot protection in hard hat zones.
- Locate and observe all warning and accident prevention signs, and confine movements only to the work area.
- Only enter restricted areas with permission from the superintendent or foreman.
- Anticipate the movement paths of all motorized equipment.
- Keep out from under raised loads, unless duties require it.
- Never ride on a load or crane hook.
- Walk around scaffolding areas. If it is necessary to work under raised platforms, be sure that the work site is covered and protected.

1.2.0 Hazard Recognition, Evaluation, and Control

The process of hazard recognition, evaluation, and control is the foundation of an effective safety program. When hazards are identified and assessed, they can be addressed quickly, reducing the hazard potential. Simply put, the more aware you are of your surroundings and the dangers in them, the less likely you are to be involved in an incident. Job sites must also have established rescue plans in place.

1.2.1 Hazard Recognition

There are many indicators of potential hazards. The best approach in determining if a situation or equipment is potentially hazardous is to ask these questions:

- How can this situation or equipment cause harm?
- What types of energy sources are present that can cause an incident?
- What is the magnitude of the energy?
- What could go wrong to release the energy?
- How can the energy be eliminated or controlled?
- Will I be exposed to any hazardous materials?

Before you can fully answer these questions, you need to know the different types of incidents and accidents that can happen and the energy sources behind them. Some of the different types of events that can cause injuries include the following:

- Falls on the same elevations or falls from elevations
- Being caught in, on, or between equipment
- Being struck by falling objects
- Contact with electricity, acid, heat, cold, radiation, pressurized liquid, gas, or toxic substances
- Being cut by tools or equipment
- Exposure to high noise levels
- Repetitive motion or excessive vibration

When equipment is the cause of an incident or accident, it is usually because there was an uncontrolled release of energy. The different types of energy sources that can be released include mechanical, pneumatic, hydraulic, electrical, chemical, thermal (heat or cold), radioactive, gravitational, and stored energy.

There are many ways to recognize hazards and potential hazards on a job site. Some techniques are more complicated than others. In order to be effective, they all must answer this question: What could go wrong with this situation or operation? No matter what hazard recognition technique you use, answering that question in advance will save lives and prevent equipment damage.

1.2.2 The Fatal Four

Twenty-one percent of the 4,693 worker fatalities in 2016 were in construction. Of those fatalities, 64 percent were caused by one of the "The Fatal Four" (*Figure 1*).

The Fatal Four include the following:

- *Falls* – Falls from elevation are incidents involving failure of, failure to provide, or failure to use appropriate fall protection.
- *Struck-by accidents* – Struck-by accidents involve unsafe operation of equipment, machinery, and vehicles, as well as improper handling of materials, such as through unsafe rigging operations.
- *Caught-in/between accidents* – Caught-in or caught-between accidents involve unsafe operation of equipment, machinery, and vehicles, as well as improper safety procedures at trench sites and in other confined spaces.

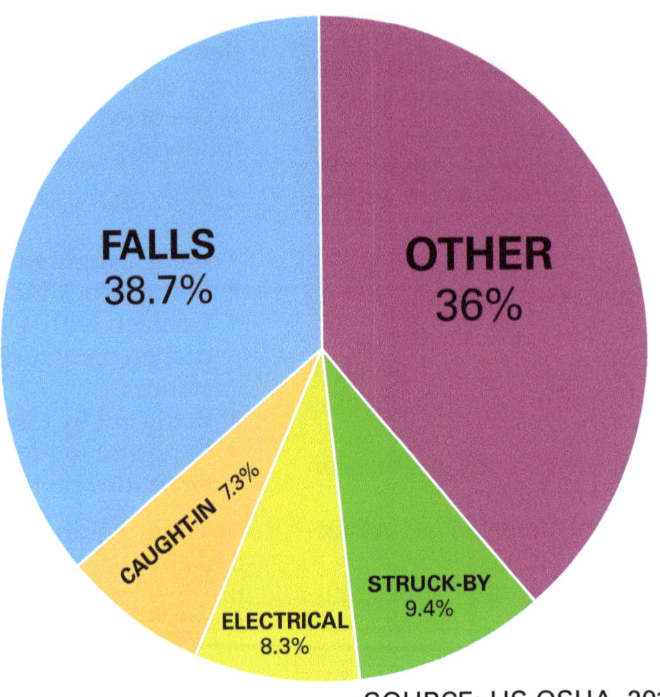

Figure 1 The four high-hazard areas.

- *Electrocution* – Electrical shock accidents involve contact with overhead wires; use of defective tools; failure to disconnect power source before repairs; or improper ground fault protection.

It is important for the craft worker to recognize that an incident or accident that initiates in one of the above hazard groups can quickly cascade into others. For instance, a worker on an elevated platform may be struck by an object falling from above and fall. In this case, situational awareness, personal protective equipment, and use of fall protection may make the difference between life and death. Safety measures which directly address each of these hazard groups have the potential to save hundreds of lives every year.

1.2.3 Job Safety Analysis (JSA) and Task Safety Analysis (TSA)

Performing a job safety analysis (JSA), also known as *job hazard analysis* (*JHA*), is one approach to hazard recognition. Another common technique is performing a task safety analysis (TSA), also called a *task hazard analysis* (*THA*).

In a JSA, the task at hand is broken down into its individual parts or steps. Each step is then analyzed for its potential hazards. Once a hazard is identified, certain actions or procedures are recommended that will correct that hazard.

For example, during a JSA, it is determined that using a chain hoist to install a pump motor in a tight space would be safer than having a worker do it manually. By using the chain hoist, the chance that the worker's hand would get crushed during installation is reduced. Using the JSA process saved the worker from injury. *Figure 2* shows an example of a form used to conduct a JSA.

JSAs can also be used as pre-planning tools. This helps to ensure that safety is planned into the job. You may be asked to take part in a JSA during job planning. When JSAs are used as pre-planning tools, they contain the following information:

- Tools, materials, and equipment needs
- Staffing or manpower requirements
- Duration of the job
- Quality concerns

A task safety analysis is similar to job safety analysis in that both require workers to identify potential hazards and needed safeguards associated with a job they are about to do. The difference is the form used to report the hazards. During a TSA, a pre-printed, fill-in-the-blank checklist is often used to document any hazard

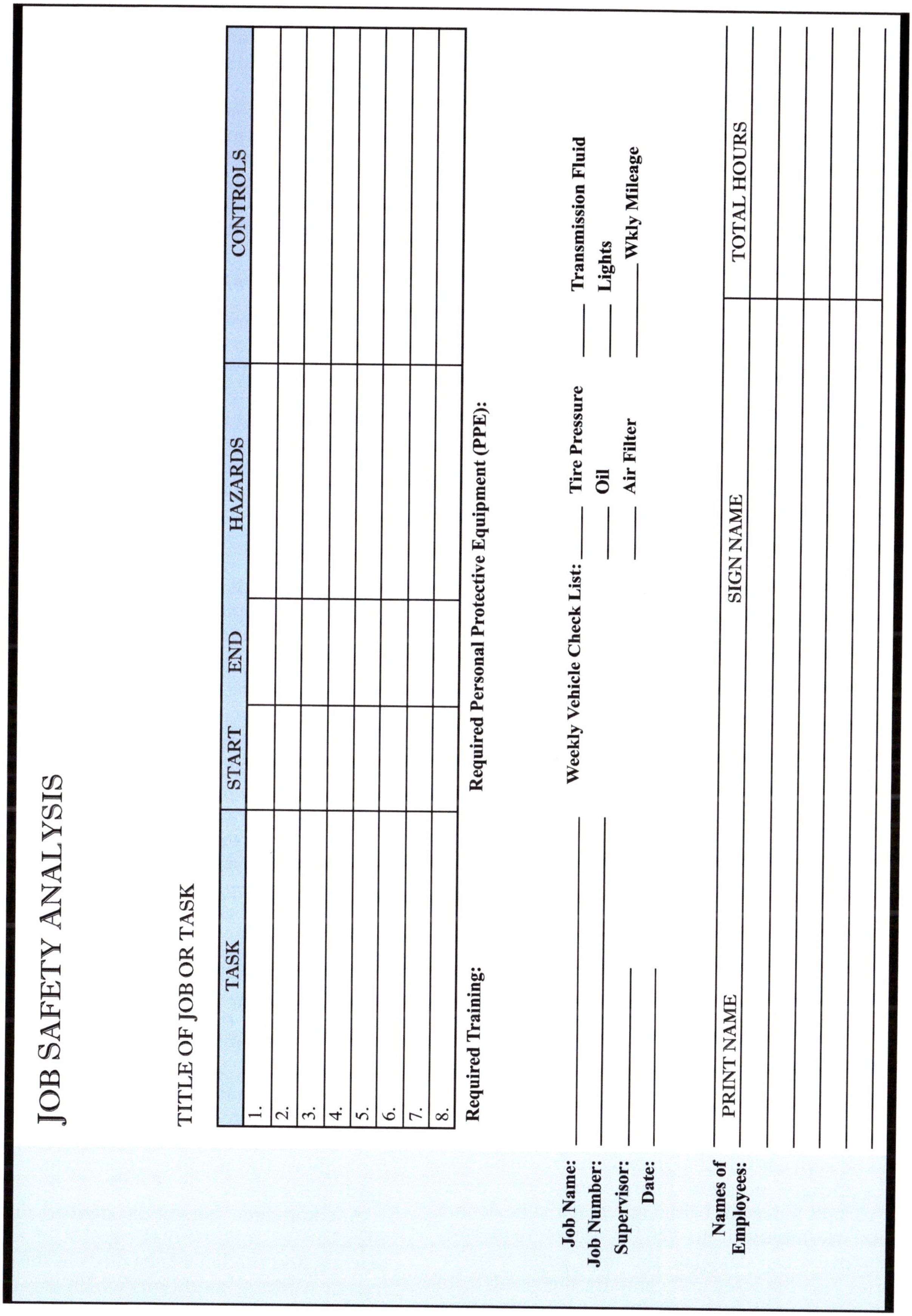

Figure 2 Job safety analysis form.

Job Safety Analysis

TITLE OF JOB OR TASK

TASK	START	END	HAZARDS	CONTROLS
1.				
2.				
3.				
4.				
5.				
6.				
7.				
8.				

Required Training:

Required Personal Protective Equipment (PPE):

Job Name: _______________________

Job Number: _______________________

Supervisor: _______________

Date: _______________

Weekly Vehicle Check List: _____ **Tire Pressure** _____ **Transmission Fluid**

_____ **Oil** _____ **Lights**

_____ **Air Filter** _________ **Wkly Mileage**

Names of Employees:

PRINT NAME	SIGN NAME	TOTAL HOURS

found during analysis. Before work begins, the first-line supervisor or team leader should discuss the conclusions found during the TSA with the crew. Some companies require workers to sign the completed TSA forms or checklists before they start work. This helps companies document the hazards and ensure that workers have been told of the potential hazards and safety procedures.

1.2.4 Risk Assessment

Whether an action is considered safe is often a matter of evaluating risk. *Risk* is a measure of the probability, consequences, and exposure related to an event. *Probability* is the chance that a given event will occur. *Consequences* are the results of an action, condition, or event. *Exposure* is the amount of time and/or the degree to which someone or something is exposed to an unsafe condition, material, or environment.

A safe operation is one in which there is an acceptable level of risk. This means there is a low probability of an incident and that the consequences and exposure risk are all acceptable. For example, climbing a ladder has risk that is considered to be acceptable if the proper ladder is being used as intended, if it is set up correctly, and if it is in good condition. The probability of exposure to a hazard and its potential consequences are all low. If any one of these conditions were different, climbing the ladder would have an unacceptable level of risk.

1.2.5 Reporting Injuries, Incidents, Near-Misses, and Damaged Equipment

All on-the-job incidents and accidents, no matter how minor, must be reported to your supervisor. Some workers think they will get in trouble if they report minor injuries. That's not true. Small injuries like cuts and scrapes can later become big problems because of infection and other complications.

US employers with more than 10 employees are required to maintain a log of significant work-related injuries and illnesses using specific OSHA forms and documents. Employee names can be kept confidential in certain circumstances. A summary of these injuries must be posted at certain intervals, although employers do not need to submit it to OSHA unless requested. Employers can calculate the total number of injuries and illnesses and compare the result with the average national rates for similar companies. By analyzing incidents and accidents, companies and OSHA can improve safety policies and procedures. By reporting an incident, you can help keep similar events from happening in the future. For details on the operation of OSHA and its important mission, refer to the *Appendix*.

All damaged equipment must be immediately reported to the job supervisor. Take steps to see that the equipment is not used until it has been checked out. Always disconnect the power source and post a sign to prevent others from using the equipment until it can be repaired.

1.2.6 Safety Training and Education

Employers are responsible for ensuring that workers are trained to perform their jobs safely within the jobsite environment. While many skills can be learned through on-the-job learning (OJL), education through a standardized training program introduces aspects of safety on the job that experience may not cover until it is too late.

For example, a worker may know how to climb scaffolding and work in an elevated environment just by watching what other workers do, but standardized training provides greater knowledge of potential hazards.

In this example, employer responsibilities for safety training include the following:

- *"The employer shall have each employee who performs work while on a scaffold trained by a person qualified in the subject matter to recognize the hazards associated with the type of scaffold being used and to understand the procedures to control or minimize those hazards."* (29 CFR 1926.454-a)
- *"The employer shall instruct each employee who is involved in erecting, disassembling, moving, operating, repairing, maintaining, or inspecting a scaffold trained by a competent person to recognize any hazards associated with the work in question."* (29 CFR 1926.54-b)

Employers are responsible for training all of their employees in safe work practices and the regulations that apply to the work they are performing. The regulations also use the terms *should* and *shall*. When dealing with regulations, *should* means that it is recommended, while *shall* means that it is required.

There are two key terms defined and used throughout the module: competent person and qualified person.

A competent person is defined in 29 *CFR* 1926.32(f) as "one who is capable of identifying existing and predictable hazards in the surroundings or working conditions which are unsanitary, hazardous, or dangerous to employees, and who has authorization to take prompt corrective measures to eliminate them."

The first part of this definition requires that the competent person have the ability to recognize hazards associated with the activity. OSHA has identified the Fatal Four leading job-site hazards. They are fall hazards, struck-by hazards, caught-in or-between hazards, and electrocution hazards. To recognize these hazards, the competent person must be thoroughly trained in the safety procedures, OSHA regulations, and safety practices associated with the activity. This person must know the safe way to do a job in order to recognize an unsafe practice.

The second part requires that the person have the authority to eliminate the hazard. This may require stopping work until the hazard can be corrected. The competent person must be given this authority by the employer.

Employers are required to provide the necessary training and are free to designate as many competent persons as necessary.

A qualified person is defined in 29 *CFR* 1926.32(m) as "one who, by possession of a recognized degree, certificate, or professional standing, or who by extensive knowledge, training, and experience, has successfully demonstrated his ability to solve or resolve problems relating to the subject matter, the work, or the project."

When specifying that all scaffolds must be designed by a qualified person, the definition above is intended. The qualified person must design and plan a scaffold project, and then pass the project off to a competent person (as defined above) to supervise the erection. It is possible for one person to fulfill both roles. Notice that the qualified person is not required to have a degree, but they must have adequate expertise for the particular project.

Regulations require that a registered engineer design any scaffold over 125 ft. (38 m) tall and determine the proper number of working levels. The number of planked and working levels is limited, based on the type of scaffold erected (*Table 1*).

1.2.7 Rescue

Every elevated job site should have an established rescue and retrieval plan. Planning is especially important in remote areas where help is not readily available. Before beginning work, make sure that you know what your employer's rescue plan calls for you to do in the event of a fall. Find out what rescue equipment is available and where it is located. Learn how to use equipment for self-rescue and the rescue of others.

If a fall occurs, any employee hanging from the fall-arrest system must be rescued safely and quickly. Your employer should have previously determined the method of rescue for fall victims, which may include equipment that lets the victim perform a self-rescue, a system of rescue by co-workers, or a way to alert a trained rescue squad. If a rescue depends on calling for outside help such as the fire department or rescue squad, all the needed phone numbers must be posted in plain view at the work site. In the event a co-worker falls, follow your employer's rescue plan. Call any special rescue service needed. Communicate with and monitor the victim constantly during the rescue. After a fall, the entire fall-protection system involved in the fall must be discarded.

Never allow a worker that has been suspended to lie down, as this can cause a sudden rush of de-oxygenated blood to the heart, triggering a cardiac arrest. This condition is known as *suspension trauma*. Proper procedure requires placing the person in a sitting position, with the person's knees pulled up tight to the chest.

Table 1 Maximum Number of Planked Levels for Each Scaffold Rating

Number of Working Levels	Maximum Number of Additional Planked Levels			Maximum Height of Scaffold (ft.)
	Light Duty	Medium Duty	Heavy Duty	
1	16	11	6	125
2	11	1	0	125
3	6	0	0	125
4	1	0	0	125

Additional Resources

Basic Construction Safety and Health, Fred Fanning. 2014. CreateSpace Independent Publishing Platform.

Construction Safety, Jimmie W. Hinze. Second Edition. 2006. New York, NY: Pearson.

DeWalt Construction Safety/OSHA Professional Reference, Paul Rosenberg, American Contractors Educational Services. 2006. DEWALT.

Online resources:

Code of Federal Regulations (CFR), **www.ecfr.gov**

Occupational Safety and Health Administration (OSHA), **www.osha.gov**

OSHA Videos, **www.osha.gov/video**

1.0.0 Section Review

1. The person primarily responsible for your safety is _____.

    a. your foreman
    b. your instructor
    c. yourself
    d. your employer

2. One approach to hazard recognition is to perform a(n) _____.

    a. JSA
    b. SDS
    c. OSHA
    d. PPE

2.0.0 REGULATORY REQUIREMENTS AND OSHA

Objective

Identify the reasons for the OSHA regulations that govern the elevated work in construction.

   a. Understand the role of OSHA in regulating the construction workplace and know where to find regulations, training, and safety requirements for elevated work.

Trade Terms

Excavations: Man-made cuts, cavities, trenches, or depressions in the earth's surface, formed by removing earth. Excavations can be made for the purpose of building anything from basements to highways.

Personal fall arrest system (PFAS): A system used to stop an employee in a fall from a working level. It consists of an anchorage, connectors, and a body harness, and may include a lanyard, deceleration device, lifeline, or suitable combinations of these.

Personal protective equipment (PPE): Equipment or clothing designed to prevent or reduce injuries.

There are many rules and procedures in place to keep employees safe, many of which are developed by OSHA. To understand how OSHA relates to the construction industry, it is important to understand the difference between codes, regulations, and standards.

A *code* is a systematically arranged and comprehensive collection of laws and rules. A *regulation* is a principle, rule, or law designed to govern behavior. A *standard* is a degree or level of performance. Standards are developed and maintained by industry groups, while codes and regulations are developed by government agencies.

2.1.0 Occupational Safety and Health Administration

The Occupational Safety and Health Act of 1970 authorized OSHA in the Department of Labor to set occupational safety and health standards for all places of employment, enforce the standards, and provide research and educational programs to support safe work practices.

OSHA requires each employer to provide a safe and hazard-free place of employment for each employee. Employees are required to comply with OSHA rules and regulations that relate to their conduct on the job. OSHA regulations that relate to ironworkers involved in steel erection are included in 29 *CFR* (*Code of Federal Regulations*) 1926, Subpart R. These regulations are available at **www.osha.gov**. According to OSHA standards, employees are entitled to on-the-job instruction.

Specifically, new employees must be properly informed and equipped as follows:

- Told what to do on the job
- Shown how to do what you have been assigned to do
- Shown how to do that job safely
- Provided with the proper tools and equipment, including appropriate personal protective equipment (PPE)
- Warned about specific hazards in the work area and in the surrounding area
- Supervised for safety while performing work

2.1.1 The Development and Intent of Regulations and Standards

The primary code, or regulation, guiding the construction industry is the OSHA code. This code is federal law. Violations of the code are punishable by fines or other similar measures. Individual states have their own codes, which are also enforceable by law. The intent of all of these laws is to provide a safe workplace. These laws apply to workers and employers equally and should be used as a basis for all rules, policies, and behaviors.

The OSHA regulations form the basic set of rules governing the construction industry. These rules cover all aspects of construction work and safety requirements, including such topics as scaffold use and misuse. The rules can be very complex and confusing. The OSHA regulations are divided into sections found in the *Code of Federal Regulations* (*CFR*). The regulatory requirements for each industry are covered in a separate section for the regulated topics, as follows:

- General Industry Standard (29 *CFR* 1910)
- Maritime Standard (29 *CFR* 1915, 1917, and 1918)
- Construction Standard (29 *CFR* 1926)

Each regulation is designed to have jurisdiction over an industry that uses scaffolds.

In addition to federal, state, and local regulations, organizations working with scaffolds have developed standards that are useful to the scaffold builder. The main sources of these standards are:

- Army Corps of Engineers (ACE) (EM-385-1.1)
- Mine Safety & Health Administration (MSHA)
- American National Standards Institute (ANSI)
- Associated Builders and Contractors (ABC)
- Scaffold and Access Industry Association (SAIA)
- Scaffold, Shoring, and Forming Institute (SSFI)
- Association of General Contractors (AGC)
- Various scaffold manufacturers' company manuals

Copies of the standards published by each of the industry associations listed are available from the individual association and are not included in this module. This training program is intended for use by the construction industry. General industry regulations are applicable to all types of scaffolds.

2.1.2 General Safety and Health Provisions

OSHA provides general safety and health regulations that apply to the construction industry as a whole. A few excerpts from these regulations that relate to this training program are as follows. The actual quotes from the regulations are in italics, and the comments and explanations are in regular print.

Accident prevention responsibilities include the following:

- *It shall be the responsibility of the employer to initiate such programs as may be necessary to comply with this part.*
- *Such programs shall provide for frequent and regular inspections of the job sites, materials, and equipment to be made by competent persons designated by the employers.*

This subpart requires the employer to develop and implement a safety program. The goal of the safety program should be to safely erect, use, and dismantle scaffold in compliance with general industry regulations. The employer is also required to designate a competent person, as defined earlier, to make regular inspections of the job site and the materials and equipment used at the site.

2.1.3 OSHA-Required Training

OSHA requires training to be provided on the proper use of each of the fall protection devices used at the work site. The employer must also provide special training for all employees engaged in installing and maintaining fall protection equipment, establishing controlled decking zones, and connecting structural members.

Did you know?

Safety Training

Safe working conditions on mine sites fall under the Mine Safety and Health Administration (MSHA), and every other job site is regulated by OSHA. However, except for a few industry-specific requirements, the regulations are the same.

2.1.4 Fall Protection Requirements

According to OSHA, falls caused 384 construction worker deaths in 2016. That is more than four times as many fatalities as the next most-common cause. OSHA regulations are designed to address this issue.

High-rise construction typically requires the use of fall protection systems. Areas or activities where fall protection is needed include, but are not limited to, scaffolds, excavations, hoist areas, holes, ramps, runways, and other walkways. It is required for formwork and rebar work, leading edge work, unprotected sides and edges, and other hazardous walking/working surfaces.

Contractors must protect their workers from fall hazards and falling objects whenever a worker is 6 ft. (1.8 m) or more above a lower level, or 4 ft. (1.2 m) above open machinery. Workers must be protected from falling into dangerous equipment. Work areas must also be protected from falling objects.

In high-rise construction, workers often find it necessary to work in places requiring wire rope or slings for access. These positioning slings and ropes should be inspected daily by a qualified person. According to OSHA regulations, no wire rope should be used when more than 10 percent of the total wires are frayed or broken in any running foot.

Three common examples of fall protection equipment are a guardrail, a personal fall arrest system (PFAS), or a safety net. These devices and their use are governed by 29 *CFR* 1926, Subpart M.

NCCER – *Fall Protection Orientation*

29 *CFR* 1926, Subpart M also covers guardrails, but the rules provided in this subpart do not apply to guardrails on scaffolds. The rules covering guardrails on scaffolds are contained in 29 *CFR* 1926, Subpart L. OSHA basically requires that all workers use guardrail systems, safety net systems, or personal fall arrest systems generally designed to protect themselves from falling more than 6 ft. (1.8 m) to the ground or a lower work level.

2.0.0 Section Review

1. The agency primarily responsible for regulating the scaffolding industry is the ______.
   a. Scaffold Access Industry Association
   b. Occupational Safety and Health Administration
   c. National Safety Council
   d. American National Standards Institute

2. Who does OSHA require to develop and implement a safety program?
   a. employers
   b. employees
   c. engineers
   d. certified inspectors

3.0.0 INCIDENTS AND ACCIDENTS

Objective

Identify incidents, accidents, causes, costs, and consequences related to falls in the workplace.

a. Identify the categories of workplace incidents and accidents.
b. Describe the costs of workplace incidents and accidents.
c. Describe the causes of workplace incidents and accidents and identify hazards that lead to falls.

Trade Terms

Body harness: Straps that may be secured about the worker in a manner that will distribute the fall-arrest forces over at least the thighs, pelvis, waist, chest, and shoulders, with means for attaching it to other components of a personal fall-arrest system.

Management system: The organization of a company's management, including reporting procedures, supervisory responsibility, and administration.

Midrail: A mid-level, horizontal board required on all open sides of scaffolds and platforms that are more than 14 in. (35 cm) from the face of the structure and more than 10 ft. (3.05 m) above the ground. It is placed halfway between the toeboard and the top rail.

Top rail: A top-level, horizontal board required on all open sides of scaffolds and platforms that are more than 14 in. (36 cm) from the face of the structure and more than 10 ft. (3 m) above the ground.

Incidents and accidents can occur at any job site. Both at-risk behavior and poor working conditions can cause these undesirable events. You can help prevent such events by using safe work habits, understanding what causes them, and learning how to prevent them.

The terms *incident* and *accident* are often used interchangeably. However, according to OSHA, an incident is an unplanned event that may or may not result in property damage.

An incident is worthy of being documented so that steps can be taken to prevent it from recurring. When an incident occurs, no personal injury has occurred.

An accident is defined as an unplanned event that results in personal injury and/or property damage. Therefore, an event that results in property damage alone could be considered an incident or an accident. An event that results in personal injury or a fatality is always considered an accident.

Note that there are varying opinions on the use of the terms *incident* and *accident*. The US National Safety Council defines an incident as an unplanned, undesired event that adversely affects the completion of a task. In this definition, there is no mention of injury or property damage. Other safety organizations across the globe are likely to have their own definitions of these terms as well, or they may use completely different terms.

The most important thing to understand is that both incidents and accidents are negative events that have a negative effect on both projects and workers. Do not be surprised when you hear the terms used interchangeably by other workers.

> **NOTE**
>
> This module uses the terms *incident* and *accident* as they are defined by OSHA.

It is within your power to spot and avoid hazardous conditions, and the lessons in this module will help you do just that. By following safety procedures and using common sense, you can protect yourself and others from injury or even death.

3.1.0 Incident and Accident Categories

Incidents and accidents are often categorized by their severity and impact, as follows:

- *Near-miss* – An unplanned event in which no one was injured and no damage to property occurred, but during which either could have happened. Near-miss incidents are warnings that should always be reported rather than overlooked or taken lightly.
- *Property damage* – An unplanned event that results in damage to tools, materials, or equipment, but no personal injuries.
- *Minor injuries* – If personnel receive minor cuts, bruises, or strains, but return to full duty on their next regularly scheduled work shift, their injuries are considered minor injuries.
- *Serious or disabling injuries* – Injuries to personnel that result in temporary or permanent disability. Included in this category would

be lost-time incidents, restricted-duty or re-stricted-motion cases, and those that resulted in partial or total disability.

- *Fatalities* – Deaths resulting from unplanned events.

Studies have shown that for every serious or disabling injury, there were 10 injuries of a less serious nature and 30 property-damage incidents. A further study showed that 600 near-miss incidents occurred for every serious or disabling injury.

3.2.0 Costs of Incidents and Accidents

Most accidents can be prevented through safe work habits and a thorough understanding of the kinds of behaviors and conditions that cause accidents. According to the National Safety Council, work accidents destroy over 90,000 lives each year in spite of increased concern from management and the federal government for human welfare on the job. These accidents include losses such as the following:

- Death or disabling injury to the person(s) involved
- Loss of job time and wages
- Medical fees and/or funeral expenses
- Loss of skills, experience, or training
- Increased insurance rates

Incidents and accidents cost billions of dollars each year and cause much needless suffering. The National Safety Council estimates that the organized safety movement has saved more than 4.2 million lives since it began in 1913. This section examines why incidents and accidents happen and how you can help prevent them.

Accidents that result in injury or death can have a lasting effect on the victims, their families, their co-workers, and their employers. An injured worker who is disabled by an accident faces potentially huge medical bills. On top of those expenses, the worker's family faces loss of the income they rely on. Insurance does not always take care of the costs. A worker who is injured because they violated established safety rules may have their insurance claims denied and may be dismissed from the company because of the violation. A worker who is injured in a fall because they were not using fall protection may not receive compensation. How these issues are handled varies dramatically among companies and countries.

On-the-job accidents also have tremendous consequences for employers. A fall from a beam, a hoisting line failure, or the improper handling of welding electrodes or torches may result in personal injury, death, and/or the loss of equipment and materials.

Employers and co-workers can be affected because many contract awards are based, in part, on a company's safety record. Therefore, incidents and accidents can also result in the loss of future jobs, which affects the company's financial position. This can result in layoffs, hiring freezes, or inability to purchase new equipment or tools. In this way, these events affect not only injured employees and their families, but everyone on the job site.

An employer may be forced to spend thousands of dollars to repair or replace equipment and materials. Additional losses that can affect an employer are as follows:

- Production slowdowns
- Fees for legal actions
- Increased insurance coverage for workers
- Loss of public confidence

There are many indirect costs of accidents, as shown in *Figure 3*. The tip of the iceberg represents direct costs, which are the visible costs. Not all of these are covered by insurance. The more numerous indirect costs aren't readily measurable, but they can represent a greater financial burden than the direct costs.

When employers lose money, so do the employees. The funds available for raises and bonuses are greatly depleted by a serious on-the-job accident. A good safety program keeps everyone in business.

3.3.0 Causes of Falls, Incidents, and Accidents

Falls can sometimes be caused by other, seemingly unrelated hazards. The following common hazards can lead to falls:

- Struck-by hazards
- Caught-in/between hazards
- Falling objects

Struck-by and caught-in/between hazards (*Figure 4*) can involve the same kinds of equipment, but there is a difference between them.

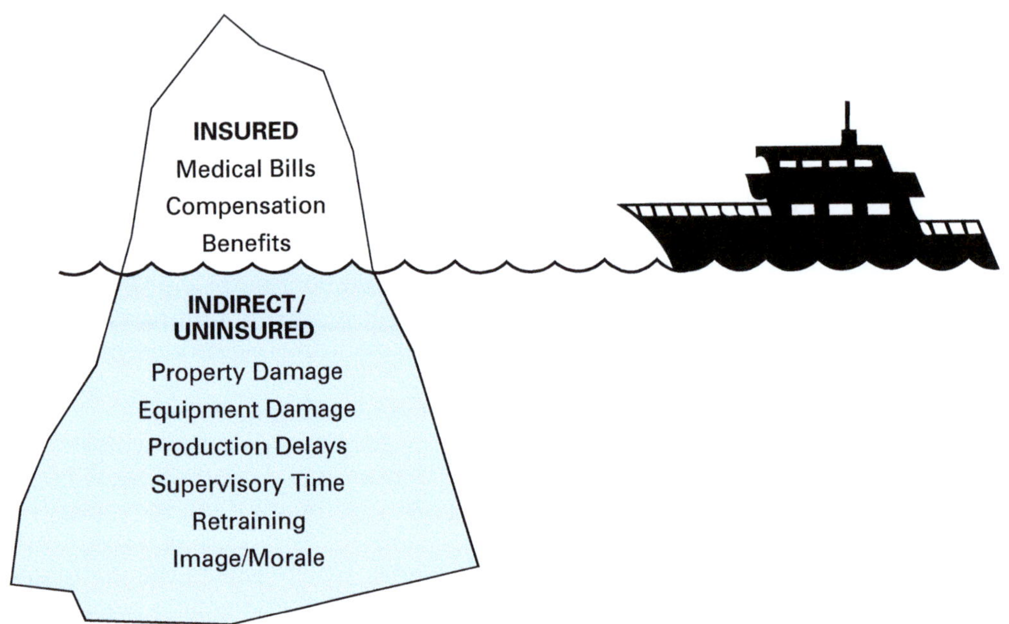

Figure 3 Costs associated with accidents.

Figure 4 Cab swing.

Struck-by means being hit by a moving object, while caught-in/between means being trapped between a moving object and a solid surface. These examples are provided for clarification. Assume a worker is working near a crane that is swinging a load and the load hits the worker. That is an example of a struck-by incident. However, if the worker is behind the crane and the crane backs up, pinning the worker against a wall, it becomes a caught-in/between incident.

On any job site, there is a risk of being struck by falling objects, such as tools dropped from above. Flying objects such as debris from grinding and chipping metal are another struck-by hazard. On any site where there is moving equipment, or where workers are near roadways, there is a further danger of being struck by a vehicle.

According to OSHA, falling objects killed 93 construction workers in 2016—9 percent of the total number of construction workers killed that year. In 2002, OSHA added two new rules to protect workers in high-rise construction from being struck by falling objects. First, all materials, equipment, and tools that are not being used must be secured in order to prevent them from being accidentally knocked or bumped off the platform. Second, the controlling contractor must bar other construction processes below steel erection unless overhead protection is provided for the employees working below. Debris netting, also called *containment netting* (*Figure 5*), is used on elevated construction sites to prevent dropped tools, materials, particulate waste such as dirt, and small items such as nails, fasteners, and screws from hitting people below. OSHA requires the installation of debris netting when a pile of materials or equipment on a scaffold is higher than the top edge of a standard toeboard.

In such cases, debris netting is installed from the top rail to the deck level so that it completely encases the work area like a protective curtain. Typical debris netting is made from a fine mesh of heavy-duty, fire-retardant polyethylene plastic, and comes in rolls in a variety of widths and lengths depending on the application. OSHA requires that debris netting maintain flexibility in extreme temperatures. Debris netting is not designed to serve as fall protection.

Figure 5 Debris netting

Keep your work area safe. Prevent injury from falling objects by following these guidelines:

- Always wear a hard hat.
- Keep openings in floors covered, secured, and properly marked.
- Do not store materials, other than masonry and mortar, within 4 feet of the working edges of a guardrail system.
- Keep the working area clear by removing excess mortar, broken or scattered masonry units, and all other material and debris on a regular basis.
- Never work or walk under loads that are being hoisted by a crane.

- Erect toeboards or guardrail systems. When guardrail systems are used to prevent material from falling from one level to another, any openings must be small enough to prevent the passage of potential falling objects.
- Erect paneling or screening from the walking/working surface or toeboard to the upper edge of the top guardrail or midrail if tools, equipment, or materials are piled higher than the top edge of the toeboard.
- Raise and lower tools and material with a rope and bucket or other lifting device. Never throw tools or material to or from a raised surface.

OSHA requires employers to investigate each work-related death, serious injury or illness, or incident having the potential to cause death or serious physical harm. The employer should document any findings from the investigation, as well as the action plan to prevent future occurrences. The employer should complete these actions immediately, with photos or video if possible. It's important that the investigation uncover the root cause of the accident to avoid similar incidents in the future. In many cases, the root cause was a flaw in the system that failed to recognize the unsafe condition or the potential for an unsafe act (*Figure 6*).

In general, many incidents and accidents can be attributed to one of the following root causes:

- Failure to communicate
- At-risk work habits
- Alcohol or drug abuse
- Lack of skill
- Intentional acts
- Unsafe acts
- Rationalizing risks
- Unsafe conditions
- Housekeeping
- Management system failure

3.3.1 Failure to Communicate

Many incidents and accidents happen because of a lack of communication. For example, you may learn how to do things one way on one job, but what happens when you go to a new job site? You need to communicate with the people at the new job site to find out whether they do things the way you have learned to do them. If you do not communicate clearly, incidents can happen. Remember that different people, companies, and job sites do things in different ways.

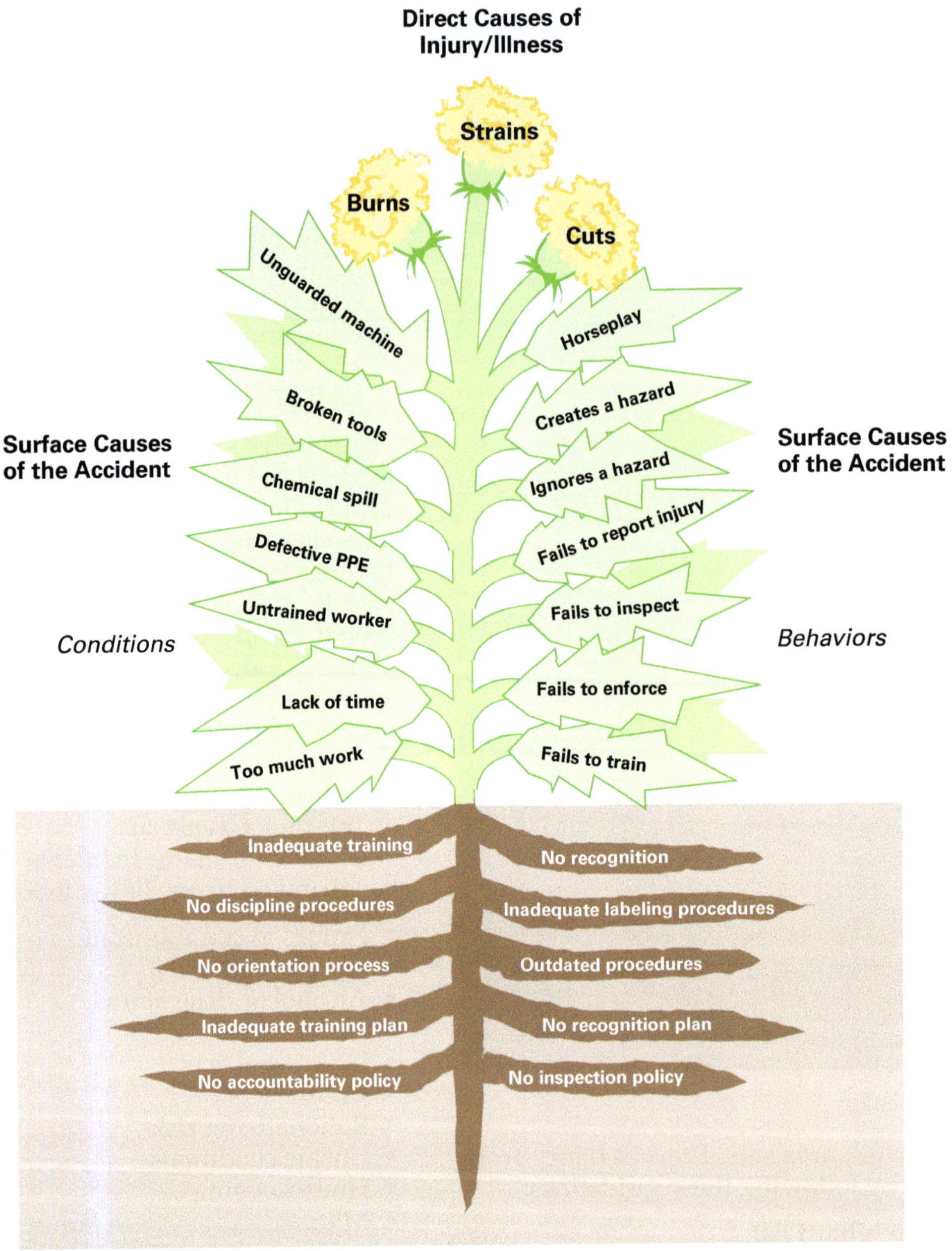

Figure 6 Root causes of accidents.

<table>
<tr><td>CAUTION</td><td>Making assumptions about what other workers know and what they will do can cause incidents and accidents. Never assume anything. For example, it is important not to use terms and jargon that people may not understand, and to be aware that some workers are native speakers of English. Another example is assuming that an electrical power source is turned off—first ask whether the power is turned off, and then check it yourself to be completely safe.</td></tr>
</table>

On Site

Toolbox Talks

Toolbox talks are one way to effectively keep all workers aware and informed of safety issues and guidelines. Toolbox talks are 5- to 10-minute meetings that review specific health and safety topics. These are very common at construction sites of all types.

One of the ways accidents can be avoided is the use of signs, safety tags, barricades and barriers, and other warning devices. When used properly, these devices not only prevent accidents, but also assist in the smooth flow of work and permit the job to be completed on schedule. Learn to recognize these types of warning signs:

- Informational signs
- Traffic and housekeeping signs
- Directional signs
- On-site traffic signs
- Safety signs
- Caution signs
- Danger signs
- Accident-prevention tags and temporary warnings

Stay alert for information and conditions that can affect everyone's safety (*Figure 7*). Do not rely on someone else to keep you safe. Think a job through before starting it. Look around for potential hazards and stay alert for changes in the work area. Remember the following:

- Accidents can kill and disable.
- Accidents cost you and your company money.
- Accidents can be prevented.

Informational Signs – Informational signs are used where it is necessary to provide general information that is not related to safety (*Figure 8*). The standard color used is blue. The background, the entire sign, or just a panel may be blue. Examples include the following:

- No Admittance
- No Trespassing
- For Employees Only

Traffic and Housekeeping Signs – Black and white informational signs are used as traffic and house-keeping markings. These signs identify things such as the following:

Figure 8 Common informational sign.

Figure 7 Stay alert for safety signs.

- Dead ends of aisles or passageways
- Location of trash cans
- Location and width of aisle-ways
- Rooms or passageways
- Stairways (risers, direction, borders)
- Drinking fountains and food-dispensing machines

Directional Signs – Directional signs help workers and visitors find locations such as restrooms, stairways, and locker rooms. Standard colors for directional signs are black and white. Exit and fire extinguisher signs are specialized directional signs that provide vital safety information and must be distinctive in color. They are red and white, as shown in *Figure 9*.

On-Site Traffic Signs – On-site traffic signs help in the safe movement of vehicles and pedestrians. Just as traffic signs must be obeyed on public highways and streets, they must also be observed in the workplace.

A slow-moving vehicle emblem is used on vehicles that move at speeds of 25 miles per hour (mph) or less. The emblem is a fluorescent yellow-orange triangle with a dark red reflective border (*Figure 10*). It is frequently seen on construction and farm equipment.

Safety Signs – Safety signs give general instructions and suggestions about safety measures (*Figure 11*). The background and lettering on these signs are typically white and green, but the colors can vary depending on the message and the location of the sign. These signs tell you where to find such important areas as the following:

- First-aid stations
- Emergency eye-wash stations
- Evacuation routes
- Safety Data Sheet (SDS) stations
- Exits (usually have white letters on a red field)

Examples of what may be written on a Safety sign include the following:

- Report All Unsafe Conditions to Your Supervisor
- Walk, Don't Run
- Help Keep This Plant Safe and Clean

Figure 9 Directional signs.

Figure 10 Slow-moving vehicle sign.

Figure 11 Common safety sign.

Caution Signs – Caution markings or signs (*Figure 12*) tell you about potential hazards or warn against unsafe acts. When you see a caution sign, protect yourself against a possible hazard. Caution signs are yellow and have a black panel with yellow letters. They may give you the following information:

* Hearing and eye protection are required
* Respirators are required
* Smoking is not allowed

When you see a caution sign, take action to protect yourself. Examples of what may be written on a Caution sign include the following:

* Caution – Do Not Operate
* Caution – Keep Aisles Clear
* Caution – Electric Fence

Yellow is the basic color used for caution. It is used to identify places where physical hazards may be caused by striking against objects, stumbling, falling, tripping, and being caught between obstacles. Solid yellow, yellow and black stripes, or yellow and black checkers caution workers against these hazards.

The caution signs on piping systems that contain dangerous materials are yellow. Yellow is also used to warn workers against starting machinery under repair. Painted barriers and flags should be located at the starting point or power source. They should be displayed so that workers will notice them easily on such things as electrical controls, ladders, scaffolds, vaults, valves, dryers, boilers, elevators, and tanks.

Danger Signs – Danger markings or signs (*Figure 13*) tell you that an immediate hazard exists and that you must take certain precautions to avoid an incident. Danger signs are red, black, and white. They may indicate the presence of the following:

* Defective equipment
* Flammable liquids and compressed gases
* Safety barriers and barricades

Figure 12 Typical caution sign.

Figure 13 Typical Danger sign.

* Emergency stop button
* High voltage

Examples of what may be written on a Danger sign include the following:

* Danger – High Voltage
* Danger – No Smoking, Matches, or Open Lights
* Danger – Keep Away

Some of the dangers that may be found on a site are radiation and biological hazards. Be on the lookout for these types of signs as well.

Accident-Prevention Tags – Accident-prevention tags, also known as *safety tags*, are used as a temporary warning to workers about immediate and potential hazards (*Figure 14*). They are similar to signs; however, they are not designed to be used in place of signs or as a permanent means of protection. For example, an Out of Order tag may be used on damaged equipment until it can be disposed of or repaired. A Do Not Start tag may be placed on machinery during lockout procedures. Tags can be an effective means of protecting workers and property. Tags and the devices used to attach them must meet specific physical requirements to ensure their durability and effectiveness. The following are some different types of tags you may encounter on a job site:

* *Do Not Start tags* – Do Not Start tags are used to prevent the unexpected energizing of equipment that could result in injury, equipment damage, or both. The tags must be placed in a conspicuous location that effectively blocks the starting mechanism. The tag is white. The square panel is red, and the letters may be white, grey, or visibly etched on the tag.
* *Danger tags* – Danger tags are used where an immediate hazard exists. They tell workers that specific precautions must be observed. Danger tags are white. White letters appear on a red oval in a black square. A danger tag may read "Danger – Unsafe – Do Not Use."

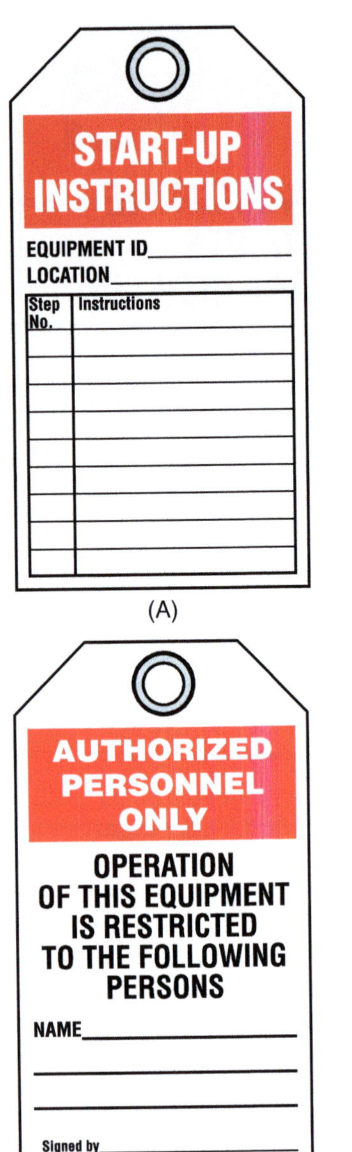

Figure 14 Examples of safety tags.

- *Caution tags* – Caution tags are used to warn workers of potential hazards and to caution workers against unsafe practices. These tags inform workers that proper precautions should be taken. For example, a tag may read "Caution—Stop Machinery to Clean, Oil, or Repair." Caution tags are yellow. They contain a black square panel with yellow letters.
- *Out of Order tags* – Out of Order tags should only be used to notify workers that a particular machine or tool is out of order and that using it may present a hazard. Out of Order tags are usually white with black letters.
- *Radiation tags* – Radiation tags are yellow with a reddish-purple panel. Any letters used against the yellow background are black. The radiation symbol should also appear on the tag. Radiation tags notify workers of actual or potential radiation hazards.
- *Biological Hazard tags* – Biological hazard tags signify the presence of an actual or potential biological hazard. Biological hazards can include any infectious agents that present a risk or potential risk to workers. Biological hazard tags are usually white or fluorescent orange with black letters. The tag also shows the biological hazard or biohazard symbol.

3.3.2 At-Risk Work Habits

Some examples of at-risk work habits are procrastination, carelessness, and horseplay. Procrastination (putting things off until the last minute) is a common cause of incidents. For example, delaying the repair, inspection, or cleaning of equipment and tools can cause incidents and even accidents. Trying to push machines and equipment beyond their operating capacities puts you and your co-workers at risk.

Work habits and work attitudes are closely related. If you resist taking orders, you may also resist listening to warnings. If you let yourself be easily distracted, you won't be able to concentrate. If you aren't concentrating, you could cause a significant problem.

Your safety is affected not only by how you do your work, but also by how you act on the job site. This is why most companies have strict policies for employee behavior. Horseplay and other inappropriate behavior are forbidden. Workers who engage in such behavior on the job site may be fired.

These strict policies are provided for the worker's protection. There are many hazards on construction sites. Each person's behavior—at work, on a break, or at lunch—must follow the principles of safety.

A person pulling a practical joke on a co-worker could consider it just having fun, but in fact, it could cause the co-worker serious, even fatal, injury. If you horse around on the job, play pranks, or don't concentrate on what you are doing, you are showing a poor work attitude that can lead to a serious incident (*Figure 15*).

3.3.3 Alcohol and Drug Abuse

Alcohol and drug abuse costs the construction industry millions of dollars a year in incidents, accidents, lost time, and lost productivity. Substance abuse can cost much more than just money—it can cost lives. Just as drunk driving kills thousands of people on our highways every year, alcohol and drug abuse kills on the construction site. Using alcohol or drugs creates a risk of injury for everyone on a job site. Many states have laws that prevent workers from collecting insurance benefits if they are injured while under the influence of alcohol or illegal drugs.

Would you trust your life to a crane operator who was under the influence of drugs? Would you bet your life on the responses of a co-worker under the influence of alcohol or drugs? Alcohol and drug abuse have no place in the workplace. A person on a job site who is under the influence of alcohol or drugs is an incident or accident waiting to happen—possibly a fatal accident (*Figure 16*).

Figure 15 Example of horseplay.

People who work while using alcohol or drugs are at risk of incident or injury; their co-workers are at risk as well. That's why most employers have a formal substance abuse policy. Be aware of and follow your employer's substance abuse policy. Avoid any substances that can affect your job performance. The life you save could be your own.

You do not have to be abusing drugs such as marijuana, cocaine, or heroin to create a job hazard. Many prescription and over-the-counter drugs, taken for legitimate reasons, can affect your ability to work safely. Amphetamines, barbiturates, and antihistamines are only a few of the prescription-controlled legal drugs that can affect the ability to work or operate machinery safely. The main thing is to understand and follow your company's substance abuse policy.

> **CAUTION**
>
> If your doctor prescribes any medication that you think might affect your job performance, ask about its effects. Your safety and the safety of your co-workers depend on everyone being alert on the job.

3.3.4 Lack of Skill

Every worker needs to learn and practice new skills under careful supervision. Never perform new tasks alone until you have been checked out by a supervisor. Lack of skill can cause incidents and accidents quickly. For example, suppose you are told to cut some boards with a circular saw, but you aren't skilled with that tool. A basic rule of circular saw operation is never to cut without a properly functioning guard. Because you haven't been trained, you don't know this. You find that the guard on the saw is slowing you down, so you jam the guard open. The result could be a serious accident. Proper training can prevent this from happening.

Never operate a power tool or machine until you have been trained to use it. You can greatly reduce the chances of incidents and accidents by learning the safety rules for each task you perform. Some power tools require that a worker be trained and certified in their use.

3.3.5 Intentional Acts

When someone purposely causes property damage or injury, it is called an *intentional act*. An angry or dissatisfied employee may purposely create a situation that leads to an incident or accident. If someone you are working with threatens to get even or pay back someone, let your supervisor know at once.

Figure 16 An impaired worker is a dangerous one!

Terrorism must also be a consideration on the job site. The level of concern is typically dependent on the site itself and the nature of the structure. Pipelines, for example, may represent a significant target due to their size, importance, location (many parts of them in remote areas), and potential for destruction. Controversial sites also tend to attract attention, perhaps from terrorist groups operating within a country that oppose the construction or what the structure represents.

As an individual worker, the most important thing you can do is to pay attention. Become familiar with people on the job site and look for strangers that do not seem to fit. A terrorist attack can come from any direction and be delivered in many different ways. However, it can also begin with a single individual gaining access to the job site.

3.3.6 Unsafe Acts

An unsafe act is a change from an accepted, normal, or correct procedure that can cause an incident or accident. It can be any conduct that causes unnecessary exposure to a job-site hazard or that makes an activity less safe than usual. Here are examples of unsafe acts:

- Failing to use required personal protective equipment (PPE)
- Failing to warn co-workers of hazards
- Lifting improperly
- Loading or placing equipment or supplies improperly
- Making safety devices (such as saw guards) inoperable
- Operating equipment at improper speeds
- Operating equipment without authority
- Servicing equipment in motion
- Taking an improper working position
- Using defective equipment
- Using equipment improperly

NCCER – *Fall Protection Orientation*

3.3.7 Rationalizing Risk

Everybody takes risks daily. When you get in your car to drive to work, you know there is a risk of being involved in an accident. Yet when you drive using all the safety practices you have learned, you know that there is a very good chance you will arrive at your destination safely. Driving is an acceptable risk because you have some control over your own safety and that of others.

Some risks are not acceptable. On the job, you must never take risks that endanger yourself or others just because you think you can justify doing so. This is called *rationalizing risk*. Rationalizing risk means ignoring safety warnings and practices. For example, because you are late for work, you might decide to run a red light. Perhaps you feel the risk is worth the time saved. This is unacceptable thinking on the job site.

The following are common examples of rationalized risks on the job:

- Crossing barricades or boundaries because there is no apparent activity
- Not wearing gloves because it's easier to do the job with bare hands
- Removing your hard hat because you are hot and you cannot see anyone working overhead
- **Not tying off your fall protection because you only have to lean over by about a foot**

Think about the job before you do it. If you think that it is unsafe, then it is unsafe. Stop working until the job can be done safely. Bring your concerns to the attention of your supervisor. Your health and safety, and that of your co-workers, make it worth taking extra care.

3.3.8 Unsafe Conditions

An unsafe condition is a physical state that is different from the acceptable, normal, or correct condition found on the job site. It usually results in an incident or accident. An unsafe condition can be anything that reduces the degree of safety normally present. The following are some examples of unsafe conditions:

- Congested workplace
- Defective tools, equipment, or supplies
- Excessive noise
- Fire and explosive hazards
- Hazardous atmospheric conditions (such as gases, dusts, fumes, and vapors)
- Inadequate supports or guards
- Inadequate warning systems
- Cluttered work area
- Poor lighting
- Poor ventilation

- Radiation exposure
- Unguarded moving parts such as pulleys, drive chains, and belts

All employees should be given the authority to stop work when an unsafe condition is observed.

3.3.9 Poor Housekeeping

Housekeeping means keeping your work area clean and free of scraps, clutter, or spills. It also means being orderly and organized. Failure to maintain good housekeeping procedures greatly increases the risk of slips and falls.

Always store your materials and supplies safely and label them properly. Arranging your tools and equipment to permit safe, efficient work practices, and easy cleaning is also important.

If the work site is indoors, make sure it is well-lit and ventilated. Don't allow aisles and exits to be blocked by materials and equipment. Make sure that flammable liquids are stored in safety cans. Oily rags must be placed only in approved, self-closing metal containers.

Remember that the major goal of housekeeping is to prevent incidents. Good housekeeping reduces the chances for slips, fires, explosions, and falling objects. Here are some good housekeeping rules:

- Remove from work areas all scrap material and lumber with nails protruding.
- Clean up spills to prevent falls.
- Remove all combustible scrap materials regularly.
- Make sure you have containers for the collection and separation of refuse. Containers for flammable or harmful refuse must have covers.
- Dispose of wastes often.
- Store all tools and equipment in the proper location and condition when you are finished with them.
- Keep all aisles and walkways clear of materials and tools.

Good housekeeping shows pride of workmanship. If you take pride in what you are doing, you won't let trash build up around you. The saying "A place for everything and everything in its place" is the right idea on the job site.

3.3.10 Management System Failure

Sometimes the cause of an incident or accident is failure of the management system. The management system should be designed to prevent or correct the acts and conditions that can create safety hazards. If the safety management system is not functioning properly, problems are likely to occur.

What traits could mean the difference between a management system that fails and one that succeeds? A company implementing a good management system will do the following:

- Put safety policies and procedures in writing
- Distribute written safety policies and procedures to each employee
- Review safety policies and procedures periodically
- Enforce all safety policies and procedures fairly and consistently
- Evaluate supplies, equipment, and services to see whether they are safe
- Provide regular, periodic safety training for employees

Additional Resources

Basic Construction Safety and Health, Fred Fanning. 2014. CreateSpace Independent Publishing Platform.

Construction Safety, Jimmie W. Hinze. Second Edition. 2006. New York, NY: Pearson.

DeWalt Construction Safety/OSHA Professional Reference, Paul Rosenberg, American Contractors Educational Services. 2006. DEWALT.

Online resources:

Code of Federal Regulations (CFR), **www.ecfr.gov**

29 *CFR* 1926, Subpart G, Accident Prevention Signs and Tags. **www.ecfr.gov**

Occupational Safety and Health Administration (OSHA), **www.osha.gov**

OSHA Videos, **www.osha.gov/video**

3.0.0 Section Review

1. The color commonly used for informational signs is _____.

    a. green
    b. red
    c. yellow
    d. blue

2. Which of the following types of signs is typically white and green?

    a. General information sign
    b. Safety sign
    c. Caution sign
    d. Danger sign

3. The sign used to inform workers that an immediate hazard exists is a(n) _____.

    a. caution sign
    b. danger sign
    c. informational sign
    d. warning sign

4.0.0 ELEVATED WORK ENVIRONMENTS

Objective

Identify the safety concerns and requirements related to elevated work.

a. Identify the work areas most likely to present fall hazards.
b. Identify safety precautions related to elevated masonry work.
c. Identify safety precautions related to elevated steel erection work.

Trade Terms

Controlled access zone (CAZ): A designated work area in which certain types of masonry work may take place without the use of conventional fall protection systems.

Controlled decking zone (CDZ): An area in which certain work (for example, initial installation and placement of metal decking) may take place without the use of guardrail systems, personal fall arrest systems, fall restraint systems, or safety net systems and where access to the zone is controlled.

Guarded: Enclosed, fenced, covered, or otherwise protected by barriers, rails, covers, or platforms to prevent dangerous contact.

Limited access zones: Restricted areas alongside a masonry wall that is under construction.

Six-foot rule: A rule stating that platforms or work surfaces with unprotected sides or edges that are 6 ft. (1.8 m) or higher than the ground or level below it require fall protection.

Falls from elevated areas are one of the leading causes of fatalities in the workplace. Falls from elevated heights account for about one-third of all deaths in the construction trade. Approximately 85 percent of the injured workers lose time from work; approximately 33 percent require hospitalization; some never return to the job.

Safety is the major consideration in organizing an elevated workstation. Keeping materials neat and organized is particularly important on an elevated workstation, as space is tight. The danger and inconvenience from dropping items increases with height. Arrange materials and equipment with the following requirements in mind:

- Check the scaffold for proper assembly and rated loads before using it.
- Keep walkways and work areas free from stored materials, tools, rope, electrical cords, and trash. Store unused tools in a contained location away from travel paths.
- Stack building materials far enough apart to leave room for work in progress. Stacks, especially of masonry materials should be no higher than 3 feet and no closer than 6 feet to any opening.
- Stack building materials directly over frame members, away from the edge of the scaffold platform, and never on the very ends of scaffold planks.

4.1.0 Hazardous Work Areas

Falls are classified into two groups: falls from an elevation and falls from the same level. Falls from an elevation can happen during work from scaffolds, work platforms, decking, concrete forms, ladders, stairs, and work near excavations. Falls from elevation often result in death unless the fall is arrested. Falls on the same level are usually caused by tripping or slipping. Sharp edges and pointed objects, such as exposed concrete reinforcing bars (rebar), could cut and otherwise harm a worker. Other bodily injuries are also common results of tripping or slipping.

In the United States, fall protection is required for platforms or work surfaces with unprotected sides or edges that are 6 ft. (1.8 m) or above the ground or the level below. This is commonly referred to as the six-foot rule (*Figure 17*). However, some international regulations and company policies may require fall protection for heights less than 6 feet (1.8 m).

4.1.1 Slippery Surfaces

Slips, trips, and falls on walking and working surfaces cause 15 percent of all incidental deaths in the construction industry. Some incidents occur due to environmental conditions, such as snow,

Figure 17 OSHA's six-foot rule.

ice, or wet surfaces. Others happen because of poor housekeeping and careless behavior, such as leaving tools, materials, and equipment out and unattended. You can avoid slips, trips, and falls by being aware of your surroundings and following the rules on your site. Remember these general walking and working surface guidelines to avoid incidents:

- Keep all walking and working areas clean and dry. If you see a spill or ice patch, clean it up, or barricade the area until it can be properly attended to.
- Keep all walking and working surfaces clear of clutter and debris.
- Run cables, extension cords, and hoses overhead or through crossover plates so that they will not become tripping hazards.
- Do not run on scaffolds, work platforms, decking, roofs, or other elevated work areas.

4.1.2 Unprotected Sides, Wall Openings, and Floor Holes

Any opening in a wall or floor is a safety hazard. There are two types of protection for these openings: (1) they can be guarded or (2) they can be covered. Cover any hole in the floor when possible. Hole covers must be clearly marked (*Figure 18*). When it is not practical to cover the hole, use barricades. If the bottom edge of a wall opening is less than 39 in. (1 m) above the floor and would allow someone to fall 6 ft. (1.83 m) or more, then place guards around the opening.

The types of barriers and barricades used vary from one job site to another. There may also be different procedures for when and how barricades are put up. Learn and follow the policies at your job site.

Barriers and guardrail systems are covered in more detail later in this module. The following are some general guidelines for working near unprotected sides, floor holes, and wall openings:

- Hole covers must be cleated, wired, or otherwise secured to prevent them from slipping sideways or horizontally beyond the hole.
- Covers must extend adequately beyond the edge of the hole.
- Hole covers must be strong enough to support twice the weight of anything that may be placed on top of them. They must also be clearly marked. Use $\frac{3}{4}$ in. (2 cm) plywood as a hole cover, provided that one dimension of the opening is less than 18 in. (46 cm); otherwise, 2 in. (5-cm) lumber is required.
- Never store material or equipment on a hole cover.
- Guard all stairway floor openings, with the exception of the entrance, with standard railing and toeboards.
- Guard all wall openings from which there is a drop of more than 6 ft. (2 m) and for which the bottom of the opening is less than 39 in. (1 m) above the working surface.
- Guard all open-sided floors and platforms 6 ft. (2 m) or more above adjacent floor or ground level, using a standard railing or the equivalent.

4.2.0 Elevated Masonry Safety

Working safely on a high-rise job means that you must work smart. Be aware of what you can do to keep your work area safe. Take precautions to prevent falls and protect yourself from falling objects. Always use personnel lifts correctly and safely.

This section provides a review of appropriate personal protective equipment (PPE) for masons, masonry access zones, and additional information that applies to working in elevated conditions.

4.2.1 Masonry PPE

Remember to take safety precautions when dressing for construction work. The following guidelines are recommended:

- Confine long hair in a ponytail or in your hard hat. Flying hair can obscure your view or get caught in machinery.
- Wear appropriate clothing and personal protective equipment.
  - Always wear a hard hat.
  - Wear goggles when cutting or grinding.
  - Wear a high-visibility vest.
  - Wear close-fitting clothing, including long-sleeved shirts (minimum 4 in. sleeves) to give extra protection if skin is sensitive.

Figure 18 Hole Covering.

- Wear gloves when working with wet mortar.
- Wear pants over boots to avoid getting mortar on legs or feet.
- Keep gloves and clothing as dry as possible.
- Wear fall protection equipment as required.

- Wear face and eye protection as required, especially if there is a risk from flying particles, debris, or other hazards such as brick dust or proprietary cleaners.
- Wear hearing protection as required.
- Wear respiratory protection as required.
- Protect any exposed skin by applying skin cream, body lotion, or petroleum jelly.
- Wear sturdy work boots or work shoes with thick soles. Never show up for work dressed in sneakers, loafers, or sport shoes.

4.2.2 Access Zones in Masonry

To prevent injury on any project where an elevated masonry wall is to be constructed, the contractor must establish controlled access zones and limited access zones before construction can begin. This section reviews the purpose of each of these access zones, and how to establish and maintain them.

A controlled access zone (CAZ) is a designated work area in which certain types of masonry work may take place without the use of conventional fall protection systems such as guardrails, personal fall arrest systems, or safety nets. Controlled access zones are established prior to the beginning of construction. They are created and clearly marked to limit entrance to authorized workers only. OSHA guidelines specify that if there are no guardrails, masons are the only workers allowed in the CAZ. The height of the CAZ is equal to the height of the wall plus 4 ft. (1.2 m). If the wall is higher than 8 ft. (2.4 m), the CAZ must remain defined until the wall is braced.

The CAZ is marked on the side of the wall without the scaffold by a rope, wire, or tape control line that runs along the entire length of the unprotected edge and is connected to each side of the guardrail or wall. The purpose of the control line is to restrict access to workers who are not authorized to work there. OSHA also requires that supporting stanchions must be clearly flagged or marked at maximum intervals of 6 ft. (1.8 m). The stanchion flagging must be rigged to ensure that the lowest point is not less than 39 in. (99 cm) and not more than 45 in. (114 cm) from the working surface. The control line must be able to withstand at least 200 pounds of force applied to it.

For block and brick wall construction, the control line is erected a minimum of 6 ft. and a maximum of 25 ft. (7.6 m) from the unprotected edge. For precast concrete wall construction, the control line should be erected a minimum of 6 ft. (1.8 m) and a maximum of 60 ft., or half the length of the precast member, whichever is less, from the edge. If overhand bricklaying techniques are used to construct the wall, the control line must be a minimum of 10 ft. (3 m) and a maximum of 15 ft. (4.6 m) from the edge.

On floors and roofs where guardrail systems are not in place before the start of overhand bricklaying operations, controlled access zones should be enlarged to enclose all points of access, materials handling areas, and storage areas. On floors and roofs where guardrail systems are in place but need to be removed to allow overhand bricklaying work or leading-edge work, remove only that portion of the guardrail necessary to accomplish that day's work.

A limited access zone (LAZ) marks a restricted area alongside a masonry wall that is under construction. Like the CAZ, the LAZ is established before construction begins. The LAZ is equal to the height of the wall plus 4 ft. (1.2 m), for the entire length of the wall on the side without scaffold. If the maximum height of the wall is 8 ft. (2.4 m), the LAZ must remain in place until the wall has been adequately supported. If the wall is higher than 8 ft. (2.4 m), the wall must be braced until permanently supported.

Barricades are used to mark off the LAZ. Only employees who are working on the construction of the wall are permitted access to the LAZ.

4.3.0 Steel Erection Safety

Steel erection is used during the construction of many types of structures including high-rise buildings, metal decking, bridges, office buildings, schools, medical facilities, and retail stores. Steel-erection jobs can be extremely dangerous.

Common types of steel-erection accidents include hand injuries, back injuries, and falling from elevations. Workers can also be struck or crushed by falling loads, electrocuted by power lines, or run over or trapped by cranes.

Hazards at the job site vary with the size and complexity of the job. However, there are certain fundamental principles for job site safety that should be observed by every worker. Most accidents in the field result from inadequate maintenance of equipment, insufficient instruction in safe practices, or failure to follow safety regulations.

Pay special attention to the hazards involved when working above the ground. Work operations and the moving of equipment and materials pose an increased risk.

4.3.1 Common Hazards

While OSHA's general fall protection requirements are 6 ft. (1.8 m) or more, 29 *CFR* 1926.760 describes the fall protection requirement during steel-erection activities. This regulation is intended to provide each worker with 100 percent fall protection above 15 ft. (4.6 m) when performing decking or detail operations and 30 ft. (9.1 m) during initial construction. This protection includes establishing perimeter safety cable systems (*Figure 19*), guardrail systems, safety nets, and personal fall arrest systems (PFAS) (*Figure 20*) for every worker who is working or walking on a surface that is 15 ft. (4.6 m) or more above the next lower level.

> Body belts can result in severe injuries during a fall and may not be used.

4.3.2 Controlled Decking Zones in Steel Erection

A controlled decking zone (CDZ) may be established for an area where metal decking is being installed more than 15 ft. (4.6 m) but less than 30 ft. (9.1 m) above the next lower level. Access to the CDZ is restricted to those employees, such as connectors, who are performing leading-edge work and who have been trained for CDZ work.

A CDZ is a substitute for fall protection in areas where decking is initially being installed. Only workers trained for CDZ work are permitted in the areas. The CDZ must be clearly marked. The zone can be no wider than 90 ft. (27.4 m) wide and 90 ft. (27.4 m) deep from any leading edge.

All openings in floors or roofs must be covered with materials capable of supporting twice the weight of workers and equipment that may be placed on the cover. These covers must be painted with high-visibility paint or marked with the word HOLE or COVER to provide warning of the hazard.

OSHA requires training to be provided on the proper use of each of the fall protection devices used at the work site. The employer must also provide special training for all employees engaged in installing and maintaining fall protection equipment, establishing CDZs, and connecting structural members.

Figure 20 Safety harness in use on test dummies.

Figure 19 Horizontal safety cable.

Additional Resources

Basic Construction Safety and Health, Fred Fanning. 2014. CreateSpace Independent Publishing Platform.

Construction Safety, Jimmie W. Hinze. Second Edition. 2006. New York, NY: Pearson.

DeWalt Construction Safety/OSHA Professional Reference, Paul Rosenberg, American Contractors Educational Services. 2006. DEWALT.

Online resources:

Code of Federal Regulations (CFR), **www.ecfr.gov**

Occupational Safety and Health Administration (OSHA), **www.osha.gov**

OSHA Videos, **www.osha.gov/video**

4.0.0 Section Review

1. The most common types of fall protection equipment include guardrails, personal fall arrest systems, and _____.

    a. large air bags
    b. scaffold cages
    c. safety nets
    d. workstation enclosures

2. The width of a limited access zone adjacent to a wall being built is equal to _____.

    a. two-thirds the wall height
    b. the wall height
    c. the wall height plus 4 ft.
    d. twice the wall height

3. A CDZ may be established where metal decking is being installed _____.

    a. up to 10 ft. (3 m) above the next lower level
    b. from 5 to 15 ft. (1.5 to 4.6 m) above the next lower level
    c. from 10 to 20 ft. (3 to 6 m) above the next lower level
    d. from 15 to 30 ft. (4.6 to 5.1 m) above the next lower level

5.0.0 FALL PROTECTION EQUIPMENT

Objective

Identify the appropriate fall protection and lifesaving equipment and describe their proper use.

  a. Describe the procedures for using fall protection equipment.
  b. Describe the requirements for anchor points.
  c. Describe the requirements for full-body harnesses and identify the steps for donning a harness.
  d. Describe the requirements for lanyards, lifelines, and deceleration devices.
  e. Identify barriers, guardrails, safety nets, and climbing devices used to prevent falls.

Trade Terms

Capacity: The total amount of weight, or load, capable of being held by a PFAS component such as an anchor point or a lanyard. This includes the weight of personnel, tools and materials, and/or equipment.

Free fall: The act of falling before a personal fall-arrest system begins to apply force to arrest the fall.

Lanyard: A short section of rope or strap, one end of which is attached to a worker's safety harness and the other to a strong anchor point above the work area.

Lifeline: A component consisting of a flexible line connected vertically to an anchorage at one end (vertical lifeline), or connected horizontally to an anchorage at both ends (horizontal lifeline), and which serves as a means for connecting other components of a personal fall-arrest system to the anchorage.

Self-retracting lanyard (SRL): A deceleration device containing a drum-wound line that can be slowly extracted from, or retracted onto, the drum under slight tension during normal employee movement, and which, after the onset of a fall, automatically locks the drum and arrests the fall.

Tensile strength: The resistance of a material to a force tending to tear it apart.

Personal fall arrest systems (PFASs) catch workers after they have fallen. These systems are designed to activate only if a fall occurs. They are rigged to prevent a worker from falling a distance of more than 6 ft. (1.8 m) to the ground or a lower work area. The following terms must be understood when describing personal fall arrest systems:

- *Free fall* – The act of falling before a personal fall arrest system begins to apply force to arrest the fall.
- *Free fall distance* – The vertical distance of a free fall.
- *Deceleration device* – A device, such as a shock absorbing *lanyard* or self-retracting *lifeline*, that brings a falling person to a stop without injury.
- *Deceleration distance* – The distance required before a person comes to a stop. The required deceleration distance for a fall arrest system is a maximum of $3\frac{1}{2}$ ft. (1.07 m).
- *Arresting force* – The force needed to stop a person from falling. The greater the free fall distance, the more force is needed to stop or arrest the fall.

Equipment used in personal fall arrest systems includes the following:

- Body harnesses and belts
- Lanyards
- Deceleration devices
- Lifelines
- Anchoring devices and equipment connectors

Remember that a PFAS (*Figure 21*) is a system, not a single piece of equipment. A complete PFAS is made up of 3 primary components: anchor points, a body harness, and connecting devices. Anchor points are related to the structure, and the type and availability of anchor points help determine what other equipment should be chosen. The body harness comprises the system of belts, rings, or hooks worn by the worker. Connecting devices or connectors are used to maintain attachment between anchor points and the PFAS and include both lanyards and various pieces of hardware.

5.1.0 Using Personal Fall Arrest Systems

When it comes to personal fall arrest systems, one size does not fit all. A proper fit is essential to avoiding injury to the worker when a fall is arrested. 29 *CFR* 1926.502, Subpart M, Section (d) lists specific

Personal Fall Arrest System

Three key components of the Personal Fall Arrest System (PFAS) must be in place and properly used to provide maximum worker protection.

Individually these components will not provide protection from a fall. However, when used properly and in conjunction with each other, they form a Personal Fall Arrest System that becomes vitally important for safety on the job site.

Figure 21 Personal fall arrest system.

requirements, such as the test strength and use of D-rings and snap hooks and the specifications for anchor points on the structure. This standard requires several important characteristics of a PFAS:

- It must limit the maximum arresting force imparted to the body to 1,800 pounds with a full body harness.
- The free fall distance (the distance it takes before a person comes to a stop when falling) must be limited to 6 ft. (1.8 m) and be rigged to prevent contact with anything below.
- It must bring the body to a stop within an additional $3\frac{1}{2}$ ft. (1.1 m).
- The system must be strong enough to withstand 5,000 pounds or twice the possible force of a body falling from a distance of 6 ft. (1.8 m).

You will often see references to American National Standards Institute (ANSI) standards regarding the performance of PFAS products. ANSI publishes standards for PFAS equipment that provide manufacturers with guidelines for their products. Its standards cover items such as design, testing, markings, and performance. Items that bear the ANSI certification markings let the user know that a product has been built and tested to ensure its integrity. Many organizations require that PFAS equipment used by its employees or agents meets ANSI standards and is marked accordingly, without exception.

A PFAS is only as good as its weakest component. They should never be used for any other task, such as lifting tools or materials to the work area. The specifications and usage instructions for components such as the full body harness (*Figure 22*) will identify the working load limit, and this limit should be strictly followed. Workers must use common sense and not overload themselves with equipment, even if the weight does fall within the limits of the PFAS specifications. Always examine fall-protection before each use and do not use it if it is worn or damaged in any way. Check with the manufacturer before mixing equipment from different manufacturers. All substitutions must be approved.

Before using fall protection equipment on the job, employers should provide employees with training in the basics of fall protection and the proper use of the equipment. All equipment supplied by employers must meet OSHA standards for strength. Before each use, always read the instructions and warnings on any fall protection equipment. Inspect the equipment using the following guidelines:

- Examine harnesses and lanyards for mildew, wear, damage, and deterioration.
- Make sure no straps are cut, broken, torn, or scraped.
- Check for damage due to fire, chemicals, or corrosives.
- Hardware should be free of cracks, sharp edges, and burns.
- Snap hooks should close and lock tightly, and buckles should work properly.
- Check ropes for wear, broken fibers, pulled stitches, and discoloration.

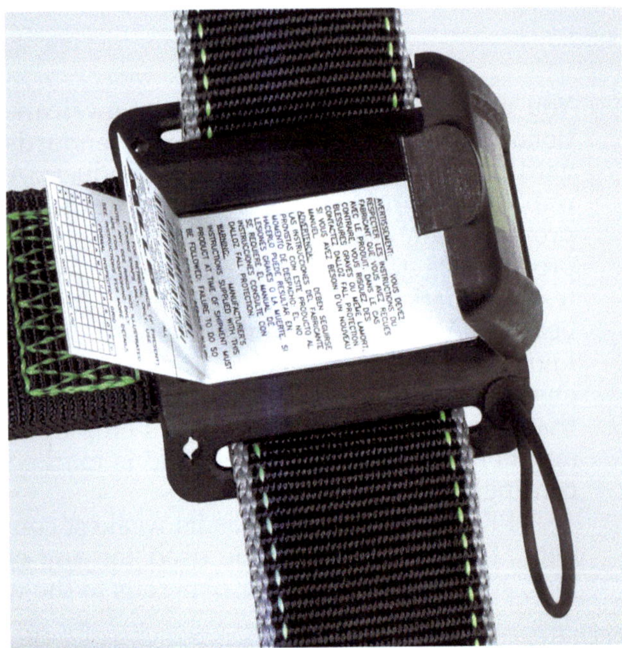

Figure 22 Body harness labeling.

- Make sure lifeline anchors and mountings are not loose or damaged.

> A supervisor must approve all equipment substitutions.

All damaged or defective parts must be immediately taken out of service and tagged as unusable or be destroyed. If the equipment has been subjected to impact from a fall, remove it from service until it can be inspected by a qualified person.

The PFAS should be inspected monthly by a competent person. This requirement should be established through your company's safety program. The competent person has the authority to impose prompt corrective measures to eliminate any hazards. If there is any question about a defect, no matter how small it may seem to be, err on the side of caution. Take the fall arrest component(s) out of service for testing or replacement.

5.2.0 Anchor Points

OSHA requires that anchor points (*Figure 23*) be rated at or equal to 5,000 pounds breaking or **tensile strength** or twice the intended load. Two workers can connect their fall-arrest lanyard to a single anchor point as long as it meets a 10,000-pound standard, or twice the intended load. Since the potential anchor load from a fall has doubled, so must the anchor point's rated **capacity**.

The ideal fall arrestance anchor point is located directly above the back D-ring (*Figure 24*) selected to minimize any swing zone hazards as well as the possible free fall distance. Swing zones are minimized when the anchor point is directly above the worker. Horizontal working situations can be far more hazardous in terms of the swing zone, as an anchor point directly above

Figure 23 Typical anchor point.

NCCER – *Fall Protection Orientation*

Figure 24 Back D-ring.

may not exist. Serious injury and damage can occur when a human body strikes an immoveable object while swinging as a pendulum. Although the PFAS may do its job by preventing the worker from falling a great distance, serious injury or death can still occur by striking an object in the swing zone.

Anchor points are often needed to secure the position of the worker, leaving the hands free to accomplish the task. Ideally, workers should select anchor points, and connect as needed, to maintain a potential fall distance of no more than 2 feet. Positioning connections are a factor in fall arrest, as the connection to a positioning strap will likely modify the swing zone and/or distance of the fall during an accident. However, they cannot be considered the primary anchor. Therefore, positioning anchor points are required by OSHA to be rated at a 3,000-pound strength instead of the 5,000-pound rating for primary fall arrest.

Remember the positioning lanyards, connected to D-rings on the harness other than the back or front chest D-ring, are fall restraints rather than fall arresting connections. A positioning lanyard does not take the place of a fall arrestance lanyard or anchor point. If you make a connection with the intention of placing weight or stress on it in any way, consider it a positioning connection.

5.2.1 *Anchoring Devices and Equipment Connectors*

Anchoring devices, commonly called *tie-off points*, support the entire weight of the fall arrest system.

The anchorage must be capable of supporting 5,000 pounds (22.2 kilonewtons) for each worker attached. Eyebolts, overhead beams, and integral parts of building structures are all types of anchorage points.

The D-rings, buckles, snap hooks, and other equipment that fasten and/or connect the parts of a personal fall arrest system are called *connectors*. Regulations specify how they are to be made and require D-rings and snap hooks to have a minimum tensile strength of 5,000 pounds. All such components should be designed for use with the attached hardware. Only locking-type snap hooks are permitted for use in personal fall arrest systems.

5.2.2 *Selecting an Anchor Point and Tying Off*

After the full-body harness has been put on, the next step is to connect it either directly or indirectly to a secure anchor point. This is called *tying off*. Tying off is always done before getting into a position where a fall could occur. Always follow the manufacturer's instructions for the best tieoff methods for the equipment. Use the following general instructions and guidelines when selecting an anchorage and tying off.

When selecting an anchorage point, ensure that it is as follows:

- Directly above the worker (when this is possible)
- Easily accessible
- Damage-free and capable of supporting 5,000 pounds per worker *or* twice the intended load *or* the safety factor of at least two for each worker attached.
- High enough so that no lower level will be struck should a fall occur

When tying off, consider the following:

- Tie-offs that use knots are weaker than other methods of attachment. Knots can reduce the lifeline or lanyard strength by 50 percent or more. A stronger lifeline or lanyard should be used to compensate for this.
- A shorter free fall can reduce the chances of falling into obstacles, being injured by the arresting force, or damaging the equipment. To limit the free fall, a shorter lanyard can be used between the lifeline and harness. Raising the tieoff point on the lifeline can reduce the amount of slack in the lanyard. The tieoff to the lifeline or anchor must always be higher than the connection to the harness. Tie off to the highest point practical.

- When using retractable systems, always be aware of the pendulum effect.

5.3.0 Harnesses

Full body harnesses like the one shown in *Figure 25* are available in sizes that are based on the height and weight of the user. The back D-ring is the only one used to connect the harness to the anchor point for primary fall arrest purposes unless you are climbing a ladder. When climbing a ladder, the front chest D-ring is the likely choice for connecting the fall arrest harness. D-rings located at the hips are used for positioning and fall restraint only. D-rings mounted to shoulders are often used for rescue situations. All of them can be used for fall restraint, but the back D-ring is the primary connection for fall arrest.

A body harness must fit correctly to ensure that it will provide proper protection. Do not place additional holes or openings in harness components under any circumstances. No field modifications to a body harness or lanyard should be attempted. Installation, maintenance, and inspection instructions are provided for every harness, and it is the responsibility of the worker to read and understand the details regarding his or her personal equipment. Workers must inspect their body harness each day that it is in use.

Harness straps are generally designed with some stretch to help absorb some of the potential force of a fall. This means good, taut installation on the body is essential so that the worker cannot fall out of the harness in the event of a fall.

Figure 26 shows the proper procedure for donning a common full body harness. The most important adjustments to be made include the chest straps, the groin straps, and the final position of the back D-ring. The related details that follow must be considered during the fitting and wearing of a full body harness:

> **NOTE**
>
> These guidelines are general in nature and the instructions provided for specific equipment by the manufacturer must always take precedence. It is the worker's responsibility to be intimately familiar with the duty of each and every ring and strap on a given harness.

- The back D-ring location is vital to proper fall arrest. Position this ring between the shoulder blades. If it is too low, it will tend to cause the body to hang in a more horizontal position during fall arrest, increasing pressure on the diaphragm and affecting breathing. If it is positioned too high or with too much slack, the D-ring may strike the worker's head at the base of the fall, and the shoulder straps may be pulled too tightly into the neck and restrict blood flow. The impact at the base of the fall arrest can be dramatic, so the force must be spread all around the body to prevent injury to any one portion.
- Chest straps generally form either an "H" pattern or an "X" pattern. Adjust "H" pattern straps to land between the bottom of the sternum and the belly button. This helps ensure the horizontal portion of the "H" does not contact the throat during a fall, choking the worker. Some harness designs may not allow for this adjustment, with the final position of the "H" being based solely on a properly sized harness.
- The position of the chest straps is also crucial for "X" pattern harnesses. Position the "X" at or just below the sternum.

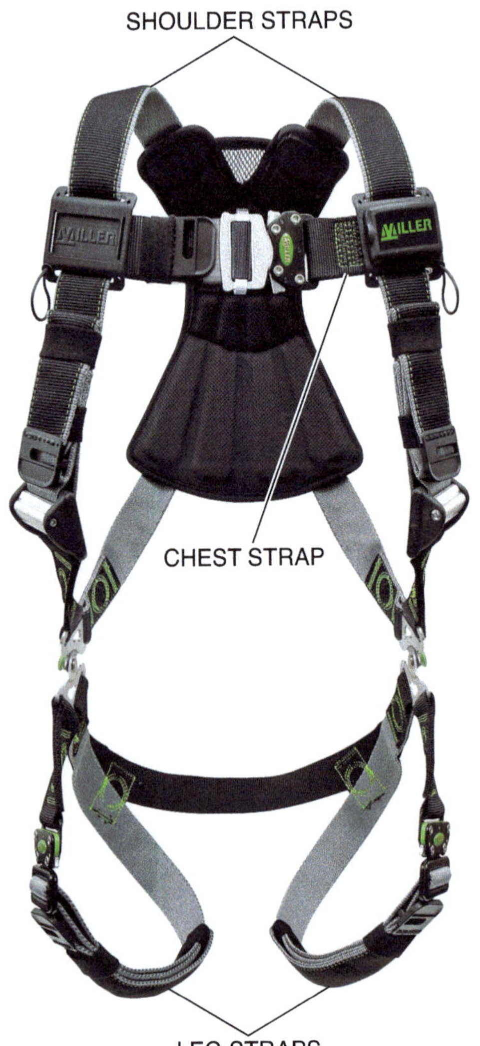

Figure 25 Full body harness.

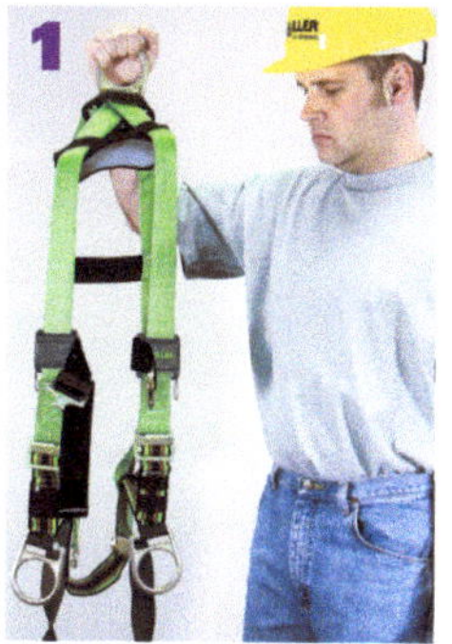	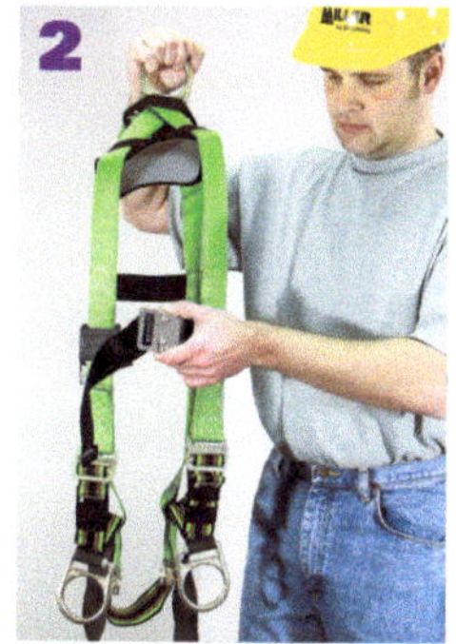	
Hold harness by back D-ring. Shake harness to allow all straps to fall in place.	If chest, leg and/or waist straps are buckled, release straps and unbuckle at this time.	Slip straps over shoulders so **D-ring is located in middle of back between shoulder blades.**

Pull leg strap between legs and connect to opposite end. Repeat with second leg strap. If belted harness, connect waist strap after leg straps.	**Connect chest strap and position in midchest area.** Tighten to keep shoulder straps taut.	After all straps have been buckled, **tighten all buckles so that harness fits snug but allows full range of movement.** Pass excess strap through loop keepers.

Figure 26 Installing the body harness.

- Leg straps are an integral and required part of a PFAS. Adjust the groin straps for a good, snug fit. Too much slack here will cause extreme discomfort in a fall, when the impact snatches them up tight and you are left suspended this way.
- A suspension trauma strap is recommended as part of the PFAS gear. The suspension trauma strap is stored in a pouch connected to the harness within easy reach. This is done by either one end of the strap being permanently sewn to the harness (by the manufacturer); one end of the strap attached to the harness with a carabiner (*Figure 27*); or by choking the pouch around a harness strap or hip D-ring. The strap

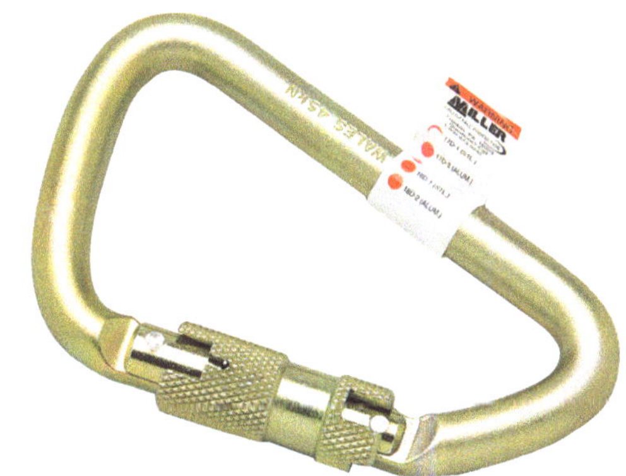

Figure 27 A carabiner.

can then be quickly removed and used without any possibility of the user dropping it. Once connected, the strap allows the worker to stand up in the harness, relieving suspended weight and pressure from the hips and groin. This helps open the path for blood flow from the legs back to the heart, preventing blood from pooling in the lower extremities.

- A separate waist or tool belt, while not considered a necessary component of the fall-arrest system, must be fitted properly. It is best used for body positioning with the D-rings precisely located at the hips, rather than in the front or rear. Do not adjust it in a way that could apply pressure to the kidneys or lower back. If the waist belt is not an integral part of the harness, it must not be worn on the outside of the harness—put it on first, then add the harness over it. This is also true of any added tool belts.
- Saddles, like waist belts, are also not considered an integral or required part of the PFAS. They are optional and often detachable. They are generally used by workers to allow a seated, suspended position when the task may require long periods in the same location.

It is important to note that not all hardware will qualify as a component of a PFAS. Some hardware is to be used only for attaching tools and equipment to the worker or to structures. Hardware used as connecting devices as part of the PFAS must be drop-forged steel and have a corrosion-resistant finish resistant to salt spray per ANSI standards. Any type of hook or carabiner must be equipped with safety gates or keepers to prevent the hooked object from being disconnected incidentally. In most cases, these safety gates will be required to be two-step, also called *double action*. Designs for these features vary, but those designed so that both movements required to open the gate can be done with a single hand are generally better. Using both hands to manipulate a single connector can be a hazard in itself.

Double-locking snap hooks (*Figure 28*) are usually curved and have an opening to allow connection to a line that can then be securely closed. They are usually not as consistent in appearance as carabiners and come in somewhat different shapes. When used as part of a PFAS, the security closure should be automatic. Snap hooks are designed to connect to D-rings primarily, not to each other. In most cases, snap hooks are already connected to a lanyard or rope to ensure the integrity of the connection and are not purchased separately.

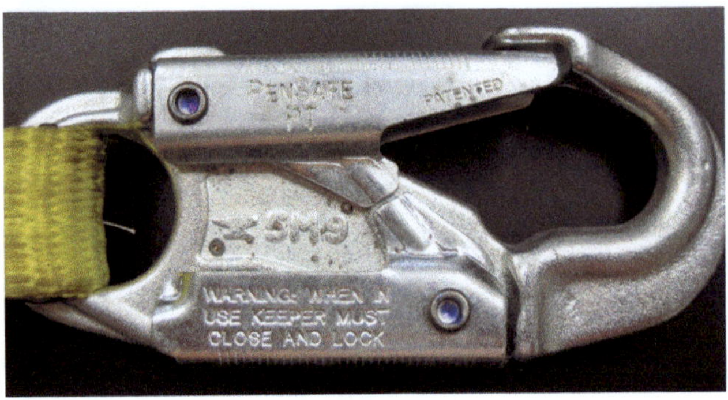

Figure 28 Double-locking snap hook.

5.4.0 Lanyards, Lifelines, and Deceleration Devices

Lanyards are an important part of the overall PFAS structure. They are used to link a body harness or body belt to a lifeline, deceleration device, or anchorage point. When properly used in conjunction with a D-ring, they can function as a fall restraint device. It is important to consider the specific function and load capacity when choosing a lanyard.

5.4.1 Lanyards

Lanyards are short, flexible lines with connectors on each end. They are used to connect a body harness or body belt to a lifeline, deceleration device, or anchorage point. There are many types of lanyards designed for different uses and climbing situations. All must have a minimum breaking strength of 5,000 pounds. Lanyards come in both fixed and adjustable lengths. They are made from steel, rope, or nylon webbing. Some have a shock absorber that absorbs up to 80 percent of the arresting force when a fall is being stopped. When choosing a lanyard for a particular job, it is always a good idea to follow the manufacturer's recommendations. Shock absorbers are required in harnesses and lanyard systems used for fall arrest.

WARNING!

When activated during the fall-arresting process, a shock-absorbing lanyard stretches as it acts to reduce the fall-arresting force. This potential increase in length must always be taken into consideration when determining the total free-fall distance from an anchor point. Failure to properly determine the free-fall distance can cause injury or death if a fall occurs.

NCCER – *Fall Protection Orientation*

Lanyards consist of some primary material of construction (rope, webbing, aircraft cable, etc.) with a connecting device attached to the ends. Lanyards used for fall arrest are either shock absorbing or self-retracting (*Figure 29*). Non-shock-absorbing lanyards, such as the one shown in *Figure 30*, are used for positioning and fall restraint. Lanyards used for positioning are not considered part of the fall-arrest system; they are fall restraints and will be attached to D-rings on the harness other than the back D-ring. Since fall restraint is all about preventing a fall from happening, lanyards used for positioning should not allow a fall or movement greater than 2 ft. (0.61 m). Two such lanyards are usually required to permit movement from one place to another.

The retractable fall arrest lanyard shown in *Figure 29 (B)* is rapidly becoming the preferred device. Since it automatically retracts or feeds lanyard as the worker moves about, there is never a great deal of slack that can pose a risk in itself.

(A) SHOCK-ABSORBING

(B) RETRACTABLE

Figure 29 Fall arrest lanyards.

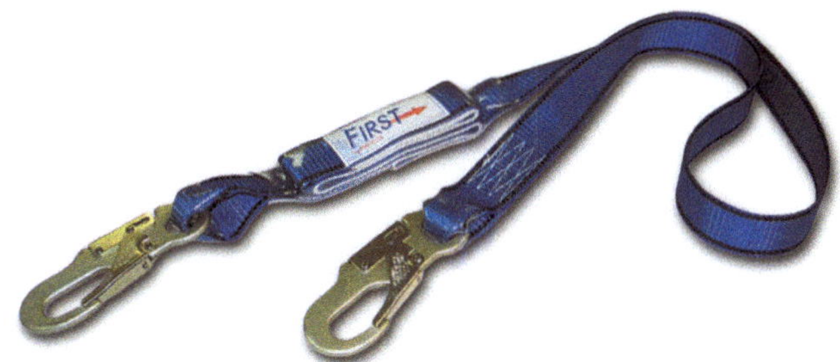

Figure 30 Non-shock-absorbing lanyard.

In addition, the potential fall distance is significantly reduced.

Lanyards for fall restraint or arrest must never be field-fabricated and must never be connected together to increase their length. Depending on the use, padding may be added during fabrication or added in the field to protect against sharp edges. Never tie a knot in a lanyard, as knots can severely reduce the load limit. Never wrap a lanyard around a structure and then choke it back into itself unless it is designed for this use. A special large D-ring is usually attached to the lanyard for this purpose. It is important to remember that whenever you are wearing a PFAS, 100 percent tie-off is required. A PFAS is absolutely useless unless you are tied off.

The D-ring used on the PFAS for fall arrest may only be connected to one live connection at a time. This can be challenging when trying to move from one point to another, especially horizontally. The user must be able to reach back and disconnect a lanyard, while maintaining one lanyard connected at all times. A Y-configured lanyard (*Figure 31*) can be used for this purpose. A Y-configured lanyard has a single point of attachment at the D-ring that is used to accommodate two lanyards. They are also referred to as *double-leg lanyards* or *tieback lanyards*.

Before using a shock-absorbing or self-retracting lanyard (SRL), the potential fall distance is determined. Then the proper equipment is selected to meet available fall clearance (see *Figure 32*). Note in the figure that the fall distance with the self-retracting lanyard is significantly shorter than it is with a shock-absorbing lanyard. This is the reason for its increasing popularity. Failure to select proper equipment and calculate fall distance may result in serious personal injury or death. These calculations must be done by experienced and qualified personnel on the job site. If a personal fall arrest system is actuated due to a fall, it must be inspected by a competent person before it can be used again. In most cases, replacement will be necessary.

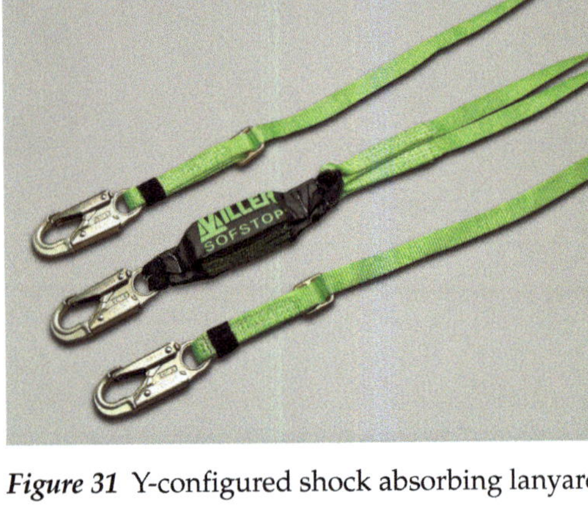

Figure 31 Y-configured shock absorbing lanyard.

> **CAUTION**
>
> When working at elevations, a dropped tool can become a serious hazard. If the tool has moving parts, like battery-powered drills, the fall will likely destroy it. Tool lanyards (*Figure 33*) are available that have been specifically designed for work in the elevated environment. Tethering tools to the tool belt or wrist can prevent injuries, save tools, prevent component damage, and prevent lost time in retrieval.

5.4.2 Lifelines

A lifeline is a flexible line such as a cable or rope connected vertically to an anchorage at one end (vertical lifeline), or horizontally to an anchorage at both ends (horizontal lifeline). It serves as a means for connecting other components of a personal fall arrest system to the available anchorage. Vertical lifelines are suspended from a fixed anchorage. A fall arrest device such as a rope grab (*Figure 34*) or retractable lanyard is attached to the lifeline. A beam grab, or beamer (*Figure 35*), is sometimes used as an anchorage for a lifeline. Vertical lifelines must have a minimum breaking strength of 5,000 pounds (2,267 kg). Workers must use separate vertical lifelines. A vertical lifeline must have a termination on the end unless it extends to the ground or to the next lower level of the structure.

Horizontal lifelines (*Figure 36*) are connected between two fixed anchorages. A lanyard is attached to the lifeline. Horizontal lifelines must be designed, installed, and used under the supervision of a competent person. The required breaking strength of any horizontal lifeline is determined by the number of workers that will be attached to it.

> **WARNING!**
>
> Keep in mind that horizontal lifelines are rated for a maximum number of connected workers. Check with your supervisor or the manufacturer before connecting to a lifeline that is being used by other workers.

5.4.3 Deceleration Devices

Deceleration devices limit the arresting force that a worker is subjected to when the fall is stopped suddenly. Rope grabs and self-retracting lifelines are two common deceleration devices. A rope grab connects to a lanyard and attaches to a lifeline. In the event of a fall, the rope grab is pulled down by the attached lanyard, causing it to grip the lifeline and lock in place. Some rope grabs have a mechanism that allows the worker to unlock the device and slowly descend down the lifeline to the ground or surface below.

Self-retracting lifelines provide for unrestricted movement and fall protection while climbing and descending ladders and similar equipment, or when working on multiple levels. Typically, they have a 25 to 100 ft. (7.6 to 30.4 m) galvanized steel cable that automatically takes up the slack in the attached lanyard, keeping the lanyard out of the worker's way. In the event of a fall, a centrifugal braking mechanism engages to limit the worker's free fall. Per OSHA requirements, self-retracting lifelines and lanyards that limit the free-fall distance to 2 ft. (0.61 m) or less must be able to support a minimum tensile load of 3,000 pounds (13.3 kilonewtons). Those that do not limit the free fall to 2 ft. (0.61 m) or less must be able to hold a tensile load of at least 5,000 pounds (22.2 kilonewtons). *Figure 37* shows a rope grab and a self-retracting lifeline.

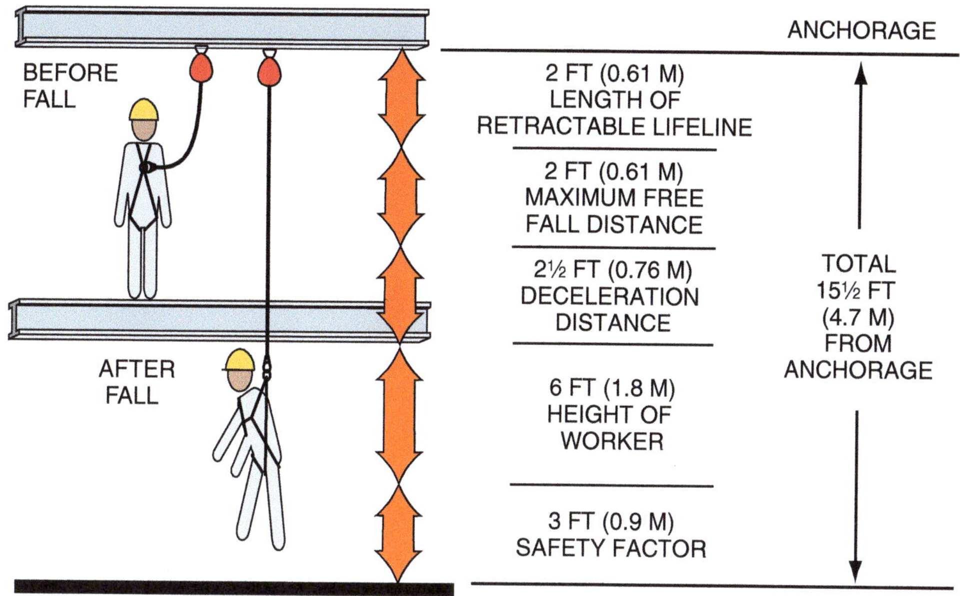

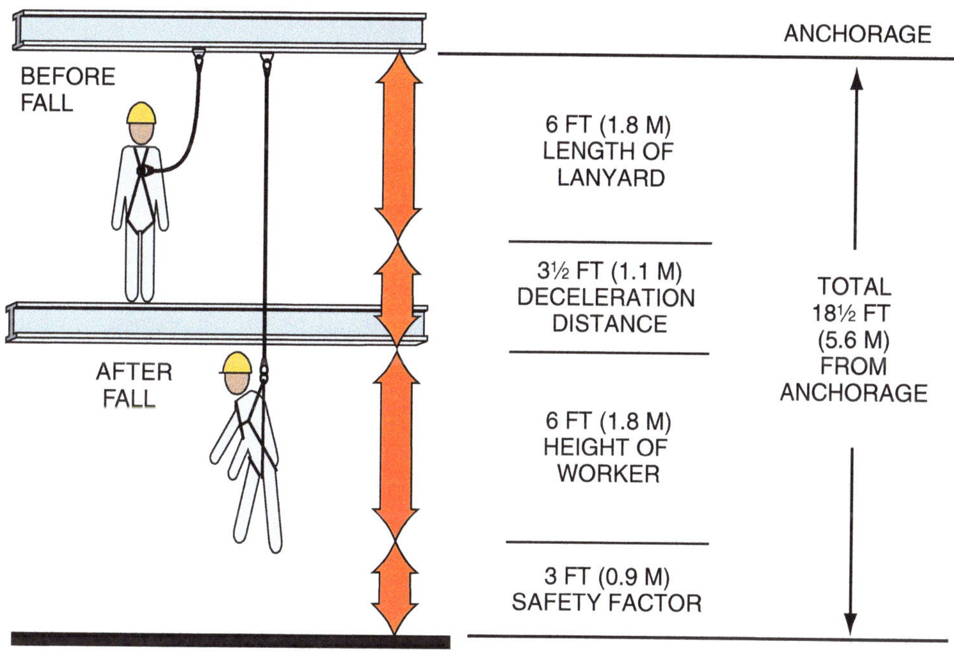

Figure 32 Potential fall distances for retractable and shock-absorbing lanyards.

Figure 33 Tool lanyard.

Figure 34 Rope grab.

5.5.0 Barriers, Guardrails, and Safety Nets

While PFAS equipment is designed to protect craft professionals in their designated work spaces, other types of devices are essential in keeping workers away from certain areas altogether. The use of barriers, guardrails, and safety nets come into play when specific areas in and around the construction site may pose a particular danger.

5.5.1 Barriers and Barricades

Barriers and barricades are used to alert workers to potential hazards and to help prevent accidents. Each work zone may have different policies

Figure 35 Beam grab.

and procedures for how and when to use them, so learn and follow the rules at the job site. Sometimes barriers will be placed in an area to prevent injury to workers, and other times they will be placed to alert heavy equipment operators of dangers. For example, a barrier may be erected to keep a bulldozer off steeply graded ground. *Figure 38* shows the various uses for barricades and barriers.

Barricades and barriers help to prevent accidents and injuries, but they are not foolproof. When you are assigned to a job site, carefully examine the area that you will be working in. Before moving heavy equipment, walk the area to ensure that there are no hidden hazards. This is especially important when there are obstructions, such as snow, mud, leaves, and other debris in the area. Never move a barricade, even in an emergency, unless certain that it is safe to enter the barricaded area.

Never remove a barricade unless you have been authorized to do so. Follow your employer's procedures for putting up and removing barricades.

5.5.2 Guardrail Systems

Guardrails (*Figure 39*) are a common type of fall prevention. They protect workers by providing a barrier between the work area and the ground or lower work areas. They may be made of wood, pipe, steel, or wire rope and must be able to support 200 pounds (90 kg) of force applied in any direction to the top rail and 150 pounds (68 kg) for the midrail. A guardrail must be 42 in. ±3 in. (106 ±8 cm) high to the top rail and have a toeboard that is a minimum of 4 in. (10 cm) high. This helps

Figure 36 Horizontal lifeline.

Figure 37 Rope grab and self-retracting lifeline.

to prevent the inadvertent loss of tools or material through the bottom rail. The toeboard must be securely fastened with not more than $\frac{1}{4}$ in. (5 mm) clearance above the floor level.

On each platform more than 10 ft. (3 m) above the ground or the next lower level, guardrails (consisting of toprails and midrails) are required on all open sides that are more than 14 inches from a solid-faced structure. Guardrail requirements can vary from state to state.

- Guardrails must be able to withstand a force of 200 pounds applied downward or horizontally.
- Minimum height for the toprail is 38 to 45 in. (11.6 to 13.7 m). (Always check regional and state requirements.)
- A midrail must be placed halfway between the platform and the toprail.
- Wire rope may be used as guardrail but must not deflect below the minimum guardrail height.

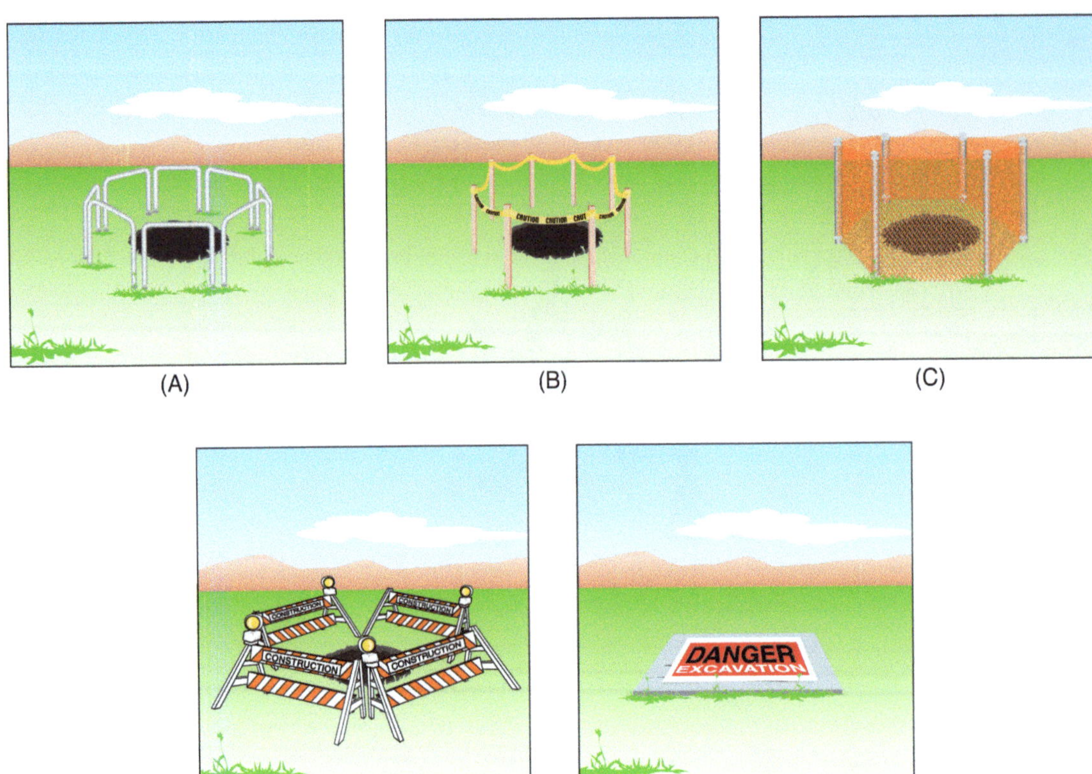

Figure 38 Common types of barricades and barriers.

Figure 39 Guardrails.

more than 30 ft. (9 m) beneath the work area. There must be enough clearance under a safety net to prevent a worker who falls into it from hitting the surface below it. There must also be no obstruction between the work area and the net. Depending on the actual vertical distance between the net and the work area, the net must extend 8 to 13 ft. (2 to 4 m) beyond the edge of the work area. Mesh openings in the net must be limited to 36 in.^{2} (232 cm^2) and 6 in. (15 cm) from the side. The border rope must have a 5,000-pound (2,267 kg) minimum breaking strength, and connections between net panels must be as strong as the nets themselves. Safety nets must be inspected at least once a week and after any event that might have damaged or weakened them. Worn or damaged nets must be removed from service.

- Guardrails must be smooth to prevent cuts. Steel or plastic banding may not be used as guardrails.
- The cross bracing may be considered as part of the toprail or midrail, depending on the height of the cross bracing.

5.5.3 Safety Nets

Safety nets are used for fall protection on bridges and similar projects. They must be installed not

5.5.4 Safe Climbing Devices

There are several different systems and devices that allow a worker to climb more confidently and safely, while helping to eliminate dangerous situations. In some cases, a worker must connect to an anchor, climb, connect to another anchor, disconnect the first anchor, climb some more, etc. Safe climbing devices and systems help to eliminate the many disconnects and allow for a smoother, safer climb.

Drop-Testing Safety Nets

Safety nets should be drop-tested at the job site after the initial installation, whenever relocated, after a repair, and at least every six months if left in one place. The drop test consists of a 400-pound (181 kg) bag of sand of 29" to 31" (74 to 79 cm) in diameter that is dropped into the net from at least 42" (107 cm) above the highest walking/working surface at which workers are exposed to fall hazards. If the net is still intact after the bag of sand is dropped, it passed the test.

Cable grabs, also called *shuttles*, pucks, rope grabs, or guided fall arresters, are mounted on the cable for the worker to use as a connection point. The cable grab locks down on the cable to help restrain a fall when movement is too rapid or aggressive. In normal use, the cable grab simply rides the cable up with the user, sliding along without resistance. Its freedom of movement upward is important to ensure that it does not suddenly snag or stop.

Like other fall-protection equipment, safe climbing devices must be inspected before use and tagged out of service if they show any damage or excessive wear.

Additional Resources

Basic Construction Safety and Health, Fred Fanning. 2014. CreateSpace Independent Publishing Platform.

Construction Safety, Jimmie W. Hinze. Second Edition. 2006. New York, NY: Pearson.

DeWalt Construction Safety/OSHA Professional Reference, Paul Rosenberg, American Contractors Educational Services. 2006. DEWALT.

Online resources:

Code of Federal Regulations (*CFR*), **www.ecfr.gov**

Occupational Safety and Health Administration (OSHA), **www.osha.gov**

OSHA Videos, **www.osha.gov/video**

5.0.0 Section Review

1. Personal fall arrest systems are designed to prevent a worker from falling more than ______.

    a. 4 ft.
    b. 6 ft.
    c. 10 ft.
    d. 12 ft.

2. If two workers connect to the same anchor point, it must be rated for twice the intended load or ______.

    a. 3,000 pounds
    b. 5,000 pounds
    c. 7,000 pounds
    d. 10,000 pounds

3. D-rings and snap hooks must have a minimum tensile strength of ______.

    a. 1,000 pounds
    b. 1,500 pounds
    c. 3,000 pounds
    d. 5,000 pounds

Figure SR01

4. The type of barricade shown in *Figure SR01* is a ______.

    a. work-zone barrier
    b. protective barricade
    c. warning barricade
    d. zone marker

6.0.0 LADDERS, STAIRS, AND SCAFFOLDS

Objective

Identify and describe the safe use of ladders, stairs, and scaffolds.

  a. Describe the hazards related to use of ladders and stairs in the workplace.
  b. Identify the safety requirements and proper use of various types of ladders and stairs.
  c. Identify types of scaffolds and describe how to use them safely.

Trade Terms

Cut: A common term for a scaffold level.

Hand line: A line attached to a tool or object so a worker can pull the tool up after climbing a ladder or scaffold.

Maximum intended load: The total weight of all people, equipment, tools, materials, and loads that a ladder can hold at one time.

Putlogs: Horizontal scaffold members on which the scaffold platform rests.

The most common accidents associated with ladders and scaffolding are falling, being struck by falling objects, and electrocution. In one instance, a worker slipped from a fixed ladder attached to a water tower and fell 40' to the ground. He died because he was not using the proper safety equipment.

Safety is your top priority on a job. It is your responsibility to learn how to set up, use, and maintain ladders. You also need to understand the construction requirements and safety hazards associated with scaffolding systems.

6.1.0 Ladder and Stair Hazards

Ladders are used daily to perform work in elevated locations. Any time work is performed above ground level, there is a risk of incidents. You can reduce this risk by carefully inspecting ladders before you use them and by using them properly.

Overloading means exceeding the maximum intended load of a ladder. Overloading can cause ladder failure, which means that the ladder could buckle, break, or topple. The maximum intended load is the total weight of all people, equipment, tools, materials, loads that are being carried, and other loads that the ladder can hold at any one time. Check the manufacturer's specifications to determine the maximum intended load. Ladders are usually given a duty rating that indicates their load capacity, as shown in *Table 2*. Note that ladders designed for the metric market are not usually direct equivalents to ladders built for the American market. Capacity ratings of 130 kg (286 lb.) and 150 kg (330 lb.) are the most common for the trades. Ladder heights will also be stated in metric units and may not be exactly the same size as those designed for the American market.

> **WARNING!**
>
> Use ladders only for their intended purposes. Ladders are not interchangeable; incorrect use of a ladder can result in injury or damage.

> **WARNING!**
>
> When using a ladder, be sure to maintain three-point contact with the ladder while ascending or descending. Three-point contact means that either two feet and one hand or one foot and two hands are always touching the ladder.

6.2.0 Ladder and Stair Types and Use

There are different types of ladders to use for different jobs (*Figure 40*). Selecting the right ladder for the job at hand is important to complete a job as safely and efficiently as possible. Ladder types include portable straight ladders, extension ladders, stepladders, and fixed ladders. Safe use of stairs is also critical on the job site.

Aluminum ladders are corrosion-resistant and can be used in situations where they might be exposed to the elements. They are also lightweight and can be used where they need to be frequently lifted and moved.

Table 2 Ladder Duty Ratings and Load Capacities

Duty Ratings	Load Capacities
Type IAA	375 lb., extra-heavy duty / professional use
Type IA	300 lb., extra-heavy duty / professional use
Type I	250 lb., heavy duty / industrial use
Type II	225 lb., medium duty / commercial use
Type III	200 lb., light duty / household use

Case History

A worker was climbing a 10 ft. (3.05 m) ladder to access a landing, which was 9 ft. (2.74 m) above the adjacent floor. The ladder slid down, and the worker fell to the floor, sustaining fatal injuries. Although the ladder had slip-resistant feet, it was not secured, and the railings did not extend 3 ft. (0.91 m) above the landing.

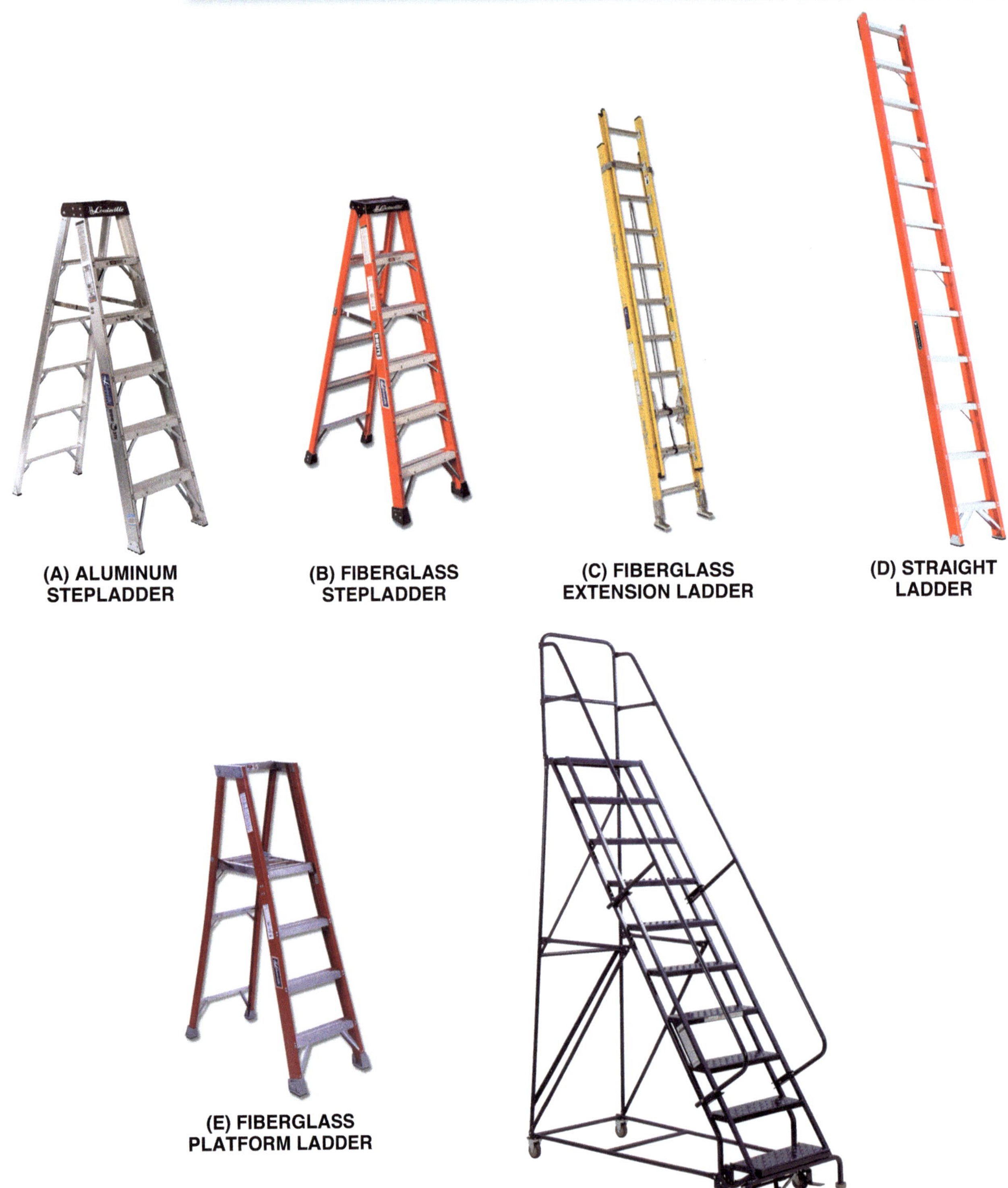

Figure 40 Different types of ladders.

Wooden ladders, which are heavier and sturdier than fiberglass or aluminum ladders, can be used when heavy loads must be moved up and down. Wooden ladders should never be painted. The paint could hide cracks in the rungs or rails. Clear varnish, shellac, or a preservative oil finish will protect the wood without hiding defects.

Fiberglass ladders are nonconductive and also very durable, so they are useful in situations involving electrical work or where some amount of rough treatment is unavoidable. Both fiberglass and aluminum are easier to clean than wood.

When selecting a ladder, consider its features and how it meets the needs of the job. Always consider the highest duty rating and weight limit needed, as well as the height requirements. A ladder that is too long or too short will not allow the work surface to be reached easily, safely, or comfortably.

When selecting a ladder for a job, it is important to choose one that will extend at least 36" above the landing surface you are trying to reach. Always place the base of the ladder so that the distance between the base and the wall is one-quarter of the ladder length from the base elevation to the point where the ladder touches the wall. It is very important to place a straight ladder at the proper angle before using it. A ladder placed at an improper angle will be unstable and could cause a fall. *Figure 41* shows a properly positioned straight ladder.

Keep the following precautions in mind when setting up and using any type of ladder:

- Do not use an aluminum ladder when performing any type of electrical work or whenever there is a possibility that you might come into contact with electrical conductors.
- Place the ladder in a stable manner.
- Place the ladder so that it leaves 6" of clearance in back of the ladder and 30" of clearance in front of the ladder.
- Place the ladder so that it leans against a solid and immovable surface. Never place a ladder against a window, door, doorway, sash, loose or movable wall, or box.
- Face the ladder when climbing up or down.
- Climb or descend the ladder one rung at a time. Never run up or slide down a ladder.
- Do not use ladders during high winds. If you must use a ladder in windy conditions, make sure you lash the ladder securely in order to prevent slippage.
- Check to make sure the soles of your shoes are free of oil, mud, and grease.

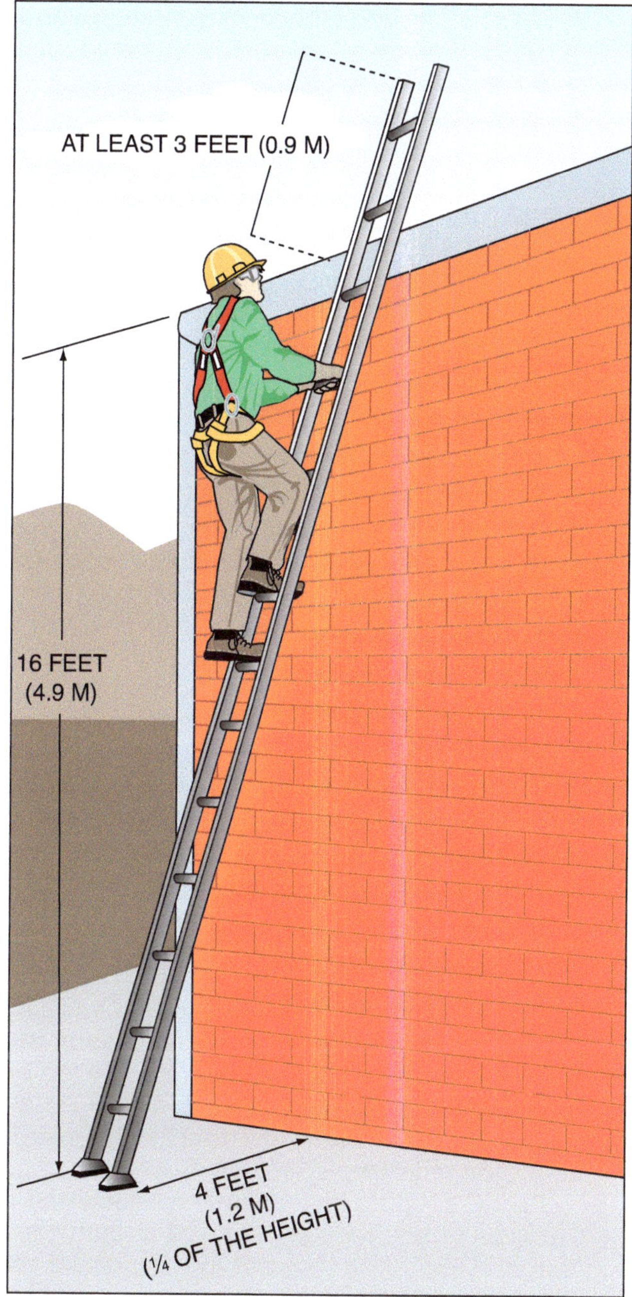

Figure 41 Proper ladder positioning

- Use a rope to raise and lower any tools and materials that you might need. This keeps both hands free to hold the ladder securely while climbing.
- Never rest any tools or materials on the top of a ladder.
- Move the ladder in line with the work to be done. Never lean sideways away from the ladder in order to reach the work area.
- Never stand on the top two rungs of a ladder.
- Use ladders only for short periods of elevated work. If you must work from a ladder for extended periods, use a personal fall-protection system.

- Lay the ladder on the ground when you have finished using it, unless it is anchored securely at the top and bottom where it is being used.
- Never use makeshift substitutes for ladders.
- Never use stepladders for straight-ladder work. Each ladder is designed for a specific purpose and climbing conditions, and they generally fall into these four categories:
  - Straight
  - Extension
  - Step
  - Fixed

6.2.1 Straight Ladders

Straight ladders consist of two rails, rungs between the rails, and safety feet on the bottom of the rails (*Figure 42*). Straight ladders are generally made of aluminum, wood, or fiberglass.

Figure 43 shows the safety feet attached to a straight ladder. Make sure the feet are securely attached and that they are not damaged or worn down. Do not use a ladder if its safety feet are not in good working order.

The distance between the foot of a ladder and the base of the structure it is leaning against must be one-fourth of the distance between the ground and the point where the ladder touches the structure. Stated another way, there should be a 4-to-1 ratio between the distances. For example, if the height of the wall shown in *Figure 41* is 16 ft. (4.9 m), the base of the ladder should be 4 ft. (1.2 m) from the base of the wall. If you are going to step off a ladder onto a platform or roof, the top of the ladder should extend at least 3 ft. (0.9 m) above the point where the ladder touches the platform, roof, side rails, etc.

Ladders should be used only on stable and level surfaces unless they are secured at both the bottom and the top to prevent any incidental movement (*Figure 44*). Never try to move a ladder while you are on it. If a ladder must be placed in front of a door that opens toward the ladder, the door should be locked or blocked open. Otherwise, the door could be opened into the ladder.

Ladders are made for vertical use only. Never use a ladder as a work platform by placing it horizontally. Make sure the ladder you are about to climb or descend is properly secure before you do so. Check to make sure the ladder's feet are solidly positioned on firm, level ground. Also check to make sure the top of the ladder is firmly positioned and in no danger of shifting once you begin your climb. Remember that your own weight will affect the ladder's steadiness once you mount

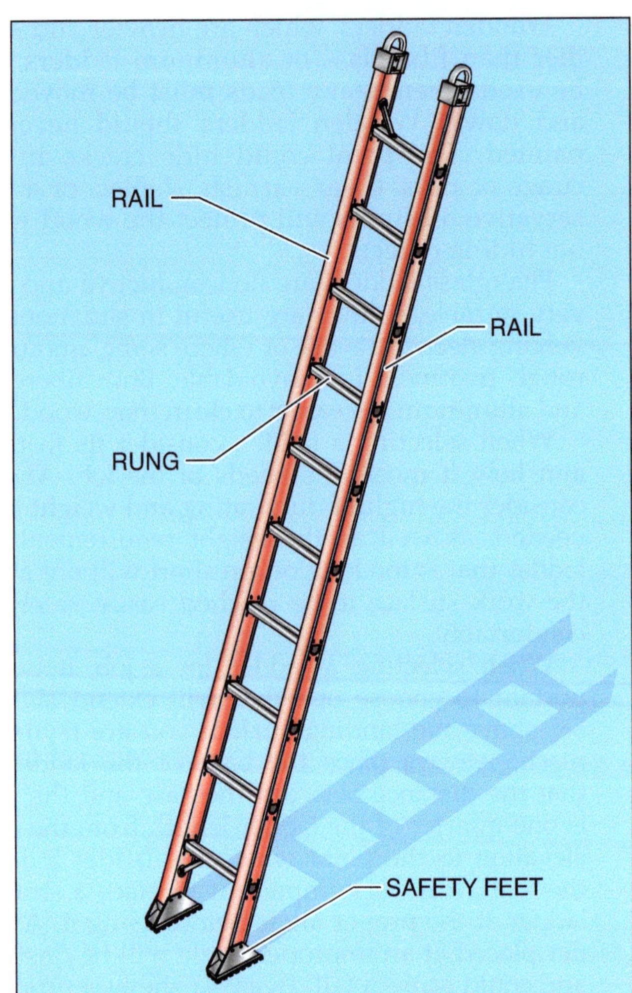

Figure 42 Portable straight ladder.

Figure 43 Ladder safety feet.

Figure 44 Securing a ladder.

it. It is important to test the ladder first by putting some of your weight on it without actually beginning to climb. This way, you can be sure that the ladder will remain steady as you climb.

When climbing a straight ladder, keep both hands on the rails or rungs. Maintain three points of contact at all times, as shown in the figure. This can be two feet and one hand, or one foot and two hands. Always keep your body's weight in the center of the ladder between the rails. Face the ladder at all times. Never go up or down a ladder while facing away from it.

> **WARNING!**
>
> Remember that the addition of your own weight will affect the ladder's steadiness once you mount it. It is important to test the ladder first by applying some of your weight to it without actually beginning to climb. This way, you will be sure that the ladder remains steady as you ascend.

To carry a tool while you are on the ladder, use a hand line or tagline attached to the tool. Climb the ladder and then pull up the tool. Don't carry tools in your hands while you are climbing a ladder.

6.2.2 Extension Ladders

An extension ladder is actually two straight ladders connected so that the overlap between them can be altered to increase or decrease the length of the ladder (*Figure 45*).

Extension ladders are positioned and secured following the same rules as straight ladders. There are, however, some safety rules that are unique to extension ladders:

- When you adjust the length of an extension ladder, always reposition the movable section from the bottom, not the top, so you can make sure the rung locks are properly engaged after you make the adjustment (*Figure 46*).
- Make sure the section locking mechanism is fully hooked over the desired rung.
- Make sure that all ropes used for raising and lowering the extension are clear and untangled.
- Make sure the extension ladder overlaps between the two sections (*Figure 47*). For ladders up to 36' (10.5 m) long, the overlap must be at least 3' (0.9 m). For ladders 36' to 48' (10.8 to 14.6 m) long, the overlap must be at least 4' (1.2 m). For ladders 48' to 60' (14.6 to 18.3 m) long, the overlap must be at least 5' (1.5 m).

ALUMINUM   **FIBERGLASS**

Figure 45 Examples of extension ladders.

Figure 46 Rung locks.

- Never stand above the highest safe standing level on a ladder. On an extension ladder, this is the fourth rung from the top. If you stand higher, you may lose your balance and fall. Some ladders have colored rungs to show where you should not stand.

> Extension ladders have a built-in extension stop mechanism. Do not remove this mechanism. If the mechanism is removed, it could cause the ladder to collapse under a load.

- Avoid carrying anything on a ladder, because it will affect your balance and may cause you to fall.
- Keep yourself centered on the ladder. Do not over-reach, lean to one side, or try to move a ladder while standing on it.

6.2.3 Stepladders

Stepladders are self-supporting ladders made of two sections hinged at the top (*Figure 48*). The section of a stepladder used for climbing consists of rails and rungs like those on straight ladders. The other section consists of rails and braces. Spreaders are the hinged arms between the sections that keep the ladder stable and prevent it from folding while in use. A stepladder may have a pail shelf to hold paint or tools.

Case History

Think Before You Act

During the construction of a building, a masonry worker was instructed by his foreman to prepare a batch of mortar on the second level and use the stairway to carry it to the third level. This worker decided it would be quicker and easier to use the top section of an extension ladder (without safety feet) instead of the stairway. He set up the ladder by placing one end of the ladder on the wet concrete floor and leaning the other end of the ladder against the wall. He then started to climb. When he was halfway up, the ladder slipped on the wet floor, causing him to fall approximately 12' to his death.

The Bottom Line: Always follow safety instructions.

Source: The National Institute for Occupational Safety and Health (NIOSH)

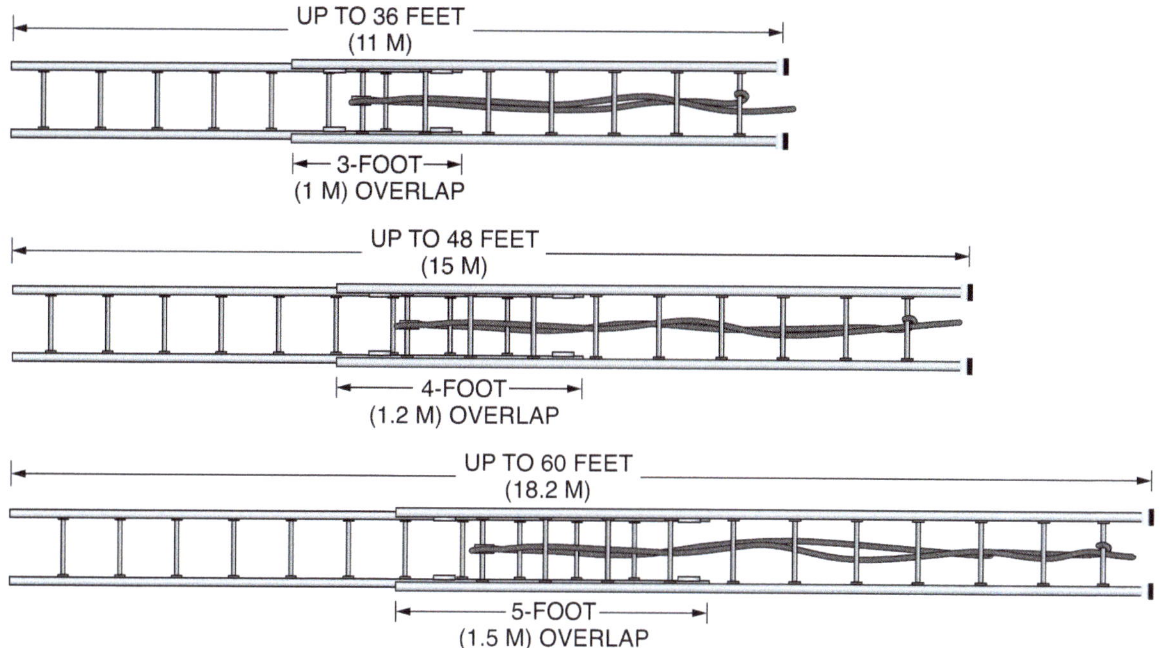

Figure 47 Overlap lengths for extension ladders.

Figure 48 Typical fiberglass stepladder.

Inspect stepladders the way you inspect straight and extension ladders. For stepladders, though, pay special attention to the hinges and spreaders to be sure they are in good repair. Also, be sure the rungs are clean. A stepladder's rungs are usually flat, so oil, grease, or dirt can easily build up on them and make them slippery.

Follow these rules when using stepladders (*Figure 49*):

- Be sure that all four feet are on a hard, even surface when you position a stepladder. If they're not, the ladder can rock from side to side or corner to corner when you climb it.
- Never stand on the top step or the top of a stepladder. Putting your weight this high will make the ladder unstable. The top of the ladder is made to support the hinges, not to be used as a step.
- Make sure the spreaders are locked in the fully open position when the ladder is in position.
- Never use the braces for climbing even though they may look like rungs. They are not designed to support your weight.

WARNING!

Never stand on a step with your knees higher than the top of a stepladder. You need to be able to hold on to the ladder with your hand. Keep your body centered between the side rails.

DOs

• Be sure your ladder has been properly set up and is used in accordance with safety instructions and warnings.
• Wear shoes with non-slip soles.

• Keep your body centered on the ladder. Hold the ladder with one hand while working with the other. Never let your belt buckle pass beyond either ladder rail.

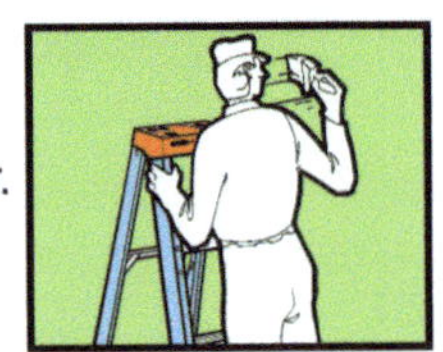

• Move materials with extreme caution. Be careful pushing or pulling anything while on a ladder. You may lose your balance or tip the ladder.

• Get help with a ladder that is too heavy to handle alone. If possible, have another person hold the ladder when you are working on it.

• Climb facing the ladder. Center your body between the rails. Maintain a firm grip.
• Always move one step at a time, firmly setting one foot before moving the other.

• Haul materials up on a line rather than carry them up an extension ladder.
• Use extra caution when carrying anything on a ladder.

Read ladder labels for additional information.

DON'Ts

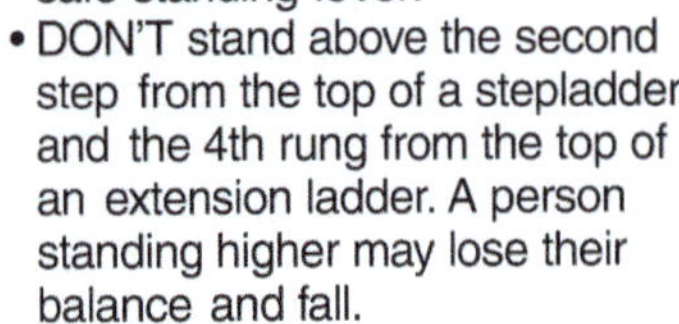

• DON'T stand above the highest safe standing level.
• DON'T stand above the second step from the top of a stepladder and the 4th rung from the top of an extension ladder. A person standing higher may lose their balance and fall.

• DON'T climb a closed stepladder. It may slip out from under you.
• DON'T climb on the back of a stepladder. It is not designed to hold a person.

• DON'T stand or sit on a step-ladder top or pail shelf. They are not designed to carry your weight.
• DON'T climb a ladder if you are not physically and mentally up to the task.

• DON'T exceed the Duty Rating, which is the maximum load capacity of the ladder. Do not permit more than one person on a single-sided stepladder or on any extension ladder.

• DON'T place the base of an extension ladder too close to the building as it may tip over backward.
• DON'T place the base of an extension ladder too far away from the building, as it may slip out at the bottom.

Please refer to the 4 to 1 Ratio Box.

• DON'T over-reach, lean to one side, or try to move a ladder while on it. You could lose your balance or tip the ladder.

Climb down and then reposition the ladder closer to your work!

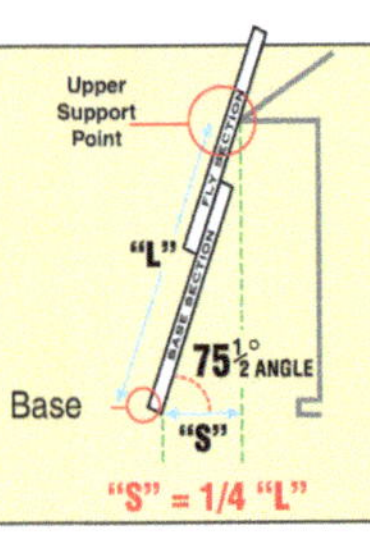

Figure 49 Ladder safety Dos and Don'ts.

6.2.4 Fixed Ladders

Fixed or stationary ladders (*Figure 50*) are ladders that cannot be readily moved or carried because they are a permanent part of a building or structure. Fixed ladders must be capable of supporting at least two loads of 200 pounds (91 kilograms) each. Each step or rung must be capable of supporting a single concentrated load of at least 200 pounds applied in the middle of the step or rung.

The angle of a fixed ladder must be no greater than 90 degrees from the horizontal, as measured at the back of the ladder. The ladder must be equipped with cages, wells, ladder safety devices, or SRLs when the climb is less than 24 ft., but the top of the ladder is a distance greater than 24 ft. above lower levels.

Ladders where the total length of the climb equals or exceeds 24 ft. must be equipped with one of the following:

- Ladder safety devices
- Self-retracting lifelines and rest platforms at intervals not to exceed 150 ft. (15.2 m)
- A cage or well and multiple ladder sections

6.2.5 Stairways

Stairways are routinely used on construction sites where there is a break in elevation of 19 in. (46 cm) or more, and no ramp, runway, sloped em-bankment, or personnel hoist is provided. Many contractors switch to a temporary stairway above the fourth cut, or level, of scaffolding. Continue the temporary stairway upward as the work progresses. It must be maintained in serviceable condition until at least one permanent stairway has been completed.

Stairways must be adequately lighted. If temporary stairways are required, they should be adequately braced, wide enough for two persons, and equipped with railings and toeboards. Ramps or runways used in place of stairways should also have railings. Stairways and ramps should be kept free of ice, snow, grease, mud, and other slipping hazards.

Observe the following regulations, based on OSHA standards, when using stairways on a job site:

- Stairways having four or more risers or rising more than 30 in. (76 cm), whichever is less, must be equipped with at least one handrail and one stair railing system along each unprotected side.
- Winding and spiral stairways must be equipped with a handrail offset sufficiently to prevent walking on those portions of the stairways where the tread width is less than 6 in. (15 cm).
- Stair railings must be not less than 36 in. (91 cm) from the upper surface of the stair railing system to the surface of the tread, in line with the face of the riser at the forward edge of the tread.

To reduce the likelihood of slips, trips, or falls, keep stairways clean and clear of debris. Do not store any tools or materials on stairways, and clean up liquid spills, rain water, or mud immediately.

Stairways must have adequate lighting. This can sometimes be a problem because permanent lighting is usually installed after stairway construction is completed. If the lighting is inadequate, temporary lighting should be installed in the stairway. Each bulb should be equipped with a protective cover and the string should be inspected daily for burned out or broken bulbs.

Whenever possible, avoid using stairways to transport materials between floors. Carrying small materials and tools is fine, as long as the materials do not block your vision. Going up or down a stairway while carrying large items is physically demanding and increases the chance of injuries and falls. Use the building elevator or crane service to transport large materials from one floor to another.

Figure 50 Typical fixed ladder.

6.3.0 Scaffolds

Scaffolds provide safe elevated work platforms for people and materials. They are designed and built to comply with high safety standards, but normal wear and tear or incidentally putting too much weight on them can weaken them and make them unsafe. That's why it is important to inspect every part of a scaffold before each use. Personnel who assemble scaffolds must be certified to do so.

Improper or careless use of scaffolding can result in accidents, injury, or death. Those who work on scaffolding can minimize their risks by being aware of the hazards involved and following the proper safety procedures and guidelines to minimize those hazards.
The main hazards involved with the use of scaffolding are as follows:

- Falls
- Workers being struck by falling objects
- Electric shock

Falls can happen because fall protection or prevention has not been provided, is not used, or is installed or used improperly. Poorly planked scaffolding causes many falls. Working on scaffolding when conditions are dangerous such as in high winds, ice, rain, and lightning also leads to accidents. Falls also happen when scaffolding collapses because of improper construction.

Fall protection is required on any scaffolding 10 ft. or more above a lower level. Note that this is the OSHA requirement and specific states or localities may have more stringent requirements. Fall-protection devices consist of guardrail systems, personal fall arrest systems, and/or safety nets. A guardrail system normally serves as adequate fall protection for most scaffolding.

> **CAUTION**
>
> Only a competent person has the authority to supervise setting up, moving, and taking down scaffolds. Only a competent person can approve the use of scaffolds on the job site after inspecting the scaffolds.

6.3.1 Types of Scaffolds

Two basic types of scaffolds—self-supporting scaffolds and suspended scaffolds—are used in the construction industry. The rules for safe use apply to both. Self-supporting scaffolds can be manufactured units or can be assembled at the site.

Manufactured scaffolds (*Figure 51*) are made of painted steel, stainless steel, or aluminum. They are stronger and more fire-resistant than wooden scaffolds. They are supplied in ready-made, individual units, which are assembled on site. A rolling scaffold has wheels on its legs so that it can be easily moved. The scaffold wheels have brakes so the scaffold will not move while workers are standing on it.

> **WARNING!**
>
> Never unlock the wheel brakes of a rolling scaffold while anyone is on it. People on a moving scaffold can lose their balance and fall.

Built-up scaffolds (*Figure 52*) are built from the ground up at a job site using steel framework sections and lumber. Swing (suspended) scaffolds, shown in *Figure 53 (A)*, are suspended by ropes or cables in a manner that allows it to be raised or lowered as needed. Another type of suspended scaffolding is a work cage, shown in *Figure 53 (B)*. A work cage is typically suspended with rigging devices that attach to I-beams with various sizes of clamps and rollers.

6.3.2 Scaffolding Hazards and Safety Requirements

A scaffold working platform should have a guardrail system that includes a top rail, midrail, toeboard, and screening. To be safe and effective, the top rail should be approximately 42 in. high, the midrail should be located halfway between the toeboard and the top rail, and the toeboard should be a minimum of 4 in. high (*Figure 54*).

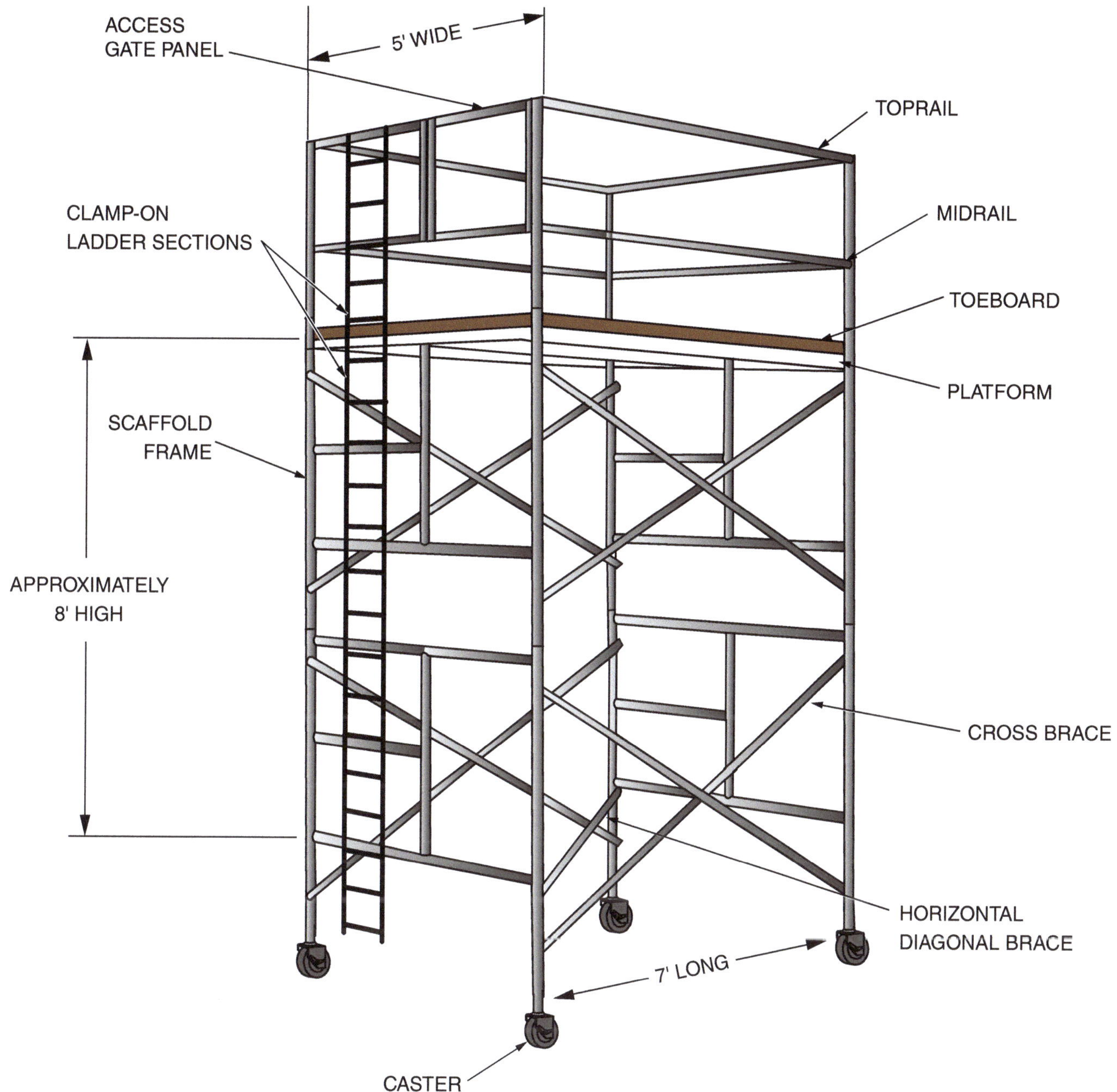

Figure 51 Typical manufactured scaffold.

If people will be passing or working under the scaffolding, the area between the top rail and the toeboard must be screened. Finally, the platform planks must be laid closely together. For safety purposes, the ends of the planks must overlap at least 6 in. and no more than 12 in.

Guardrail systems must extend around all open sides of the scaffolding. The side facing the work surface need not have a guardrail if it is located less than 14 in. away from the work surface. Any opening on a scaffolding platform must be protected by a guardrail system, including the access opening(s) and platforms that do not extend across the entire width of the scaffolding. People who work or pass under scaffolding may be hit by falling objects. Tools, materials, debris, and scaffolding parts may fall to the surface below. Those working on scaffolding may also be injured if there are others working above them, or if the structure or workpiece extends above the work level of the scaffolding. Any worker who is exposed to the danger of falling objects is required to wear a hard hat. Depending on the situation, additional protection may be needed such as debris nets, screens or mesh, canopy structures, and toeboards. Barricades that prevent access under the scaffolding can also be used to protect workers and others.

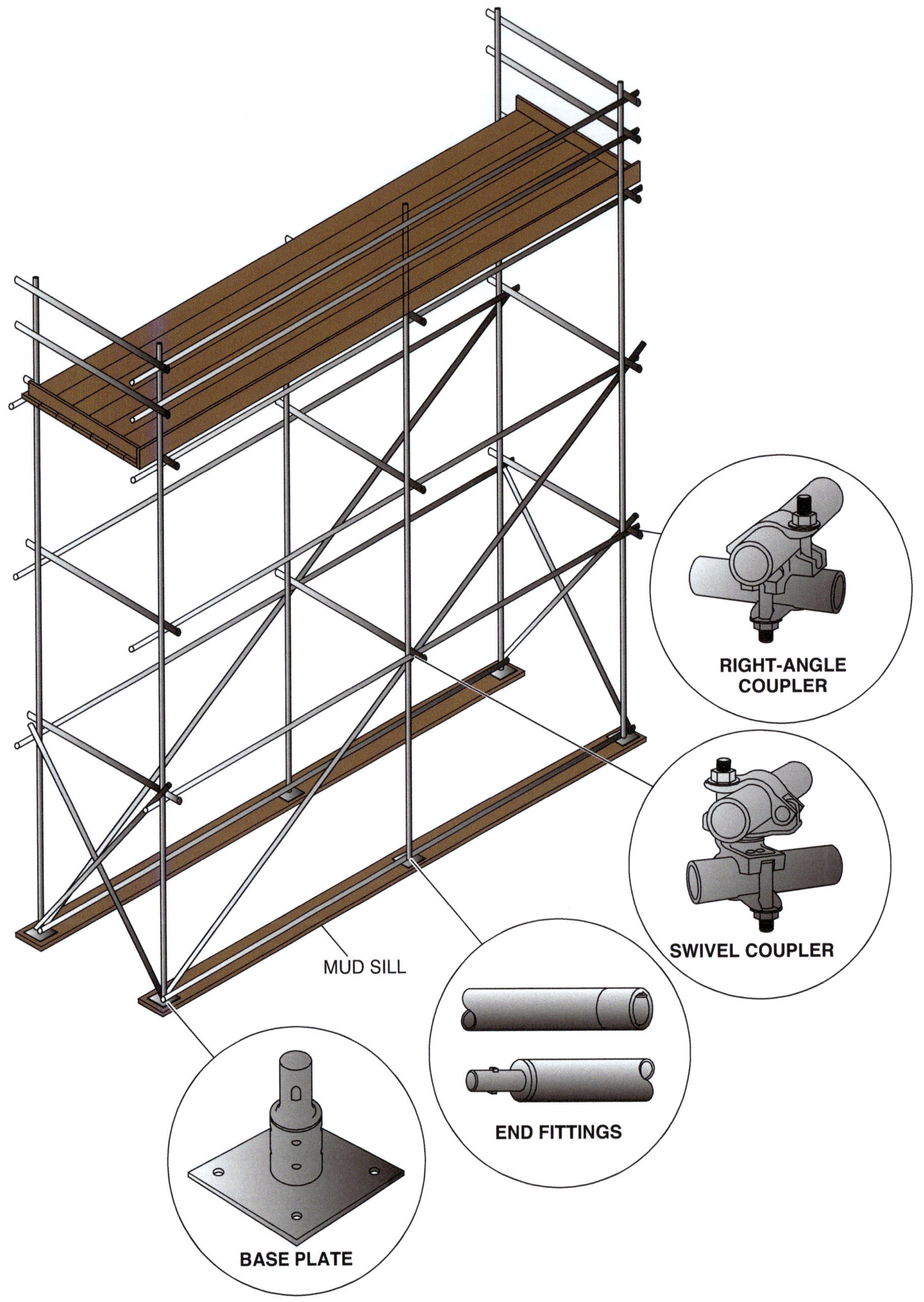

Figure 52 Built-up scaffold.

NCCER – *Fall Protection Orientation*

(A) SUSPENDED SCAFFOLD

(B) WORK CAGE

Figure 53 Suspended scaffold and work cage.

Because most scaffolding is made of metal, the chance of electric shock is always a hazard. Never assume that you can work around high-voltage wires just by avoiding contact. High voltages can arc through the air and cause electrocution without direct contact. When scaffolding must be erected close to power lines, the utility company must be called in to de-energize, move, and/or cover the lines with insulating protective barriers.

Safety begins by getting training in the proper use of scaffolding. It is equally important to always use the right safety equipment, including a hard hat and personal fall-protection systems. Never work on scaffolding, in a work cage, or on a platform if you have any of the following conditions:

- Are subject to seizures
- Become dizzy or lightheaded when working at an elevation
- Take medication that might affect your stability and/or performance
- Are under the influence of drugs and/or alcohol

When working on scaffolding, always follow these guidelines:

- Erect and use scaffolding according to the manufacturer's instructions. Scaffolding must also be erected and used in accordance with all local, state, and federal/OSHA requirements.
- An industry best practice is to attach a green, red, or yellow tag as needed to any scaffolding that is assembled and erected to alert users of its current mechanical and/or safety status. Do not rely solely on the tag. Inspect all parts of scaffolding before each use.

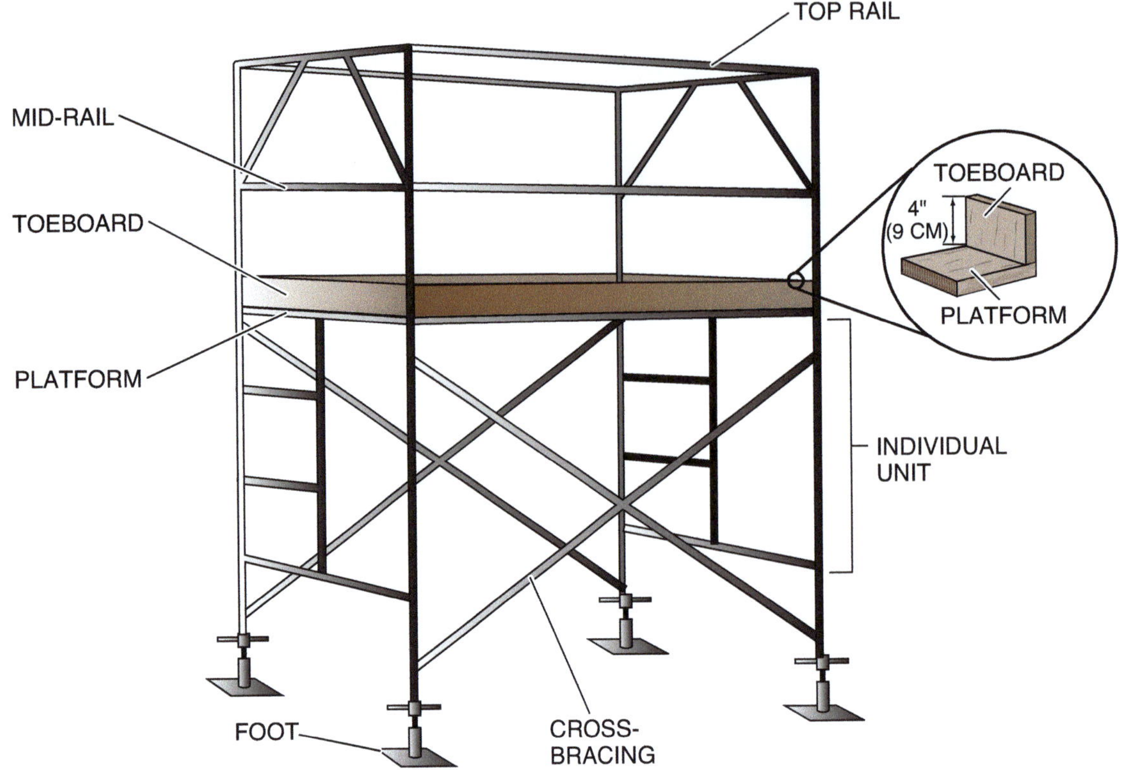

Figure 54 Scaffold toeboard distance.

- If the scaffolding shifts, exit the scaffolding immediately.

Safety Guidelines for Built-Up Scaffolding – Use the following guidelines when erecting and using tubular built-up scaffolding:

- Inspect all scaffolding parts before assembly.
- Never use parts that are broken, damaged, or deteriorated. Be cautious of rusted materials.
- Follow the manufacturer's recommendations for the proper methods of erecting and using scaffolding.
- Do not interchange parts from different manufacturers unless permitted by the manufacturer.
- Do not force braces or other parts to fit. Adjust the level of the scaffolding until the connections can be made easily.
- Provide adequate sills with baseplates for all scaffolding built on filled or soft ground. Compensate for uneven ground by using adjusting screws or leveling jacks.
- Do not use boxes, concrete blocks, bricks, or other similar objects to support scaffolding.
- Keep scaffolding free of clutter and slippery material.

- Be sure scaffolding is plumb and level at all times. Follow the prescribed spacing and positioning requirements for the parts of the scaffolding. Anchor or tie-in scaffolding to the building at prescribed intervals.
- Use ladders rather than cross braces to climb the scaffolding. Position ladders with caution to prevent the scaffolding tower from tipping.
- Do not work on scaffolding that is more than 10 ft. high without guardrails, midrails, and toeboards on open sides and ends.
- Lock or remove the casters of mobile scaffolding when it is positioned for use.
- Avoid building scaffolding near power lines.

Safety Guidelines for Swing and Other Suspended Scaffolding – Use the following guidelines when erecting and using swing and other suspended scaffolding:

- Follow the manufacturer's recommendations for installation, use, and maintenance of the equipment. Before installation, inspect all parts of a structure to which rigging and tieback lines will be secured to ensure that they can support the load.

NCCER – *Fall Protection Orientation*

- Be sure rigging devices are of the proper size and design to support the scaffolding and that they are installed properly. If counterweights are used to secure the inner end of outrigger beams, they should be fastened to the outrigger. Roofing materials and sand bags are not appropriate counterweights.
- Check that tieback lines are installed perpendicular to the face of the structure and are secured to a solid support.
- Check for power lines or electric service wires on the job site. If they pose a hazard, contact the utility company to have them temporarily de-energized.
- Observe the scaffolding's load capacity; never overload the equipment.
- Inspect all scaffolding equipment each day. Check ropes and cables thoroughly for wear, fraying, corrosion, brittleness, damage, or other conditions that may weaken them. Have them replaced as necessary by qualified personnel.
- Keep suspension ropes and cables straight and perpendicular to the platform during use. Do not affix them to anything to change the line of travel.
- Lash the scaffolding to the building or structure to prevent it from swaying.
- Stay off scaffolding during storms or high winds; watch for icy or slippery platforms.
- Guarantee safe access to the swing stage at all times.
- Use two-way radios for communication between workers on the scaffolding and on the ground.
- Do not combine two or more two-point swing scaffolding units to form one unit.
- Raise and lash the scaffolding in a safe position when not in use.

Case History

Inspect All Materials

A crew laying bricks on the upper floor of a three-story building built a 6' platform to connect two scaffolds. The platform was correctly constructed of two 2" x 12" planks with standard guardrails. One of the planks however, was not scaffolding-grade lumber. It also had extensive dry rot in the center. When a bricklayer stepped on the plank, it broke and he fell 30' to his death.

The Bottom Line: Make sure that all planking is sound and secure. Your life depends on it.

Source: OSHA

6.3.3 Inspecting Scaffolds

Any scaffold that is assembled on the job site must be tagged to indicate whether the scaffold meets OSHA standards and is safe to use. Three colors of tags are used: green, yellow, and red (*Figure 55*).

- A green tag means the scaffold meets all OSHA standards and is safe to use.
- A yellow tag means the scaffold does not meet all OSHA standards. An example is a scaffold on which a railing cannot be installed because of equipment interference. To use a yellow-tagged scaffold, you must wear a safety harness attached to a lanyard. You may have to take other safety measures as well.

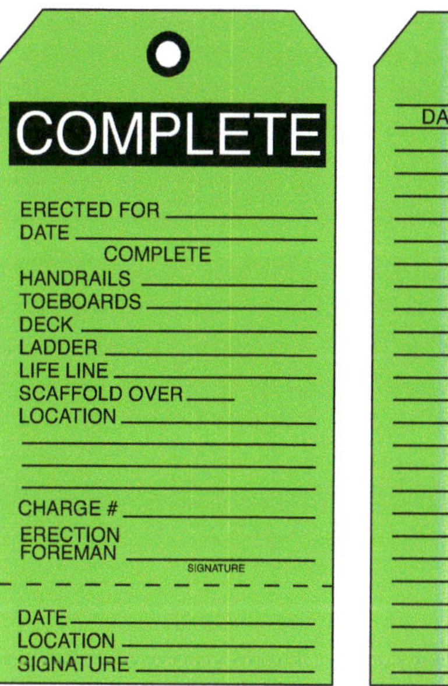

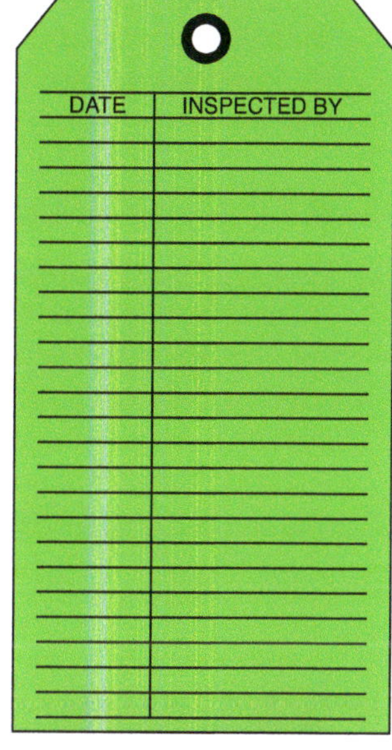

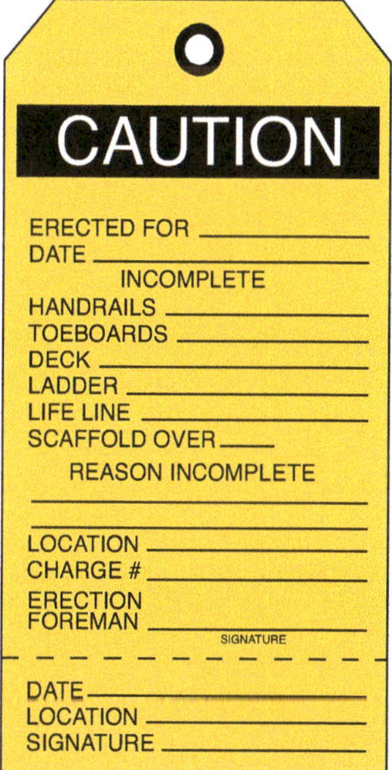

Figure 55 Typical scaffold tags.

- A red tag means a scaffold is being put up or taken down. Never use a red-tagged scaffold.

Don't rely on the tags alone; inspect all scaffolds before you use them. Check for bent, broken, or badly rusted tubes. Also check for loose joints where the tubes are connected. Any of these problems must be corrected before the scaffold is used.

Make sure you know the weight limit of any scaffold you will be using. Compare this weight limit to the total weight of the people, tools, equipment, and material you expect to put on the scaffold. Scaffold weight limits must never be exceeded.

When you examine a rolling scaffold, check the condition of the wheels and brakes. Be sure the brakes are working properly and can stop the scaffold from moving while work is in progress. Be sure all brakes are locked before you use the scaffold.

6.3.4 Using Scaffolds

Be sure that a competent person inspects the scaffold before you use it. There should be firm footing under each leg of a scaffold before putting any weight on it. If you are working on loose or soft soil, you can put planks or matting under the scaffold's legs or wheels. When moving a rolling scaffold, first unlock the brakes and then move the scaffold. Once the scaffold is repositioned, don't forget to relock the brakes.

> **WARNING!**
>
> Keep scaffolds a minimum of 3 ft. (91 cm) from power lines up to 300 volts and a minimum of 10 ft. (3.1 m) from power lines above 300 volts. Never move a scaffold while someone is on it.

This list is a brief list of precautions to be considered while working on scaffolds. It is not intended to be a substitute for proper scaffold use. Scaffold loading must be constantly monitored. In addition to the final inspection by a competent person, scaffold users should always perform a visual inspection prior to use.

- Do not use a scaffold that has not been inspected during your work shift. Check for a tag indicating that the scaffold is safe and when the inspection was done.
- Do not use a scaffold that does not have proper access.

> **WARNING!**
>
> Do not climb guardrails or bracing. Use only approved ladders, stairs, or access from an existing structure. Guardrails and braces could fail, causing personal injury or death.

- Do not work from a platform that is not fully planked, unless fall protection is used.
- Do not use the scaffold if the planking is not scaffold grade, was not made specifically for decking, or is not in good condition.
- Do not use a scaffold if the planking is bowing more than $\frac{1}{60}$ of the span. That calculates to 2 inches on a 10-foot span.
- Do not use a scaffold that is not rigid, plumb, and square.
- Do not use a scaffold that is taller than four times the minimum base dimension unless it is properly braced.
- Do not work on a scaffold if you feel sick, dizzy, or weak.
- Do not climb scaffold ladders carrying tools or materials. Use both hands on the side-rails.
- Do not jump on the planking.
- Do not use a ladder to extend the height of the scaffold.
- Do not use a scaffold if you see damaged parts.
- Do not allow material, debris, or even tools to create a hazard on a platform.

> **WARNING!**
>
> Due to fire hazard, do not use gas-powered tools, equipment, or hoists when working from a suspended scaffold. If a fire occurs there is nowhere to escape its effects.

- Do not work on a platform that is covered with snow or ice, unless you are involved in removing it.
- Do not work on a scaffold during high winds or storms.
- Do not use a scaffold as a platform for a hoist unless it has been designed and built for it.
- Do not load a platform beyond the designed capacity.
- Do not concentrate loads in a single spot on a frame or platform.
- Do not violate clearances from electrical or process lines.
- Do not use the scaffold unless proper falling-object protection is in place.
- Do not ride on rolling scaffold towers.

- Do not use flame-producing equipment unless proper safeguards are protecting the flammable portions of the scaffolds.
- Do not attach fall arrest equipment to any guardrail system unless it is recommended by the scaffold manufacturer.

Scaffold users are not required to be trained in scaffold construction. However, training for the user should cover the hazards that the user might encounter, such as electrical hazards, fall hazards, falling-object hazards, and caught-in or -between hazards. The training of scaffold builders and any employees who might modify, repair, maintain, or inspect scaffolds is required, not just recommended. Employees must be trained when conditions change and retrained as necessary. Industry practices state the following:

- The employer shall have each employee who performs work while on a scaffold trained by a person qualified in the subject matter to recognize the hazards associated with the type of scaffold being used and to understand the procedures to control or minimize those hazards. The training shall include the following areas as applicable:

  - The nature of any electrical hazards, fall hazards, and falling-object hazards in the work area;
  - The correct procedures for dealing with electrical hazards for erecting, maintaining, and disassembling the fall protection system and falling-object protection system being used;
  - The proper use of the scaffold, and the proper handling of materials on the scaffold;
  - The maximum intended load and the load-carrying capacity of the scaffold used; and
  - Any other pertinent requirements of this subpart.

- The employer shall have each employee who is involved in erecting, disassembling, moving, operating, repairing, maintaining, or inspecting a scaffold trained by a competent person to recognize any hazards associated with the work in question. The training shall include the following topics as applicable:

  - The nature of scaffold hazards;
  - The correct procedures for erecting, disassembling, moving, operating, repairing, inspecting, and maintaining the type of scaffolds in question;

  - The design-criteria maximum intended load-carrying capacity and use of the scaffold;
  - Any other pertinent requirements of this subpart.

- When the employer has reason to believe that an employee lacks the skill or understanding needed for safe work involving the erection, use, or dismantling of scaffolds, the employer shall retrain each employee so that the requisite proficiency is regained. Retraining is required in at least the following situations:

  - Where changes at the work site present a hazard about which the employee has not been previously trained; or
  - Where changes in the types of scaffolds, fall protection, falling-object protection, or other equipment present a hazard about which an employee has not been previously trained; or
  - Where inadequacies in an affected employee's work involving scaffolds indicate that the employee has not retained the requisite proficiency.

6.3.5 Sway Prevention

After the scaffold is erected, measures must be taken to ensure that the scaffold will not sway or become displaced.

- Legs, poles, or uprights must be plumbed, secured, and rigidly braced to prevent swaying or displacement.
- Putlogs must satisfy the following:

  - Putlogs are for support of personnel only. Do not place materials on platforms supported by putlogs, unless designed by a qualified person.
  - Lengths greater than 10 feet long require knee bracing.
  - Depending on the expected loading, putlogs may require close spacing.

> **NOTE**
>
> Follow the recommendations of the manufacturer.

Additional Resources

Basic Construction Safety and Health, Fred Fanning. 2014. CreateSpace Independent Publishing Platform.

Construction Safety, Jimmie W. Hinze. Second Edition. 2006. New York, NY: Pearson.

DeWalt Construction Safety/OSHA Professional Reference, Paul Rosenberg, American Contractors Educational Services. 2006. DEWALT.

Fall Protection and Scaffolding Safety: An Illustrated Guide, Grace Drennan Gagnet, CSP. 2000. Government Institutes.

Online resources:

Code of Federal Regulations (CFR), **www.ecfr.gov**

Occupational Safety and Health Administration (OSHA), **www.osha.gov**

OSHA Videos, **www.osha.gov/video**

6.0.0 Section Review

1. Metal ladders should not be used near _____.

   a. stairways
   b. scaffolds
   c. electrical equipment
   d. windows

2. A fixed ladder must be able to support at least two loads of _____.

   a. 150 pounds each
   b. 200 pounds each
   c. 250 pounds each
   d. 300 pounds each

3. The scaffolding side facing the work surface need not have a guardrail if it is located less than _____.

   a. 14" away from the work surface
   b. 16" away from the work surface
   c. 18" away from the work surface
   d. 24" away from the work surface

7.0.0 AERIAL LIFTS

Objective

State the guidelines for the safe operation of aerial lifts.

  a. Identify the types of aerial lifts and precautions for safe use.

Trade Terms

Aerial lifts: Mobile work platforms designed to transport and raise personnel, tools, and materials to overhead work areas.

As the work level becomes higher, ladders and stairways are discarded and replaced with aerial lifts (also called *personnel lifts*) and materials hoists (*Figure 56*). Lifts should be plumb, securely braced, and enclosed their full height with expanded metal or wire mesh. The doors or gates should be 6 feet tall. They should lock and unlock only from the inside. The locking mechanism should operate only when the cage is stopped at the landing level. Hoisting equipment should be thoroughly inspected each day and tested whenever subject to a new use.

Lifts must be equipped with guardrails, toeboards, and hand controls that operate from the ground or the platform. Only persons riding the lift should be able to lock the doors. The lifts must also have safety brakes to prevent free fall if the cabling fails.

OSHA requires that personnel lifts be given a trial lift and proof testing before they are used to lift people. This is accomplished by loading the lift to 125 percent of capacity, operating it, and inspecting the cabling. The lift capacity must be posted so that it will not be overloaded by accident. All personnel lift towers (*Figure 57*) that are outside the structure must be enclosed their full height on the side or sides used for entrance and exit to the structure. At the lowest landing, the enclosure on the sides not used for exit or entrance to the structure should reach a height of at least 10 feet. Other sides of the tower structure adjacent to floors or scaffold platforms should be enclosed to a height of 10 feet above the level of such floors or scaffolds. Towers inside structures should be enclosed on all four sides throughout their full height.

Figure 56 Aerial lifts.

Figure 57 Personnel lift tower.

Typically, materials hoists do not have the safety features of lifts. On some jobs, however, personnel hoists do double duty because they have enclosed cabs, internal controls, and safety brakes. Be sure that materials hoists are clearly marked so that everyone knows that they do not have the safety equipment required for carrying workers.

Although using cranes to lift people was common in the past, OSHA regulations, as spelled out in 29 *CFR* 1926.550, now discourage the practice. Using a crane to lift personnel is not specifically prohibited by OSHA, but the restrictions are such that it is only permitted in special situations where no other method is suitable. When it is allowed, certain controls must be in place:

- The rope design factor is doubled.
- The lift capacity is cut in half.
- Free falling is prohibited.
- Devices are required that provide warnings and prevent the lower load block (pulley) or hook from coming into contact with the upper load block, boom point, or boom-point machinery, likely causing the failure of the rope or release of the load or hook block.
- The platform must be specifically designed for lifting personnel.
- Before the lift is used, it must be tested with a comparable weight and then inspected.
- Every intended use must undergo a trial run with weights rather than people.

Safety precautions unique to aerial lifts must also be followed. Remember that the other safety precautions already discussed in this module also apply to aerial lifts. Each manufacturer provides specific safety precautions in the operator's manual that comes with the equipment. Specifically, 29 *CFR* 1926.453 defines and governs the use of aerial lifts as expressed in the following precautions:

- Avoid using the lift outdoors in stormy weather or in strong winds.
- Prevent people from walking beneath the work area of the platform.
- Use personal fall arrest equipment (body harness and lanyard) as required for the type of lift being used. Use approved anchorage points.
- Do not use an aerial lift on uneven ground.
- Lower the lift and lock it into place before moving the equipment. Also, lower the lift, shut off the engine, set the parking brake, and remove the key before leaving it unattended.
- Stand firmly on the floor of the basket or platform.

Do not lean over the guardrails of the platform, and never stand on the guardrails. Do not sit or climb on the edge of the basket or use planks, ladders, or other devices to gain additional height.

7.1.0 Aerial Lift Types and Precautions

There are two main types of lifts: boom lifts and scissor lifts. Both types are available in various models. Some are transported on a vehicle to a job site, where they are unloaded. Others are trailer-mounted and towed to the job site by a vehicle. Still others are permanently mounted on a vehicle. Depending on their design, aerial lifts can be used for indoor work, outdoor work, or both.

7.1.1 Boom Lifts

Boom lifts (*Figure 58*) are work platforms mounted on the end of an arm that can be raised and lowered and may be able to extend beyond the mobile base. They are designed for both indoor and outdoor use and are capable of holding one or two workers. Some models have a jointed (articulated) arm that allows the work platform to be positioned both horizontally and vertically; thus the boom section may consist of two or more telescoping sections. The machines usually can be driven by the operator from the platform. However, base-mounted controls are also provided and are typically used only in an emergency.

7.1.2 Scissor Lifts

Scissor lifts (*Figure 59*) are powered personnel lifts which raise a work enclosure vertically by means of crisscrossed supports. They use hydrau-

NCCER – *Fall Protection Orientation*

(A) ARTICULATING BOOM LIFT

(B) TELESCOPIC BOOM LIFT

Figure 58 Boom lifts.

lic rams to raise and lower the platform. Power is supplied to the hydraulic system by a battery, a gasoline engine, or an LPG (liquefied petroleum gas) engine. The controls for the lift are located at an operator's station on the platform itself. Some models also include an operator's station at the ground and the platform level. In addition to con-

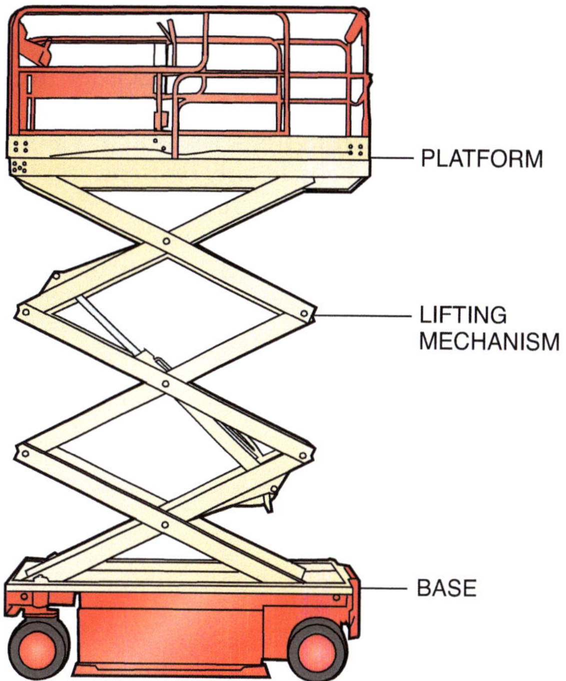

Figure 59 Scissor Lift.

trolling the height of the platform, the operator is also able to move the unit to the front or rear and to steer it as it moves. Lift heights and capacities vary by model and manufacturer. Lift heights may be as high as 41 feet and capacities may reach 2,000 pounds. Both the height and capacity of the lift can usually be found on a decal conspicuously placed on the lift. Some manufacturers offer optional manual outriggers for added stability.

No one is permitted to operate a scissor lift without training in the proper operating procedures for the particular model. Keep the following safety guidelines in mind when operating a scissor lift:

- Read and observe the warnings and cautions placed on the unit.
- Make sure your safety harness is properly anchored to an appropriate place when you are in the basket.
- Keep the vehicle and work platform at least 10 feet from electrical power lines. Scissor lifts are not insulated to protect against arcing or electrical contact.
- Check for loose or missing parts before operating the lift. Anything that is not working correctly must be taken out of service and repaired before use.
- Do not remove any parts or safety devices.
- Make certain that the immediate area is clear of all personnel and obstructions before raising or lowering the lift.

- Keep the guardrails and chains in place when operating the lift.
- Do not bridge between two scissor lifts or between a scissor lift and a building.
- Never crawl into the scissors at any point.
- Operate the vehicle only on firm, solid ground. Do not operate the vehicle near a drop-off.
- Do not impose side loads on the work platform.
- Never exceed the rated capacity of the lift.
- Do not operate the vehicle in an explosive atmosphere.
- Never climb on railings.
- Only operate per manufacturer's instructions.

Most models of aerial lifts are self-propelled, allowing workers to move the platform as work is performed. The power to move these lifts is provided by several means, including electric motors, gasoline, propane, or diesel engines, and hydraulic motors.

Additional Resources

Basic Construction Safety and Health, Fred Fanning. 2014. CreateSpace Independent Publishing Platform.

Construction Safety, Jimmie W. Hinze. Second Edition. 2006. New York, NY: Pearson.

DeWalt Construction Safety/OSHA Professional Reference, Paul Rosenberg, American Contractors Educational Services. 2006. DEWALT.

Online resources:

Code of Federal Regulations (CFR), **www.ecfr.gov**

Occupational Safety and Health Administration (OSHA), **www.osha.gov**

OSHA Videos, **www.osha.gov/video**

7.0.0 Section Review

1. Before it can be used, a personnel lift must be tested with a load equal to ______.

    a. 90 percent of its capacity
    b. 125 percent of its capacity
    c. 150 percent of its capacity
    d. 200 percent of its capacity

2. Boom lifts are capable of holding up to ______.

    a. two workers
    b. three workers
    c. four workers
    d. five workers

SUMMARY

Although the typical job site has many hazards, it does not have to be a dangerous place to work. Your employer has programs to deal with potential hazards. Basic rules and regulations help protect you and your co-workers from unnecessary risks.

The basic approach to safety is to eliminate hazards in the equipment and the workplace; to learn the rules and procedures for working safely with and around the remaining hazards; and to apply those rules and procedures.

Employees are responsible for maintaining a safety awareness at all times. 29 *CFR* 1926, Subpart R provides safety guidelines for construction work in general. All forms of construction require close attention to the appropriate safety procedures for that aspect of the work.

Falls are a very common and serious type of accident on construction sites. You have a responsibility to yourself and others to work safely and use fall-protection equipment. Falls can cause serious injuries or death when no fall protection equipment or the wrong kind of equipment is used. The three types of fall protection commonly used when working at elevated levels are guardrails, personal fall arrest systems, and safety nets.

Falls from ladders, scaffolds, lifts, or unprotected work areas can cause serious injuries or death when no fall-protection equipment or the wrong kind of equipment is used. Similar mishaps can occur when the right type of equipment is improperly used.

Aerial work platforms and scissors lifts are used to raise and lower workers to and from elevated work areas. 29 *CFR* 1926, Subpart N defines and governs the use of aerial lifts.

1. The four leading causes of death in the construction industry include electrical accidents, struck-by accidents, caught-in/between accidents, and _____.

   a. asphyxiation
   b. falls
   c. radiation exposure
   d. chemical burns

2. Which of the following must be reported to your supervisor?

   a. Only major injuries
   b. Only incidents and major injuries
   c. All injuries, incidents, and incidents
   d. Only incidents in which a death occurred

3. Which of the following is true regarding terminology used in OSHA regulations?

   a. "Should" means something is recommended, and "shall" means it is required
   b. "Should" and "shall" mean the same thing
   c. A "qualified person" supervises job activities
   d. An "incident" always involves injury to personnel

4. Someone who, by possession of a recognized degree, certificate, or professional standing, or who by extensive knowledge, training, and experience, has successfully demonstrated the ability to solve or resolve problems related to the subject matter, the work, or the project is defined by OSHA as a(n) _____.

   a. designated person
   b. competent person
   c. authorized person
   d. qualified person

5. The agency primarily responsible for regulating the scaffolding industry is the _____.

   a. Scaffold Access Industry Association
   b. Occupational Safety and Health Administration
   c. National Safety Council
   d. American National Standards Institute

6. Who does OSHA require to develop and implement a safety program?

   a. employers
   b. employees
   c. engineers
   d. certified inspectors

7. Debris netting must be installed when _____.

   a. scaffold reaches a third-story level
   b. an injury involving falling objects occurs
   c. material or equipment piles are higher than a standard toeboard
   d. masonry materials are being lifted onto scaffolds

8. The sign used to inform workers that an immediate hazard exists is a(n) _____.

   a. caution sign
   b. danger sign
   c. informational sign
   d. warning sign

9. To inform workers about potential hazards or unsafe practices, you would see a(n) _____.

   a. caution sign or tag
   b. informational sign or tag
   c. red flag
   d. Stop/Slow paddle

10. A danger sign warns workers that _____.

    a. an immediate hazard exists and specific precautions must be observed
    b. equipment is out of order
    c. radiation is present
    d. general safety rules apply

11. Accident-prevention tags are used _____.

    a. where it is necessary to tell workers about information not related to safety
    b. where it is necessary to inform workers about general instructions related to safety measures
    c. on landmarks where traffic must slow or stop completely
    d. as a temporary way of warning workers about immediate and potential hazards

12. OSHA requires that supporting stanchions in a controlled access zone be clearly flagged or marked at maximum intervals of _____.

    a. 3 ft.
    b. 6 ft.
    c. 9 ft.
    d. 12 ft.

13. OSHA requires that positioning anchor points be rated at _____.

    a. 1,800 pounds
    b. 2,400 pounds
    c. 3,000 pounds
    d. 5,000 pounds

14. A barricade that alerts workers to hazards but provides no real protection is called a(n) _____.

    a. hole cover
    b. temporary barricade
    c. railing
    d. warning barricade

15. When positioning a straight ladder against a wall, how far from the wall should the base of the ladder be?

    a. 4 ft. (1.2 m)
    b. One-fourth the distance from the ground to the point where the ladder touches the wall
    c. The height of the wall minus 4 ft. (1.2 m)
    d. One-half the distance from the ground to the point where the ladder touches the wall

16. The two basic types of scaffolds are _____.

    a. self-supporting and suspended scaffolds
    b. fixed and portable scaffolds
    c. metal and wooden scaffolds
    d. assembled and deliverable scaffolds

17. If a scaffold has a yellow tag, it means _____.

    a. the scaffold may only be used by one person at a time
    b. the scaffold is condemned and cannot be used
    c. a safety harness must be worn when using it
    d. the scaffold is under assembly and cannot be used

18. An unacceptable means of access to scaffolds is _____.

    a. stairs
    b. bracing
    c. ladder
    d. from an existing structure

19. The base-mounted controls on a boom lift are used _____.

    a. for maintenance purposes
    b. mainly in emergency situations
    c. for primary control
    d. for raising the boom only

20. Which of the following is true regarding scissor lifts?

    a. They are limited to a lift height of 15 feet.
    b. Their capacity can reach 2,000 pounds.
    c. The operation of scissor lifts is the same across all manufacturers.
    d. Scissor lifts are insulated against electrical contact.

OSHA MISSION AND STANDARDS

The mission of the Occupational Safety and Health Administration (OSHA) is to save lives, prevent injuries, and protect the health of America's workers. To accomplish this, federal and state governments work in partnership with millions of working men and women who are covered by the Occupational Safety and Health Act (OSH Act) of 1970.

Nearly every worker in the US comes under OSHA's jurisdiction. There are some exceptions, such as miners, transportation workers, many public employees, and the self-employed. These specific groups are not covered by OSHA standards, but most are covered by standards developed by the specific industry.

The Code of Federal Regulations

The *Code of Federal Regulations* (*CFR*) Part 1910 covers the OSHA standards for general industry. 29 *CFR* 1926 covers the OSHA standards for the construction industry. Either or both may apply to you, depending on where you are working and what you are doing. If a job-site condition is covered in the *CFR* book, then that standard must be used. However, if a more stringent requirement is listed in 29 *CFR* 1910, it should also be met. Check with your supervisor to find out which standards apply to your job.

29 *CFR* 1926 is divided into subparts A through Z. As you progress in task-specific training, you will learn about all the subparts applicable to your work. Subpart C of 29 *CFR* 1926 applies to all construction and maintenance work. It outlines the general safety and health provisions for the construction industry. It covers the following topics:

- Safety training and education
- Injury reporting and recording
- First aid and medical attention
- Housekeeping
- Illumination
- Sanitation
- Personal protective equipment (PPE)
- Standards incorporated by reference
- Definitions
- Access to employee exposure and medical records

- Means of egress
- Employee emergency action plans

For assistance in identifying parts, sections, paragraphs, and subparagraphs of an OSHA standard, refer to *Figure A01*.

All of OSHA's safety requirements in the Code of Federal Regulations apply to residential as well as commercial construction. In the past, OSHA enforced safety only at commercial sites. The increasing rate of incidents at residential sites led OSHA to enforce safety guidelines for the building of houses and townhomes. Today, however, OSHA still focuses its enforcement efforts on commercial construction.

The General Duty Clause

If a standard does not specifically address a hazard, the general duty clause must be invoked. Failing to adhere to the general duty clause can result in heavy fines for your employer. The general duty clause reads as follows:

In practice, OSHA, court precedent, and the review commission have established that if the following elements are present, a general duty clause citation may be issued:

- The employers failed to keep the workplace free of a hazard to which employees of that employer were exposed.

An OSHA Standard reference may look like this:

29 CFR 1926.501 (a)(1)(i)(A)

and breaks down like this:

29	=	Title (Labor)
CFR	=	Code of Federal Regulations
1926	=	Part (Construction)
.501	=	Section
(a)	=	Paragraph
(1)	=	Subparagraph
(i)	=	Subparagraph
(A)	=	Subparagraph

Figure A01 Reading OSHA standards.

- The hazard was recognized. (Examples might include: through your safety personnel, employees, organization, trade organization, or industry customs.)
- The hazard was causing or was likely to cause death or serious physical harm.
- There was a feasible and useful method to correct the hazard.

Employee Rights and Responsibilities

While it is the employer's responsibility to keep workers safe by complying with the General Duty Clause and all other OSHA regulations, workers have certain rights and responsibilities on the job site as well. First and foremost, workers must follow their employers' safety rules. While workers cannot be cited or fined by OSHA, they can be disciplined for violating their employer's safety rules. Workers must also wear the provided personal protective equipment. Workers should also inform their foreman about health and safety concerns on the job.

Section 11(c) of the OSH Act prohibits employers from disciplining or discriminating against any worker for practicing their rights under OSHA, including filing a complaint. You have the right to file a complaint if you do not think that your employer is protecting your health and safety at work. You may submit a written request to OSHA asking for an inspection of your worksite. Workers who file a complaint have the right to have their names withheld from their employers, and OSHA will not reveal this information.

Workers who would like an on-site inspection must submit a written request. You have the following rights when job site inspection is conducted:

- You must be informed of imminent dangers. An OSHA inspector must tell you if you are exposed to an imminent danger. An imminent danger is one that could cause death or serious injury now or in the near future. The inspector will also ask your employer to stop any dangerous activity.
- You have the right to accompany the OSHA inspector in the walk-around inspection. Walk-around activities include all opening and closing conferences related to the conduct of the inspection.
- You have the right to be told about citations issued at your workplace. Notices of OSHA citations must be posted in the workplace near the site where the violation occurred and must remain posted for three days or until the hazard is corrected, whichever is longer.

After an inspection has been performed, OSHA will give the employer a date by which any hazards cited must be fixed. Employers can appeal these dates, and appeals must be filed within 15 days of the citation. Workers have the right to meet privately with the OSHA inspector to discuss the results of the inspection.

If you have been discriminated against for asserting your OSHA rights, you have the right to file a complaint with the OSHA area office within 30 days of the incident. Make sure you file your complaint as soon as possible, as the time limit is strictly enforced.

You also have the right to see and copy any medical records about you that the employer has obtained. Your employer is required by 29 *CFR* 1926.33 and 29 *CFR* 1910.1020 to maintain your medical records for 30 years after you leave employment. If you are employed for less than one year, the employer can maintain your records or give them to you when you leave the job.

Inspections

OSHA conducts six types of inspections to determine if employers are in compliance with standards:

- *Imminent danger inspections* – OSHA's top priority for inspection, conducted when workers face an immediate risk of death or serious physical harm.
- *Catastrophe inspections* – Performed after an incident that requires hospitalization of three or more workers. Employers are required to report fatalities and catastrophes to OSHA within eight hours.
- *Worker complaint and referral inspections* – Conducted due to complaints by workers or a worker representative, or a referral from a recognized professional.
- *Programmed inspection* – Aimed at high-risk areas based on OSHA's targeting and priority methods.
- *Follow-up inspection* – Completed after citations to assure employer has corrected violations.
- *Monitoring inspection* – Used for long-term abatement follow-up or to assure compliance with variances.

Before beginning an inspection, OSHA staff must be able to determine from the complaint that there are reasonable grounds to believe that a violation of an OSHA standard or a safety or health hazard exists. If OSHA has information indicating the employer is aware of the hazard and is correcting it, the agency may not conduct an inspection after obtaining the necessary documentation from the employer.

Complaint inspections are typically limited to the hazards listed in the complaint, although other violations in plain sight may be cited as well. The inspector may decide to expand the inspection based on professional judgment or conversations with workers.

Complaints are not necessarily inspected in first-come, first-served order. OSHA ranks complaints based on the severity of the alleged hazard and the number of workers exposed. That is why lower-priority complaints can often be handled more quickly using the phone/fax method than through on-site inspections.

Inspections are typically performed by conducting a walk-around. During a walk-around inspection, the inspector typically does the following:

- Observes conditions of the job site.
- Talks to workers.
- Inspects records.
- Examines posted hazard warnings and signs.
- Points out hazards and suggests ways to reduce or eliminate them.

After the walk-around inspection, there is typically a closing conference held between the inspector and the site contractor or company managers. During this conference, inspectors discuss their findings, citing specific violations and suggested abatement methods. Inspectors may also conduct interviews with the employers, workers, and representatives at this point.

Violations

Employers who violate OSHA regulations can be fined. The fines are not always high, but they can harm a company's reputation for safety. Fines for serious safety violations can cost up to $7,000. Fines for each violation that was done willfully can cost up to $70,000. In 2012, roughly $260,000,000 in fines were issued against employers. In the decade prior to 2012, annual fines averaged approximately $150,000,000. Significant increases and decreases in annual fines tend to be dependent on the current federal administration and OSHAs budget provided by federal lawmakers.

Compliance

Just as employers are responsible to OSHA for compliance, employees must comply with their company's safety policies and rules. Employers are required to identify hazards and potential hazards within the workplace and eliminate them, control them, or provide protection from them. This can only be done through the combined efforts of the employer and employees. Employers must provide written programs and training on hazards, and employees must follow the procedures. You, as the employee, must read and understand the OSHA poster at your job site explaining your rights and responsibilities. If you are unsure where the OSHA poster is, ask your supervisor.

To help employers provide a safe workplace, OSHA requires companies to provide a competent person to ensure the safety of the employees. In 29 *CFR* 1926, OSHA defines a competent person as "one who is capable of identifying existing and predictable hazards in the surroundings or working conditions which employees, and who has authorization to take prompt corrective measures to eliminate them."

In comparison, 29 *CFR* 1926 defines a qualified person as "one who, by possession of a recognized degree, certificate, or professional standing, or who by extensive knowledge, training, and experience, has successfully demonstrated his ability to solve or resolve problems relating to the subject matter, work, or the project."

In other words, a competent person is experienced and knowledgeable about the specific operation and has the authority from the employer to correct the problem or shut down the operation until it is safe. A qualified person has the knowledge and experience to handle problems. A competent person is not necessarily a qualified person.

These terms will be an important part of your career. It is important for you to know who the competent person is on your job site. OSHA requires a competent person for many of the tasks you may be assigned to perform, such as confined space entry, ladder use, and trenching. Different individuals may be assigned as a competent person for different tasks, according to their expertise. To ensure safety for you and your co-workers, work closely with your competent person and supervisor.

Accident: As defined by OSHA, an unplanned event that results in personal injury or property damage.

Aerial lifts: Mobile work platforms designed to transport and raise personnel, tools, and materials to overhead work areas.

Body harness: Straps that may be secured about the worker in a manner that will distribute the fall-arrest forces over at least the thighs, pelvis, waist, chest, and shoulders, with means for attaching it to other components of a personal fall-arrest system.

Capacity: The total amount of weight, or load, capable of being held by a PFAS component such as an anchor point or a lanyard. This includes the weight of personnel, tools and materials, and/or equipment.

Competent person: As defined by OSHA, one who is capable of identifying existing and predictable hazards in the surroundings or working conditions which are unsanitary, hazardous, or dangerous to employees, and who has authorization to take prompt corrective measures to eliminate them.

Controlled access zone (CAZ): A designated work area in which certain types of masonry work may take place without the use of conventional fall protection systems.

Controlled decking zone (CDZ): An area in which certain work (for example, initial installation and placement of metal decking) may take place without the use of guardrail systems, personal fall arrest systems, fall restraint systems, or safety net systems and where access to the zone is controlled.

Cut: A common term for a scaffold level.

Excavations: Man-made cuts, cavities, trenches, or depressions in the earth's surface, formed by removing earth. Excavations can be made for the purpose of building anything from basements to highways.

Free fall: The act of falling before a personal fall-arrest system begins to apply force to arrest the fall.

Guarded: Enclosed, fenced, covered, or otherwise protected by barriers, rails, covers, or platforms to prevent dangerous contact.

Hand line: A line attached to a tool or object so a worker can pull the tool up after climbing a ladder or scaffold.

Incident: As defined by OSHA, an unplanned event that does not result in personal injury but may result in property damage or is worthy of recording.

Lanyard: A short section of rope or strap, one end of which is attached to a worker's safety harness and the other to a strong anchor point above the work area.

Lifeline: A component consisting of a flexible line connected vertically to an anchorage at one end (vertical lifeline), or connected horizontally to an anchorage at both ends (horizontal lifeline), and which serves as a means for connecting other components of a personal fall-arrest system to the anchorage.

Limited access zones: Restricted areas alongside a masonry wall that is under construction.

Management system: The organization of a company's management, including reporting procedures, supervisory responsibility, and administration.

Maximum intended load: The total weight of all people, equipment, tools, materials, and loads that a ladder can hold at one time.

Midrail: A mid-level, horizontal board required on all open sides of scaffolds and platforms that are more than 14 in. (35 cm) from the face of the structure and more than 10 ft. (3.05 m) above the ground. It is placed halfway between the toeboard and the top rail.

Occupational Safety and Health Administration (OSHA): An agency of the US Department of Labor whose mission is to assure safe and healthful working conditions for employees by setting and enforcing standards and by providing training, outreach, education and assistance.

On-the-job learning (OJL): The learning an apprentice acquires while working on the job under the supervision of journey workers. Commonly called *on-the-job training (OJT)*.

Personal fall arrest system (PFAS): A system used to stop an employee in a fall from a working level. It consists of an anchorage, connectors, and a body harness, and may include a lanyard, deceleration device, lifeline, or suitable combinations of these.

Personal protective equipment (PPE): Equipment or clothing designed to prevent or reduce injuries.

Planked: Having pieces of material 2 in. (5 cm) thick or greater and 6 in. (15 cm) wide or greater used as flooring, decking, or scaffold decks.

Putlogs: Horizontal scaffold members on which the scaffold platform rests.

Qualified person: As defined by OSHA, one who, by possession of a recognized degree, certificate, or professional standing, or who by extensive knowledge, training, and experience, has successfully demonstrated his ability to solve or resolve problems relating to the subject matter, the work, or the project.

Safety culture: The culture created when the whole company sees the value of a safe work environment.

Scaffolding: A temporary built-up framework or suspended platform or work area designed to support workers, materials, and equipment at elevated or otherwise inaccessible job sites.

Self-retracting lanyard (SRL): A deceleration device containing a drum-wound line that can be slowly extracted from, or retracted onto, the drum under slight tension during normal employee movement, and which, after the onset of a fall, automatically locks the drum and arrests the fall.

Six-foot rule: A rule stating that platforms or work surfaces with unprotected sides or edges that are 6 ft. (1.8 m) or higher than the ground or level below it require fall protection.

Tensile strength: The resistance of a material to a force tending to tear it apart.

Top rail: A top-level, horizontal board required on all open sides of scaffolds and platforms that are more than 14 in. (36 cm) from the face of the structure and more than 10 ft. (3 m) above the ground.

Additional Resources

This module presents thorough resources for task training. The following reference material is recommended for further study.

Basic Construction Safety and Health, Fred Fanning. 2014. CreateSpace Independent Publishing Platform.

Construction Safety, Jimmie W. Hinze. Second Edition. 2006. New York, NY: Pearson.

DeWalt Construction Safety/OSHA Professional Reference, Paul Rosenberg, American Contractors Educational Services. 2006. DEWALT.

Fall Protection and Scaffolding Safety: An Illustrated Guide, Grace Drennan Gagnet, CSP. 2000. Government Institutes.

Online resources:

> *Code of Federal Regulations (CFR)*, **www.ecfr.gov**
>
> 29 *CFR* 1926, Subpart G, Accident Prevention Signs and Tags. **www.ecfr.gov**
>
> Occupational Safety and Health Administration (OSHA), **www.osha.gov**
>
> OSHA Videos, **www.osha.gov/video**

Figure Credits

JLG Industries, Inc., Figures 56, 58A, 58B

Courtesy of Louisville Ladders, Figures 40, 45 ,48

LPR Construction, MO, Figures 19, 20

Fall protection materials provided courtesy of Miller Fall Protection, Franklin, PA, Figures 21-27, 29, 31, 34, 37

Courtesy of PERI Formwork Systems, Inc., Figure 5

Courtesy of Snap-on Industrial - Tools at Height Program, Figure 33

Spider Staging, Figures 53A, 53B

Werner Co., Figure 49

Section Review Answer Key

Answer	Section Reference	Objective
Section One		
1. c	1.0.0	1
2. a	1.2.3	1b
Section Two		
1. b	2.1.1	2a
2. a	2.1.2	2a
Section Three		
1. d	3.3.1	3c
2. b	3.3.1	3c
3. b	3.3.1	3c
Section Four		
1. c	4.2.0	4b
2. c	4.2.2	4b
3. d	4.3.2	4c
Section Five		
1. b	5.0.0	5
2. d	5.2.0	5b
3. b	5.2.1	5b
4. b	5.5.1	5e
Section Six		
1. c	6.2.0	6b
2. b	6.2.4	6b
3. a	6.3.2	6c
Section Seven		
1. b	7.0.0	7
2. a	7.1.1	7a

NCCER CURRICULA — USER UPDATE

NCCER makes every effort to keep its textbooks up-to-date and free of technical errors. We appreciate your help in this process. If you find an error, a typographical mistake, or an inaccuracy in NCCER's curricula, please fill out this form (or a photocopy), or complete the online form at **www.nccer.org/olf**. Be sure to include the exact module ID number, page number, a detailed description, and your recommended correction. Your input will be brought to the attention of the Authoring Team. Thank you for your assistance.

Instructors – If you have an idea for improving this textbook, or have found that additional materials were necessary to teach this module effectively, please let us know so that we may present your suggestions to the Authoring Team.

NCCER Product Development and Revision

13614 Progress Blvd., Alachua, FL 32615

Email: curriculum@nccer.org
Online: www.nccer.org/olf

❑ Trainee Guide   ❑ Lesson Plans   ❑ Exam   ❑ PowerPoints   Other______________________

Craft / Level: ____ Copyright Date: ____________

Module ID Number / Title: ____

Section Number(s): ____

Description: ____

____

____

____

Recommended Correction: ____

____

____

____

____

Your Name: ____

Address: ____

____

Email: ____ Phone: _________________

This page is intentionally left blank.

Drawings in Roofing

OVERVIEW

Construction drawings are used to represent a building and its components. Drawings provide specific information about the layout and parts of a structure and types of materials to be used. Roofers must be able to interpret construction drawings and documents and identify information needed to install a roof system.

Module 16103

Trainees with successful module completions may be eligible for credentialing through the NCCER Registry. To learn more, go to **www.nccer.org** or contact us at 1.888.622.3720. Our website, **www.nccer.org**, has information on the latest product releases and training.

Your feedback is welcome. You may email your comments to **curriculum@nccer.org**, send general comments and inquiries to **info@nccer.org**, or fill in the User Update form at the back of this module.

This information is general in nature and intended for training purposes only. Actual performance of activities described in this manual requires compliance with all applicable operating, service, maintenance, and safety procedures under the direction of qualified personnel. References in this manual to patented or proprietary devices do not constitute a recommendation of their use.

16103 V1.0

From *Roofing, Trainee Guide*. NCCER.
Copyright © 2021 by NCCER. Published by Pearson. All rights reserved.

DRAWINGS IN ROOFING

Objective

Successful completion of this module prepares you to do the following:

1. Identify drawings found in a drawing set, including their fundamental components and features, and describe how construction documents are used in a roofing project.
   a. Identify various types of construction drawings.
   b. Identify and describe the basic components of construction drawings.
   c. Identify and describe various drawing elements and explain the use of dimensions and drawing scales.
   d. Explain how to find roof details and information about a roofing project using construction documents.

Performance Tasks

This is a knowledge-based module. There are no Performance Tasks.

Trade Terms

As-built drawings
Building information modeling (BIM)
Computer-aided drafting (CAD)
Copings
Detail drawings
Drawing set
Flashings
Legend

Nominal size
Project manual
Purlins
Record drawings
Reroofing
Riser diagram
Schedules
Takeoff

Industry Recognized Credentials

If you are training through an NCCER-accredited sponsor, you may be eligible for credentials from NCCER's Registry. The ID number for this module is 16103. Note that this module may have been used in other NCCER curricula and may apply to other level completions. Contact NCCER's Registry at 1.888.622.3720 or go to **www.nccer.org** for more information.

> **NOTE**
>
> The figures and construction drawings shown in this module are not to scale (NTS).

Contents

This page is intentionally left blank.

1.0.0 CONSTRUCTION DOCUMENTS AND DRAWINGS

Objective

Identify drawings found in a drawing set, including their fundamental components and features, and describe how construction documents are used in a roofing project.

a. Identify various types of construction drawings.
b. Identify and describe the basic components of construction drawings.
c. Identify and describe various drawing elements and explain the use of dimensions and drawing scales.
d. Explain how to find roof details and information about a roofing project using construction documents.

Trade Terms

As-built drawings: Drawings of a completed installation that reflect the structure as it was constructed and show any unexpected changes during construction.

Building information modeling (BIM): An intelligent process that uses computer databases and computer-aided drafting (CAD) modeling to create a dynamic, interactive model of an entire structure and track its lifecycle.

Computer-aided drafting (CAD): The making of a set of construction drawings with the aid of a computer.

Copings: Coverings on top of a wall exposed to the weather, usually made of metal, brick, or stone. Copings are usually sloped to shed water.

Detail drawings: Drawings shown at a larger scale in order to show specific features or connections.

Drawing set: The set of detailed drawings or plans drawn to scale by an architect and/or engineer, showing all information and dimensions necessary to build or remodel a structure.

Flashings: Components used to seal the edges of a roof system at perimeters, penetrations, walls, expansion joints, valleys, drains, and other places where the roof covering or membrane is interrupted or terminated.

Legend: In maps, plans, and diagrams, an explanatory table defining all symbolic information contained in the document.

Nominal size: Approximate or rough size by which materials are known and sold. Nominal size is usually slightly larger than the actual size.

Project manual: Documents prepared by the architect/engineer for the owner in order to execute a project. Documents in a project manual include bidding documents, contracts, project details, and a list of construction drawings. Also called a *specification book* or *spec book*.

Purlins: Horizontal secondary structural members that transfer loads to the primary structural framing.

Record drawings: Drawings that show a record of any unexpected changes made during the construction of a structure.

Reroofing: The process of re-covering or tearing off and replacing an existing roof system.

Riser diagram: A schematic drawing that depicts the layout, components, and connections of a piping system.

Schedules: Tables that describe and specify the various types and sizes of construction materials used in a building. Door schedules, window schedules, and finish schedules are the most common types. Other types include equipment schedules, beam schedules, column schedules, and footing schedules.

Takeoff: The process of surveying, measuring, itemizing, and accounting for all materials and equipment needed for a construction project.

Construction drawings are used to represent a structure or system before it is built. Most construction drawings are created by architects, engineers, and designers in computer-aided drafting (CAD) software. CAD software allows designers to create a virtual 3D model of a building and all of its components. These 3D models are then used to generate and print the drawings (sometimes called *blueprints* or *prints*) in a drawing set. CAD may also be used with building information modeling (BIM), enabling construction models and data to be viewed from any location and used to track the lifecycle of the building or structure.

Drawings convey to craft professionals the information needed to construct or repair a building. Drawings and documents needed for a roofing project vary depending on the project.

Roofing projects fall into the following general categories:

- *New Construction* — When applying a roof system to a new structure, workers must reference the drawing set and project manual for the project. Together, these are referred to as *construction documents*, and they convey all the information needed to complete a construction project.
- Reroofing — When re-covering or replacing an existing roof system, workers reference a roof replacement plan. They may also consult the original drawing set, as-built drawings, or record drawings, if available.

Manufacturers of roofing materials often have plans and drawings showing how their products should be installed, including insulation fastening patterns and panel layout.

1.1.0 Types of Drawings

There are several types of construction drawings (*Figure 1*). These drawings are assembled into the drawing set (*Figure 2*). The drawings in the set illustrate a variety of different views in order to convey complete information about the structure.

Each type of drawing in a set is assigned a letter. For example, since electrical drawings are assigned the letter E, the first few electrical drawings in a set are numbered E1, E2, E3, and so on. A complete drawing set typically includes the following types of drawings:

- General — G
  - Cover sheet
  - Index of drawings
  - Abbreviations
- Civil — C

Did You Know?

History of the Blueprint

In 1861, French chemist Alphonse Louis Poitevin discovered that ferro-gallate, a chemical found in gum, turns blue when exposed to strong light. This chemical was used to coat paper for the blueprinting process. The original drawing was placed over the light-sensitive paper and then exposed to sunlight, allowing ultraviolet light to shine through the drawing and producing a copy.

Today, most drawings are created on computers and printed. However, the term *blueprint* is still widely used when referring to copies made from original drawings.

Figure Credit: iStock@Michael Burrell

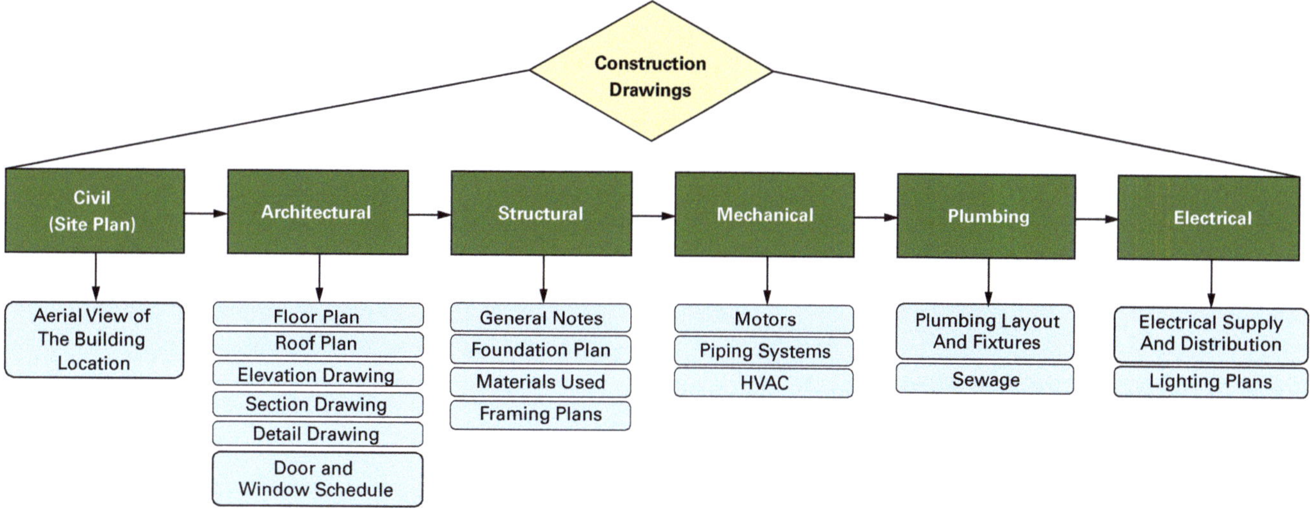

Figure 1 Types of construction drawings.

- Architectural — A
  - Plan views (roof, floor, and reflected ceiling plans)
  - Elevation views
  - Wall sections
  - Schedules
  - Detail drawings
- Structural — S
  - Foundation plan
  - Framing plans
- Mechanical — M
  - Heating Ventilation, and Air Conditioning (HVAC) plans
- Plumbing/Piping — P
- Electrical — E
- Fire Protection — F or FP

1.1.1 Civil Plans

A civil plan (*Figure 3*), sometimes called a *site plan* or *plot plan*, shows the location of the building on the site from an overhead view. A civil plan may also show the contours of the ground surrounding the building site. Other information that may be shown on the civil plan includes the following:

- Water features (ponds, lakes, rivers, and storm retention ponds)
- Sidewalks, driveways, curbs, and gutters
- Utilities
- Property dimensions
- Notes regarding grading
- A legal description of the property
- Site drainage

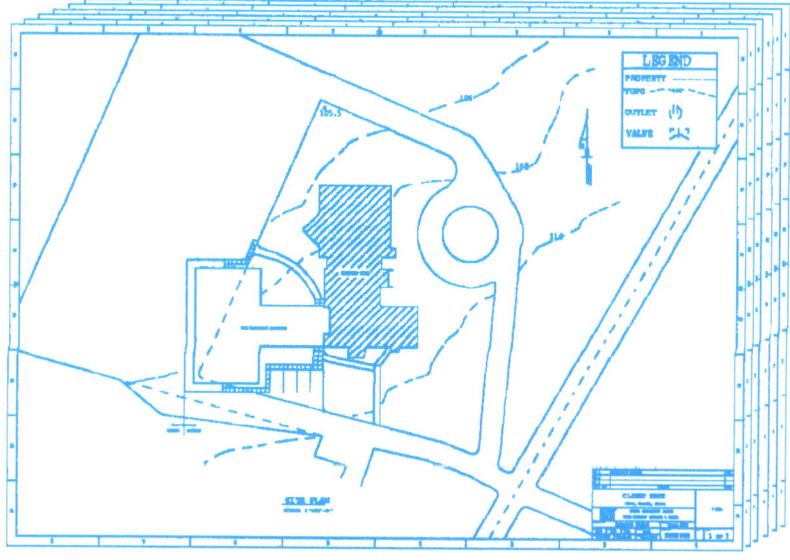

Figure 2 Typical organization of a construction drawing set.

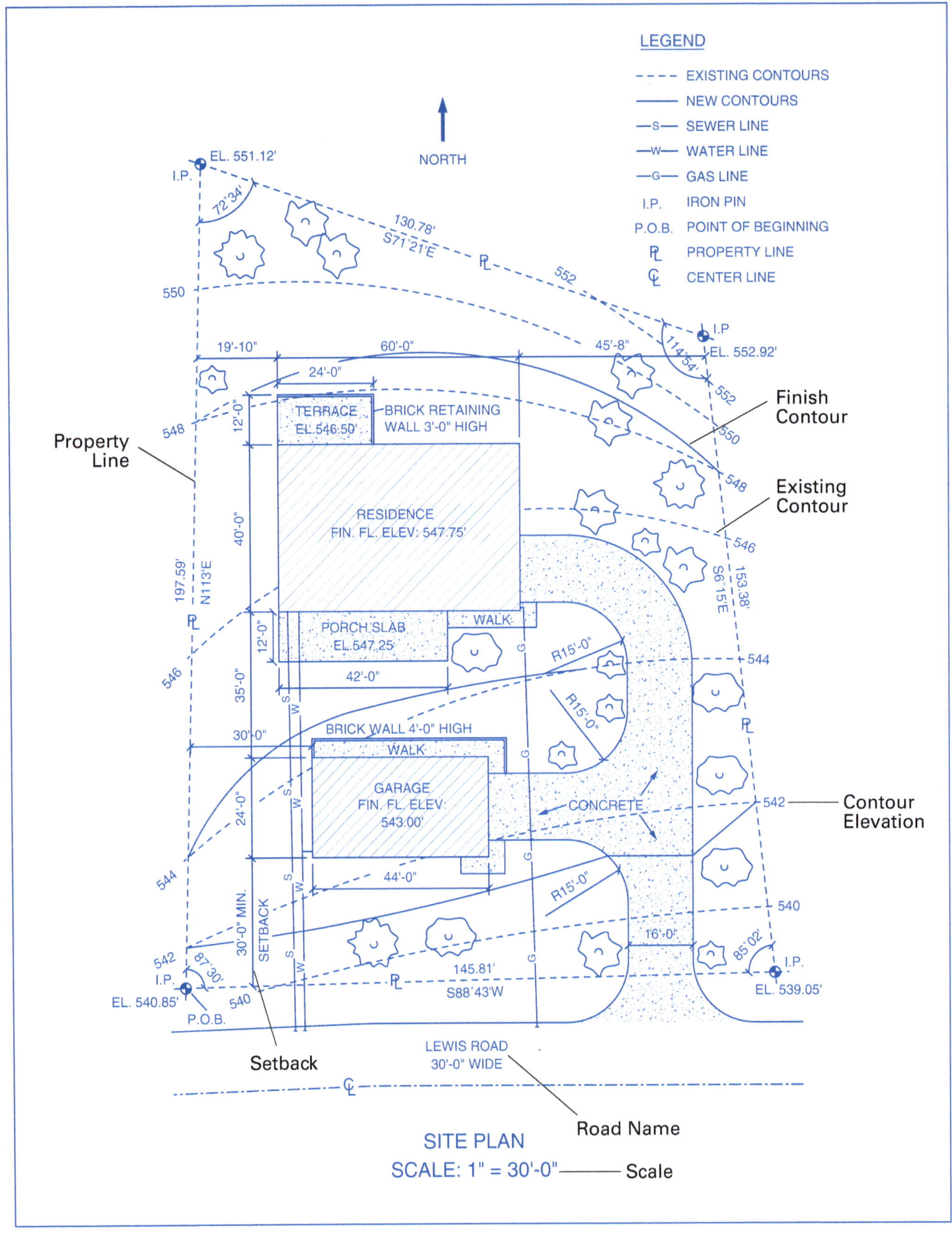

Figure 3 Typical civil plan showing topographical features.

Contour lines spaced at regular intervals (usually 1') are used to show the rise and fall of the ground. When contour lines are close together on the drawing, it indicates the land grades steeply. The farther apart the lines are, the more gradual the grade being represented.

1.1.2 Architectural Plans

An architectural plan includes multiple drawings showing how each part of the building should look from various views. These views include elevation drawings and floor, roof, and ceiling plans. Architectural plans also include schedules that describe the quantities, types, and sizes of items such as windows and doors. They also indicate ceiling type and height of specific rooms.

Floor Plans, Roof Plans, and Ceiling Plans

Floor plans show the layout of a structure. For a floor plan, an imaginary line is cut horizontally across the structure at varying heights so all the important features, such as windows, doors, and plumbing fixtures, can be shown. For multi-story buildings, separate floor plans are typically provided for each floor. *Figure 4* shows an example of a basic floor plan.

Floor plans show types of construction materials to be used, as well as the locations of the following parts of a structure:

- Outside walls, including exterior openings
- Interior walls and partitions
- Doors and windows
- Stairways
- Cabinets, electrical and mechanical equipment, and fixtures
- Cutting plane lines

Roof plans provide information about the roof slope, roof drain placement, and other information, such as gutters, downspouts, and ornamental sheet metal. Where applicable, the roof plan may also indicate the location of air conditioning units, exhaust fans, and other ventilation equipment. Examples of roof plans are shown in *Figure 5*, *Figure 6*, and *Figure 7*.

Some drawings sets may include a reflected ceiling plan that shows the location of supply diffusers, exhaust grilles, access panels, and the location of structural and mechanical components.

Elevation Drawings

An elevation drawing (*Figure 8*) shows a face-on view of a building, room, or part of a building, such as a stairwell or elevator shaft. They are called *elevation drawings* because they show height.

A set of plans typically includes two types of elevation drawings: a presentation view and multiple working views. The presentation view gives a complete picture of how the building will look. Working views may be at floor level, roof level, or ceiling level and contain information such as roof slope.

A Museum for Architecture

The Athenaeum of Philadelphia is a museum of American architecture. The museum, which has collected the work of about 1,000 American architects, has 150,000 drawings, 50,000 photographs, and many architectural documents.

Figure Credit: File: "The Athenaeum from north.jpg" is licensed under the Creative Commons Attribution-Share Alike 4.0 International, 3.0 Unported, 2.5 Generic, 2.0 Generic and 1.0 Generic license.

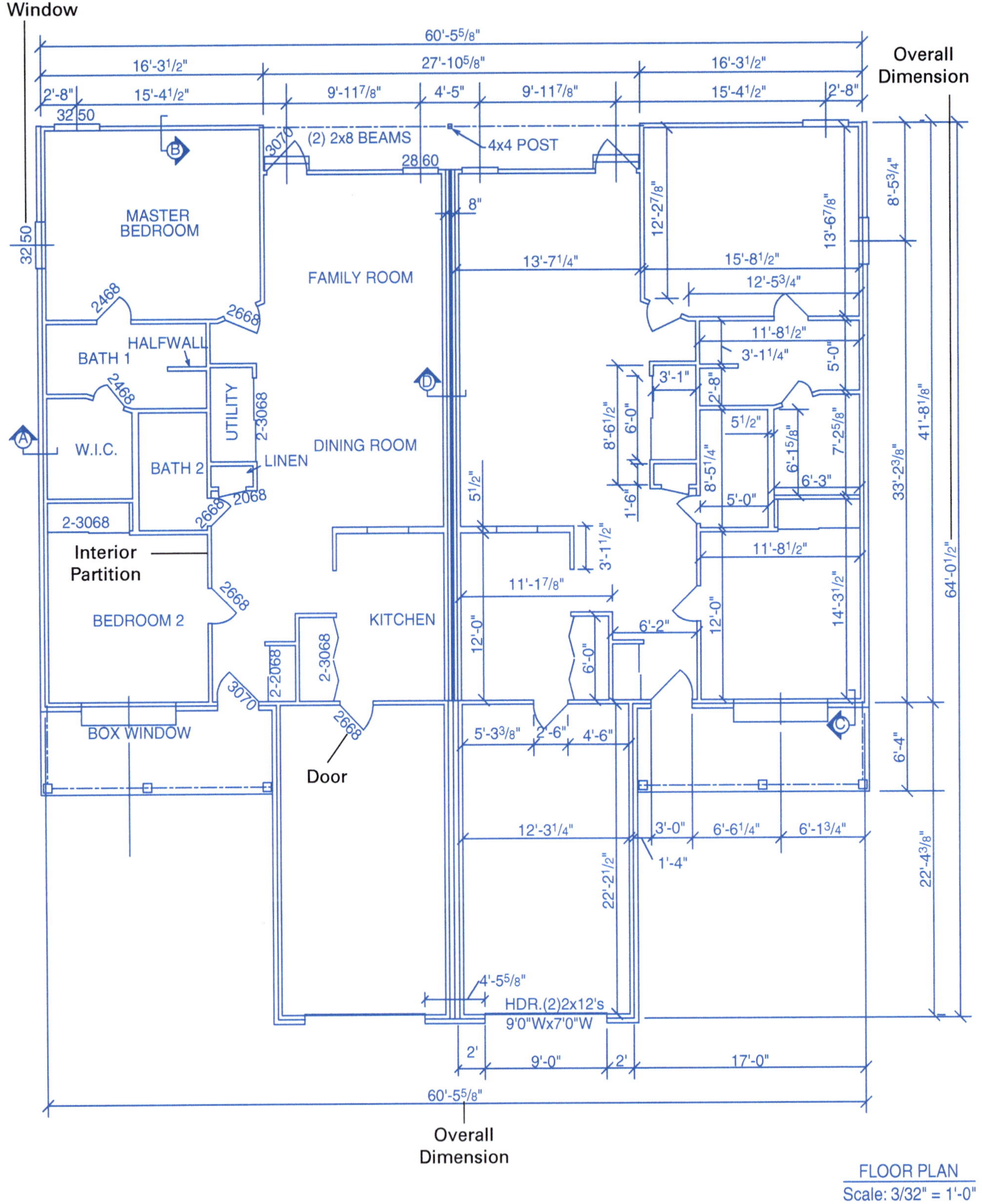

Figure 4 Basic floor plan.

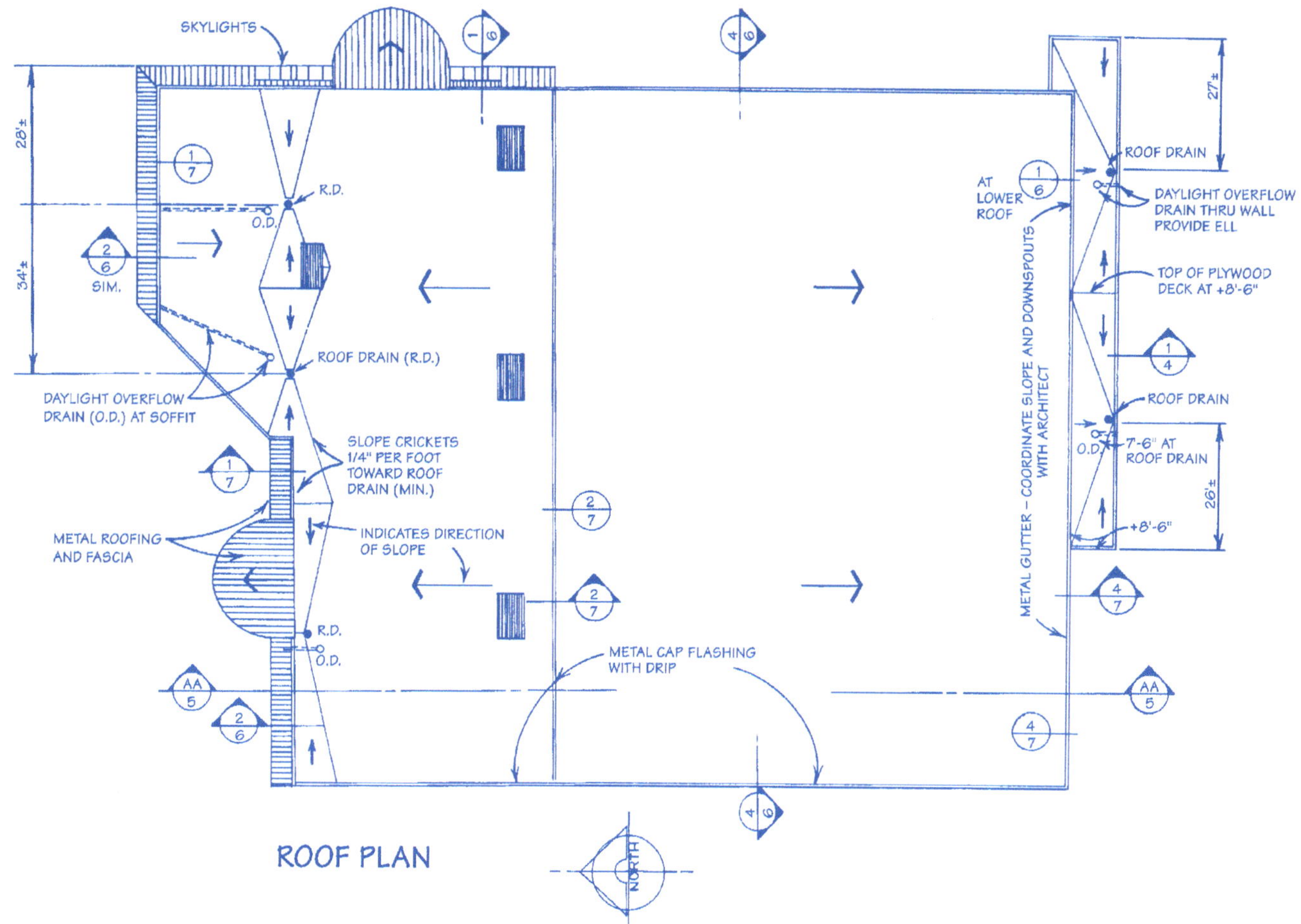

Figure 5 Roof plan (example 1).

Elevation views include the following information:

- Exterior style of the building
- Height of building
- Locations of doors, windows, chimneys, covered entryways, and decorative trim
- Exterior finish materials

1.1.3 Structural Plans

Structural plans are created by a structural engineer and accompany the architect's plans. Structural drawings show requirements for structural elements of the building, including floor and roof systems, columns, beams, stairs, canopies, and loadbearing walls. They also show information pertinent to roofing work, including the following:

- Type of roof deck (important when selecting fasteners)
- Amount of slope (if any) built into the deck

- Where deck changes direction and expansion joints are needed
- Locations of purlins

1.1.4 Mechanical, Electrical, and Plumbing Plans

For many construction jobs, mechanical, electrical, and plumbing (MEP) information is included on the basic floor plans (*Figure 9*). However, for complex commercial projects, MEP information is generally shown on separate plan drawings.

> **NOTE**
> While roofers usually work with architectural drawings, useful notes and views are included on drawings in other sections, especially the mechanical and plumbing sections. For example, mechanical plans show penetrations associated with rooftop HVAC units, and plumbing plans show drainage system penetrations.

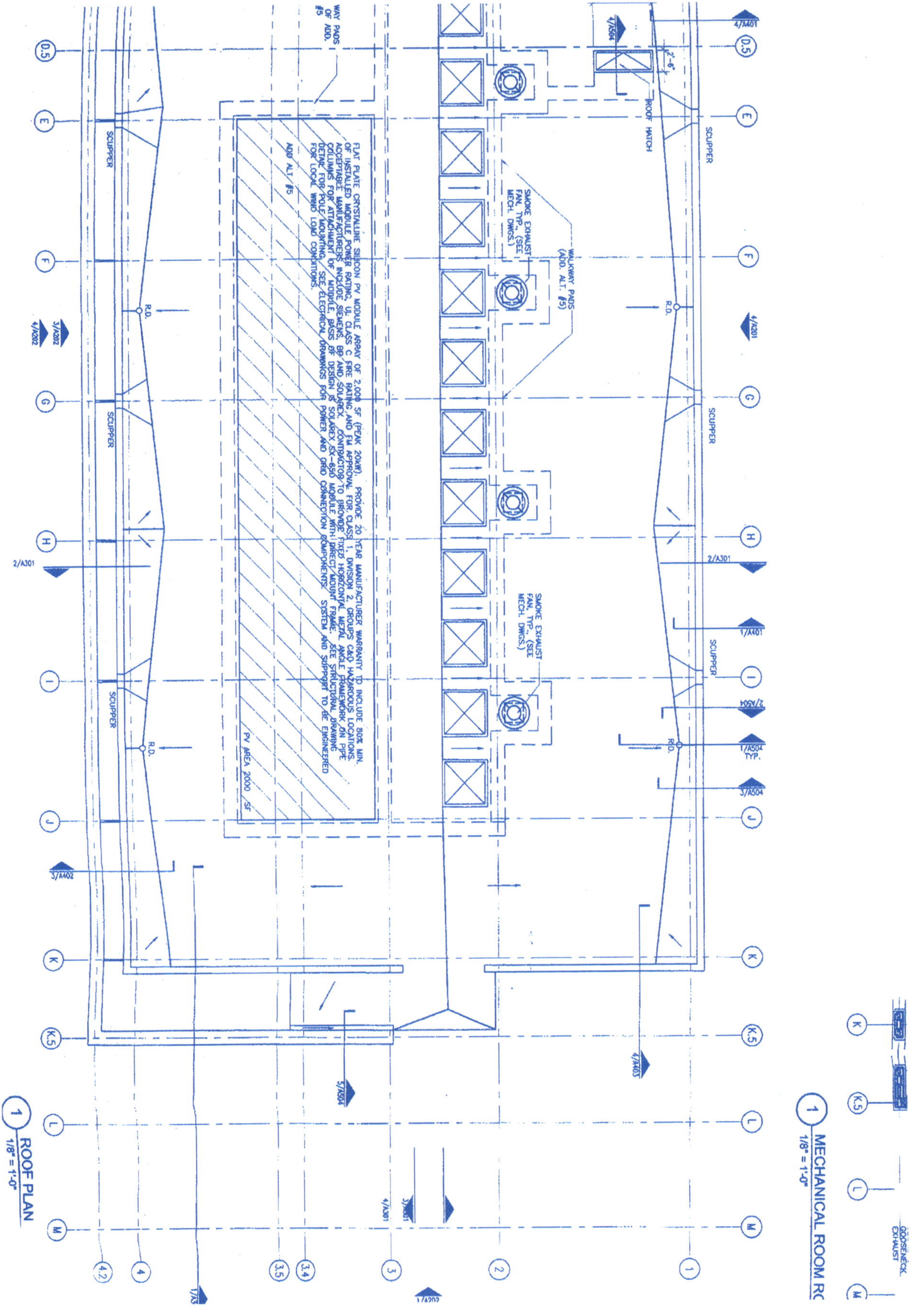

Figure 6 Roof plan (example 2).

NCCER – *Roofing*

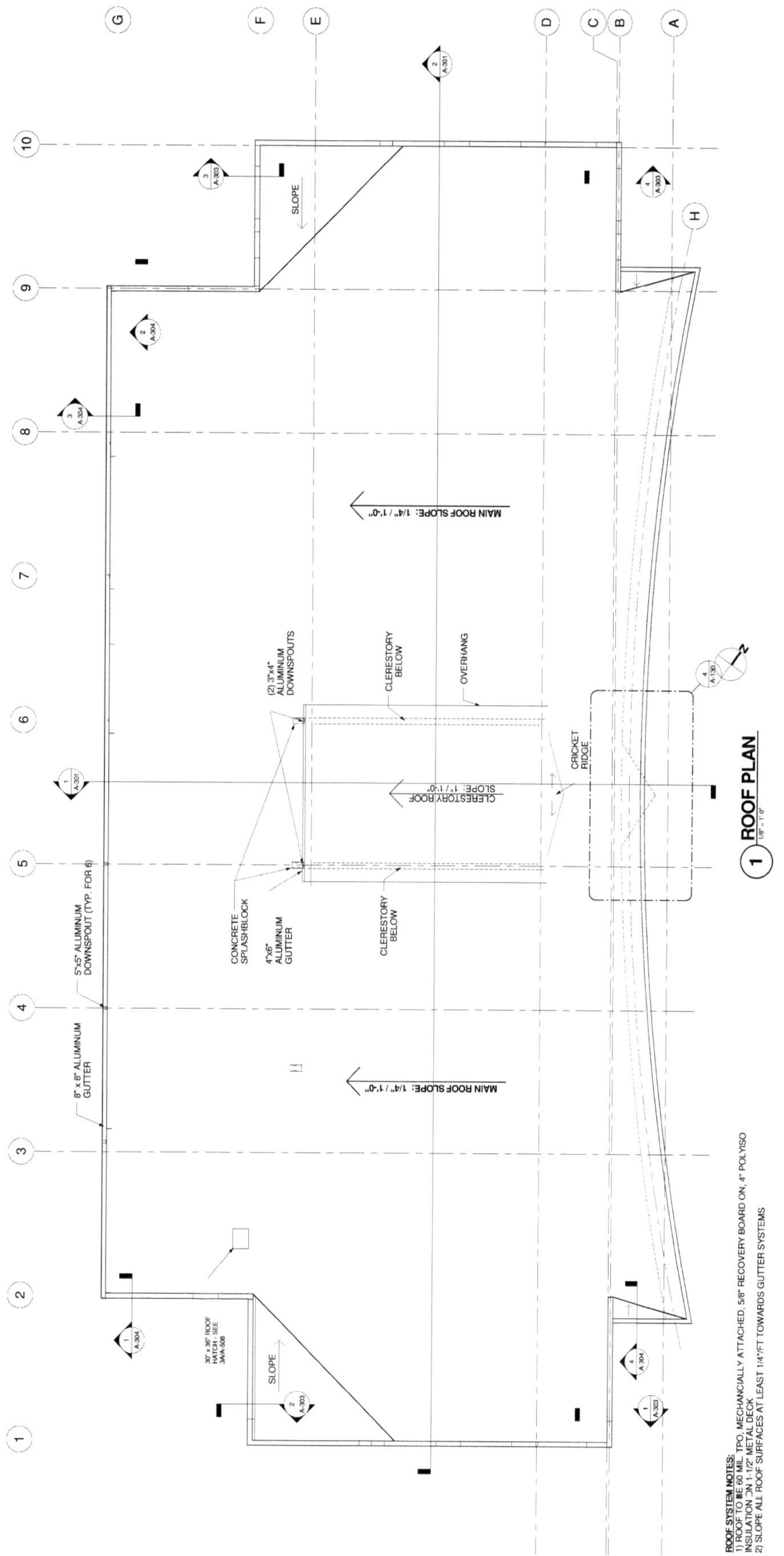

Figure 7 Roof plan (example 3).

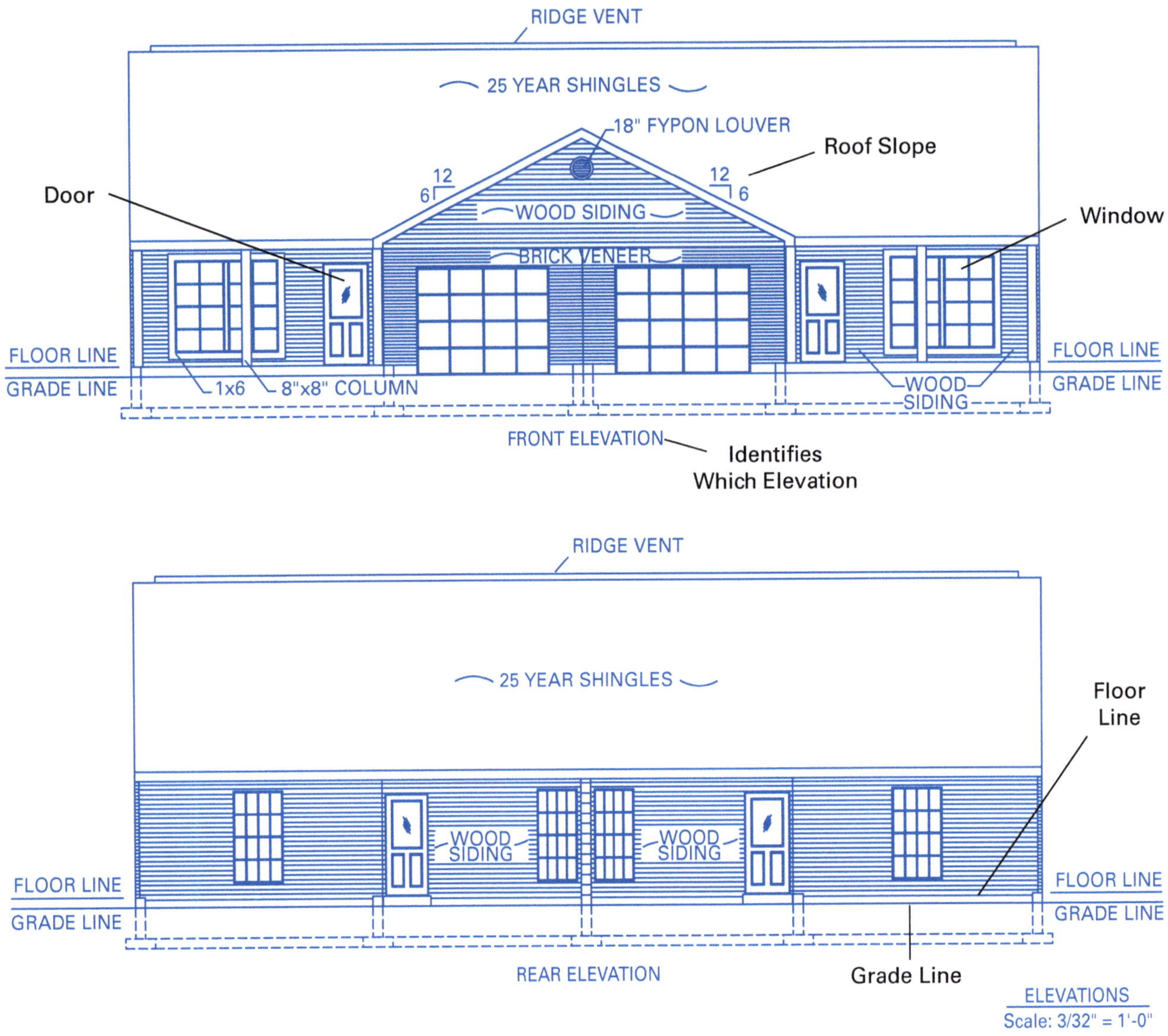

Figure 8 Example elevation views.

Presentation Drawings

Presentation drawings are rendered from 3D models to show a building from a desirable vantage point and display its most interesting features. Presentation drawings do not provide detailed information for construction purposes. They are used mainly by an architect or contractor to sell the design of a house or building to a prospective customer.

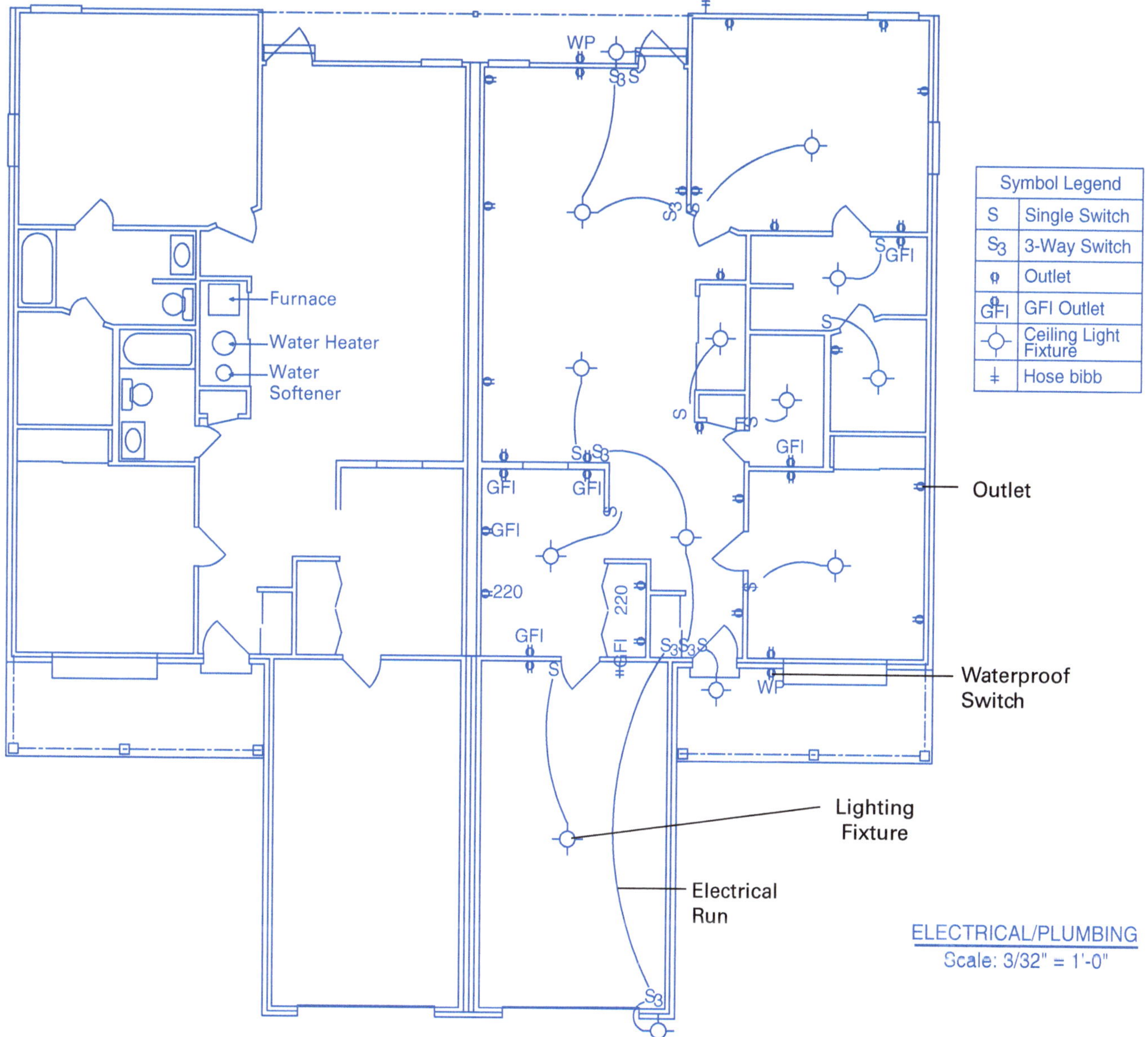

Figure 9 Example of an electrical/plumbing plan.

Mechanical plans show details for installation of the HVAC and other equipment. Depending on regional climate, they may include a refrigerant piping schematic that shows the types and sizes of the piping and identification of the fittings. They may also include a legend defining the HVAC symbols used on the drawing.

Electrical plans show the details for installation of the electrical system and equipment. They show the location of the meter, distribution panel, fixtures, switches, and other special items, as well as specifications for load capacities and wire sizes. The plan may also include a legend that defines the symbols used in the drawing.

Plumbing plans show the layout of water supply and sewage disposal systems and fixtures. Plumbing system plans generally have a separate plumbing riser diagram that shows the layout and identification of the piping and fixtures. A plumbing legend is usually located on the plan.

1.1.5 Fire Protection Plans

Fire protection plans show the piping, valves, sprinkler heads, switches, and other components that are part of the structure's sprinkler system. A list of fire sprinkler symbols is usually included on a separate sheet along with specifications, details and assembly drawings, and riser diagrams. For completeness, the sprinkler system is often included as part of the general mechanical plans to show how its installation relates to other mechanical systems.

1.2.0 Basic Components of Drawings

Once you can navigate a drawing set to find the sheets you need, you must be able to interpret the sheets themselves. Title sheets show generic information about a plan, and section and detail drawings show more detailed, "zoomed in" views within large systems. Some components that may be found on a sheet include title blocks, revision blocks, notes, and legends.

1.2.1 Title Sheets, Title Blocks, and Revision Blocks

A title sheet is typically placed at the beginning of a set of drawings or at the beginning of a major section of drawings. The title sheet provides an index to other drawings, a list of abbreviations used on the other drawings and their meanings, and various other information such as the project location, size of the land parcel, and building size.

The title sheet(s) should be carefully reviewed to understand the specific symbols and abbreviations used throughout the set of drawings. These symbols and abbreviations may vary from plan to plan.

A title block and a revision block are commonly placed on each sheet in a set of drawings, usually in the lower-right corner of the sheet (*Figure 10*). It is also common for the title block to go up the entire right side of the sheet.

A title block typically contains the name of the of the company that prepared the drawings, the owner's name, and the address and name of the project. It also includes the title of the sheet, the sheet number, the date the sheet was prepared, the scale, and the initials or names of the people who prepared and checked the drawing.

A revision block is typically shown on each sheet in a set of drawings, usually in the upper- or lower-right corner of the sheet near or within the title block. A revision block is used to record any changes (revisions) to the drawing. When a revision is made, an entry is added in the revision block. An entry typically contains the revision number or letter (keyed to a location on the plan), a brief description of the change, the date, and the initials of the person making the changes.

When using drawings, it is essential to note the revision on each drawing and ensure it is the latest one. Failure to do so could result in costly mistakes. If you are not sure about the revision status of a drawing, check with your supervisor to ensure the most recent version of the drawing is being used.

1.2.2 Section and Detail Drawings

Section drawings essentially show a slice of an object or structure. Showing what the finished structure would look like if it were cut in half makes it easier to see the materials and understand how it is put together. Most section views are vertical, but horizontal section views may also be included.

Section drawings are usually drawn to a larger scale than other drawings, so details can be easily seen. Detail drawings, which are often included in section drawings, are zoomed in to an even larger scale. *Figure 11* shows an example of a section drawing, and *Figure 12* shows an example of a detail drawing.

1.2.3 Schedules

Schedules are tables that describe and specify the various types and sizes of construction materials used in a building. Door schedules, window schedules, and finish schedules are the most common. Drawing sets for commercial projects also include schedules for mechanical equipment and controls, plumbing fixtures, lighting fixtures, and any other equipment.

1.2.4 Notes, Abbreviations, and Legends

General notes for each plan are included in a drawing set, as well as a legend that lists symbols and abbreviations used on the plans. *Figure 13* shows general notes for a mechanical plan.

Sheets may have notes with information needed for construction or installation of specific components. If present, these notes are usually listed near the top or bottom of the sheet.

1.3.0 Drawing Elements, Dimensions, and Scale

Construction drawings use lines, symbols, abbreviations, and notes to deliver the architect's message as clearly and concisely as possible. Although drawings must be created smaller than the building they represent, they maintain the

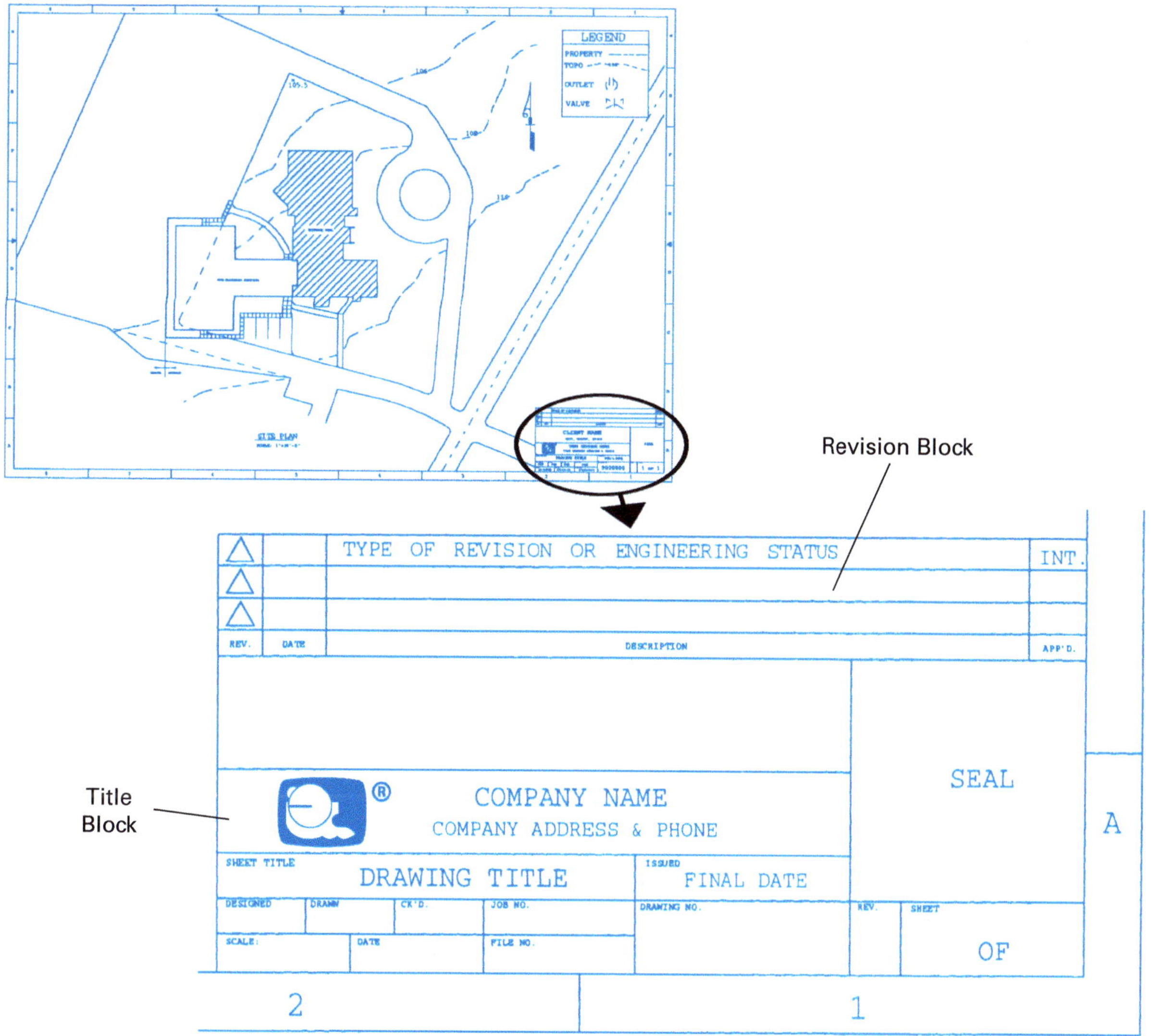

Figure 10 Title and revision blocks.

exact proportions of the building to ensure they are an accurate representation of the intended structure.

1.3.1 Dimensions and Drawing Scale

Dimensions provided on drawings indicate actual sizes, distances, and measurements of the objects and spaces being represented. As shown in *Figure 14*, dimensions may be indicated from outside to center, center to center, wall to wall, or outside to outside.

Drawings are created using a specified scale to ensure all dimensions are proportional, which allows the drawings to be used to calculate the actual dimensions of the structure.

Inches or fractions of an inch on the drawing are used to represent feet in the actual measurement of a building. For example, in a plan drawn to $\frac{1}{4}$" scale, $\frac{1}{4}$" on the drawing represents 1' of the building. The scale would be written on the drawing as $\frac{1}{4}$" = 1'.

In general, plans are drawn at a $\frac{1}{4}$" scale, and details are drawn at a $\frac{3}{4}$" scale. The scale of a drawing is usually shown directly below or above the drawing.

> **NOTE**
> In cases where all the details on a sheet are drawn to the same scale, the scale may be noted in the title block of the sheet. The same scale may not be used for all the drawings that make up a complete set of plans.

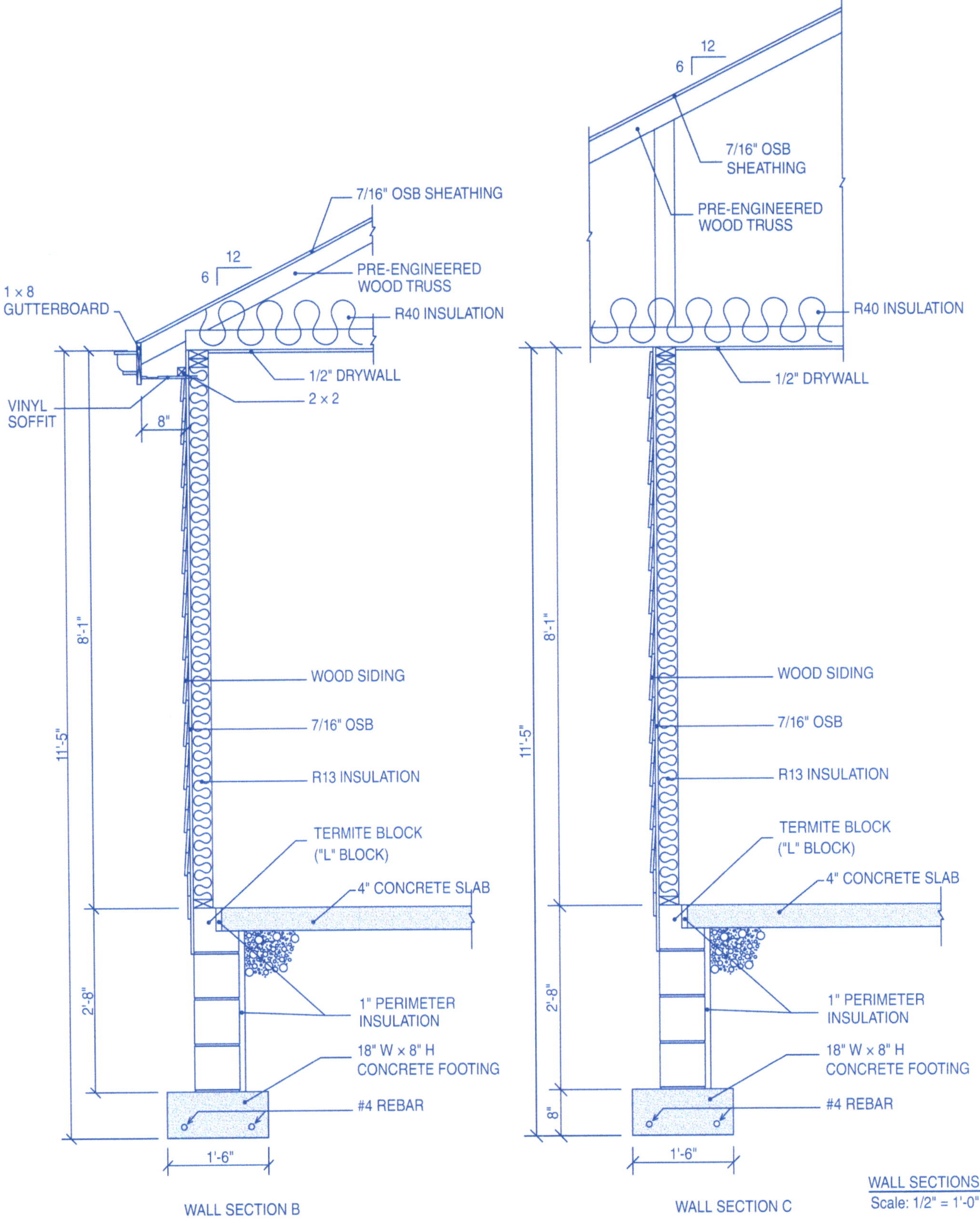

Figure 11 Example of a section drawing.

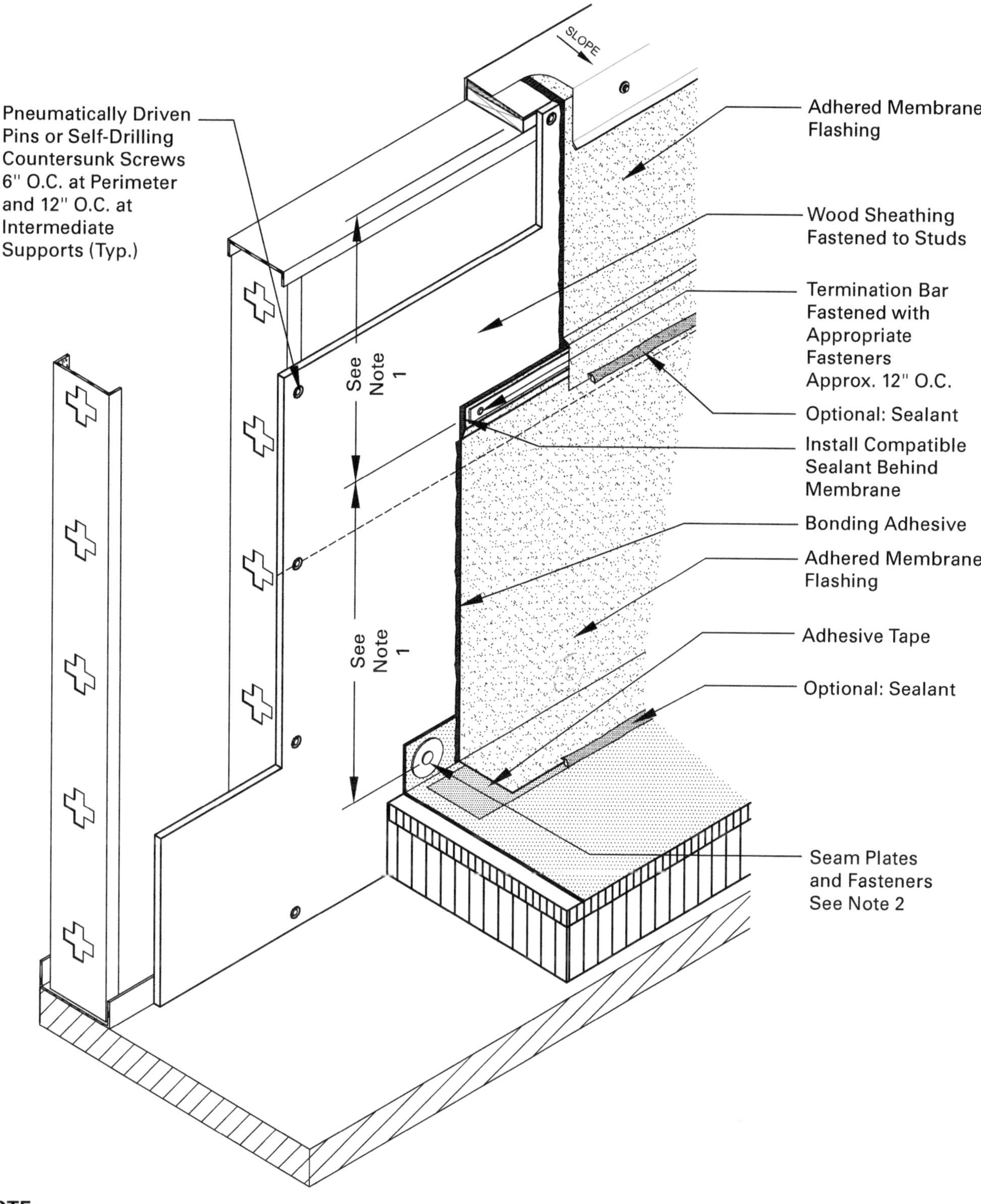

NOTE:

1. Max spacing for membrane flashing fastening as recommended by manufacturer.

2. Refer to the introduction of the construction details chapter for alternative base securement options.

Figure 12 Example of a roofing detail.

GENERAL NOTES

1. VERIFICATION OF EXISTING CONDITIONS

1.1 Dimensions indicated on drawings, which have a direct relationship with the existing construction, must be verified in the field prior to the construction, fabrication or installation of any related work. Such dimensions and elevations include, but are not limited to, building dimensions, parapet heights and thickness, etc. If field verifications of dimensions at existing construction differ from that indicated on the drawing, the contractor should document the actual conditions and submit to the architect. If necessary a field directives (Change Order) will be issued to clarify requirements of the work.

2. VERIFICATION OF EXISTING ROOF CONSTRUCTION

2.1 Existing roof construction including type of deck, insulation, wood nailer, and roofing membrane is indicated of the drawings for the convenience of the contractor. This information is not warranted as to accuracy or completeness. The contractor is responsible for verification prior to bid of existing roof construction relative to its impact on demolition and disposal operations. No allowance will be made for failure of the contractor to obtain such existing conditions information prior to bidding except as indicated below.

3. The Contractor shall take special care to insure all work complies with all federal, state and local codes and regulatory agencies having jurisdiction.

4. The scope of work which includes but is not limited to, the specified work to all parapets, flashings, wood blocking, metal coping caps, metal fascia, roof insulation, curbs, plumbing stacks, roof drains, tapered components, saddles, crickets, membrane, equipment supports, penetrations, heater stacks, vent stacks, roof hatches, etc.

5. Proposed contractor storage area, staging area, contract limits and parking area locations are show on Sheet 'SP.1' and shall be discussed @ the Pre-Bid Conference.

6. All work/setup areas shall be barricaded off with caution tape, orange cones, orange plastic fencing, warning signs e.g. "Caution - Work Area", "Over head work in progress", "No Admittance - Work Area", etc.

7. Proposed contractor storage and staging area, all contract limits and employee parking area locations shall be verified with owner prior to the start of work.

8. Where existing rooftop mechanical equipment is to remain, raise equipment, ductwork, etc., and flashing as necessary and/ or called for on the drawings, to meet roof warranty requirements.

9. Remove all existing roofing membrane, flashing system, existing cant, metal counter flashing, where indicated, down to existing roof substrate and as required for new roofing installation.

10. Roof Drains/ Downspouts: Contractor shall test all roof drains & downspouts to verify water flow. Submit in writing notice that all roof drains &/ or downspouts have been inspected and tested and are functioning properly. Notify owner in writing of all roof drains &/ or downspouts that are clogged and/or if the water flow is very slow.

11. Where existing rooftop mechanical equipment, exhaust fans, gravity vents, roof hatch, skylights, etc., are to remain, disconnect and raise equipment and provide new wood blocking and flashings as necessary to meet roof warranty requirements. Reset and reconnect unit(s) as required. Disconnect and reconnect of roof top equipment shall be coordinated with the Owner and the responsibility of the Contractor.

11.1 Contractor shall advise Owner a minimum of seventy-six (76) hours in advance, in writing, prior to disconnect/reconnect or any shut down of any fans, HVAC units, condensers or other mechanical equipment.

11.2 Shop drawings (reviewed and approved by manufacturer) shall be submitted as per specifications for final review/approval prior to any construction.

12. Where gas lines, conduits, cables, wiring, etc., remain on roof surface, mount on pre-manufacturered rubber pipe supports @ 4'-0" o.c. Rubber pipe supports to have s.s. unistruts & s.s. clip(s) to anchor pipe, conduit, cable, wiring, and etc., to top of supports.

13. Where wiring/piping/conduit/unit support post, etc., entrance into the building is made through the roof; the Contractor shall provide liquid applied flashings or penetration hood at the roof penetration.

14. Maintain watertight roof condition at all times, before and during roof construction.

- NO phased construction allowed. Remove only the amount of existing roofing and insulation that can be covered over with new roof insulation and membrane in the same day.

15. Contractor shall verify, on site, all dimensions and equipment locations prior to ordering materials.

16. In the event of any discrepancies and/or errors found in the drawings or if problems are encountered during construction that would cause hardships on the contractor, the contractor shall be required to notify the Owner before proceeding with the work.

17. All areas requiring patching, it shall be done in a manner to match adjacent surfaces and finishes.

18. Contractor shall be responsible for closing, patching and repairing of all holes, openings, damage to all surfaces, e.g., metal coping caps, masonry walls, roof decks, etc., resulting during construction by Contractor and/or Subcontractor.

19. Contractor shall be responsible for maintaining security of the work place, setup areas, storage areas, parking areas, equipment, tools, etc.

20. Care shall be taken to protect all utilities and equipment.

Figure 13 Mechanical plan general notes.

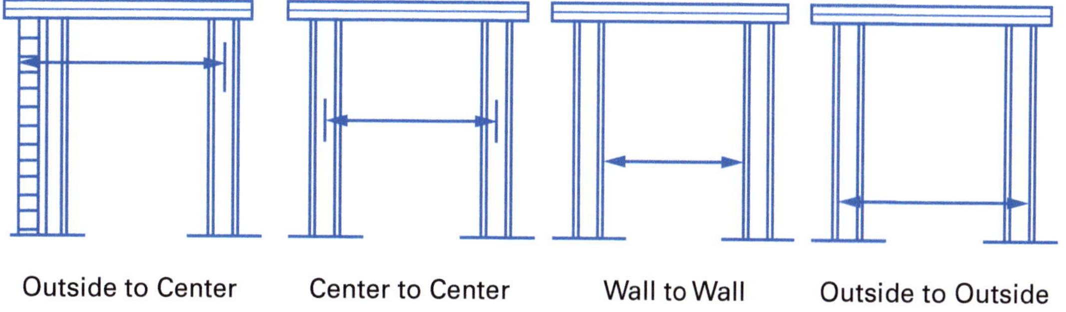

Figure 14 Common methods of indicating dimensions on drawings.

Think About It

Scaling Drawings

Measuring an object on a drawing and converting the measurement to the true length of the object is called scaling a drawing. If a line on a floor plan measures 6" and the object being represented is drawn to a scale of $\frac{1}{4}$" = 1', what is the true length of the object?

NOTE

When a dimension is not found on a drawing, it is sometimes measured with an architect's scale. However, this is not a reliable practice. If you have questions about dimensions, your foreman can contact the architect to confirm the missing dimensions.

Section and detail views may use a nominal size in labeling. Nominal sizes or dimensions are approximate or rough sizes by which supplies, such as lumber, pipes, etc., are commonly sold. For example, a 2 × 4 is named after its nominal dimensions (2" × 4"), but it actually measures $1\frac{1}{2}$" × $3\frac{1}{2}$".

1.3.2 Drawing Symbols

Symbols are used in construction drawings to indicate different kinds of materials, fixtures, and structural members. Some symbols are standard, and others can vary in different regions. A set of drawings generally includes a sheet identifying symbols used in the set and their meanings. The following types of symbols may be in a drawing set:

- Material symbols (*Figure 15*)
- Electrical symbols (*Figure 16*)
- Plumbing symbols (*Figure 17*)
- HVAC symbols (*Figure 18*)

Measuring Roofs

Drawings are used to measure and calculate how much roofing material is needed. When measuring a two-dimensional (or flat) plan that shows a sloped roof from above, the slope factor (sometimes called *pitch factor*) is used to account for a roof not being flat. The table shown here includes slope factors for various roof slopes. For example, if a section of a 5:12 roof measures 200 ft^2 on the plan, this value must be multiplied by the corresponding slope factor of 1.083 to calculate the actual surface area of the roof (200 ft^2 × 1.083 = 216.6 ft).

Roof Slope	Slope Factor	Valley and Hip Factor
1:12	1.0035	1.4167
2:12	1.0138	1.4240
3:12	1.0308	1.4362
4:12	1.0541	1.4530
5:12	1.0833	1.4743
6:12	1.1180	1.5000
7:12	1.1577	1.5298

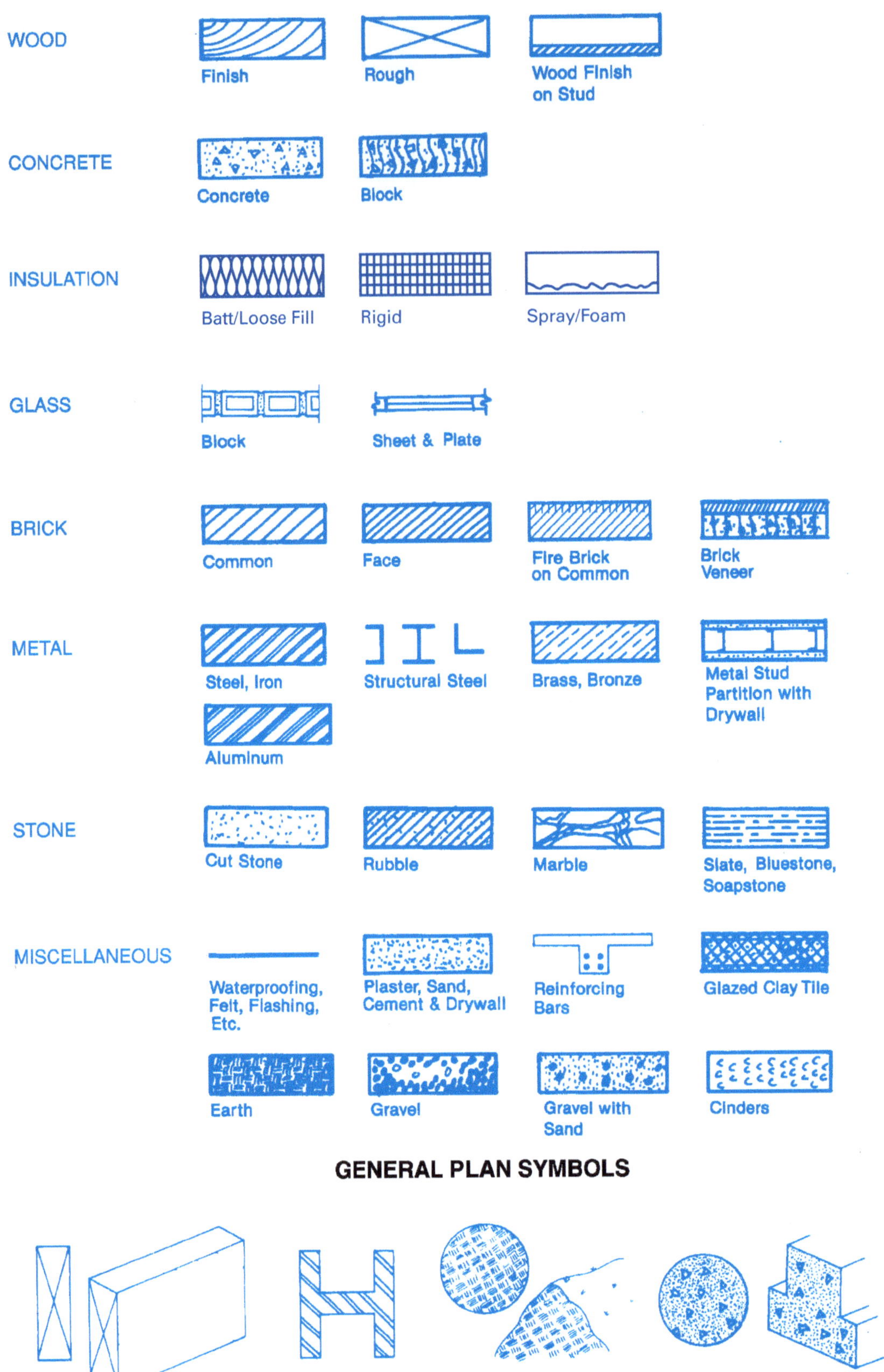

Figure 15 Example material symbols.

 NCCER – *Roofing*

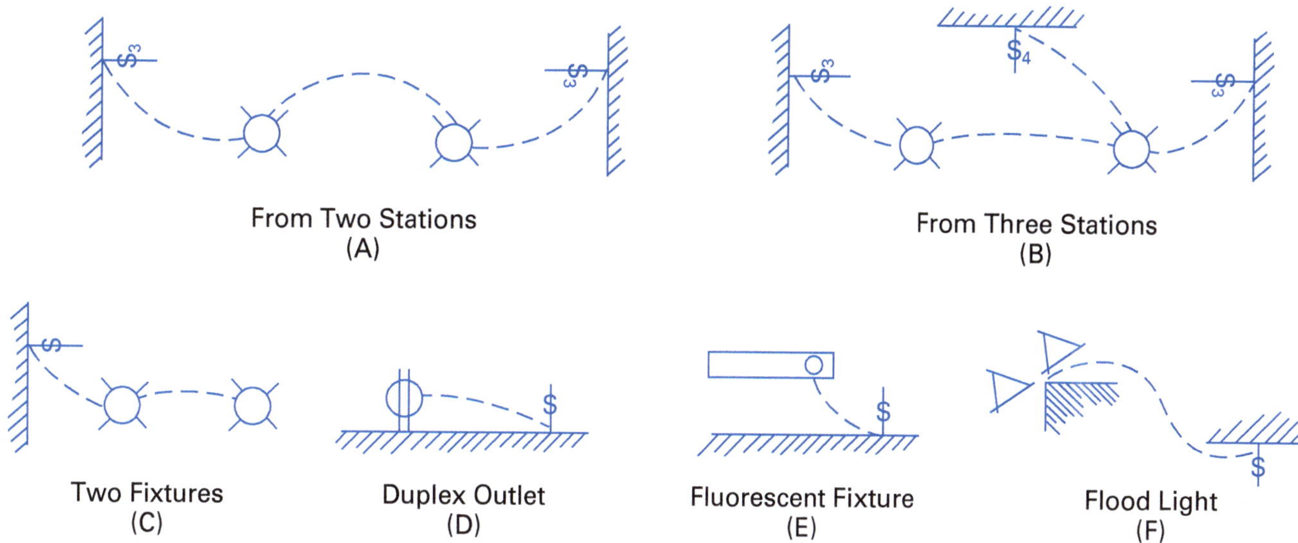

Figure 16 Electrical symbols showing control of light fixtures or an outlet.

- Site plan and topographical survey symbols (*Figure 19*)
- Structural member symbols (*Figure 20*)

> **NOTE**
>
> All workers on the project must be able to accurately locate where their work is to be done. To help workers orient themselves to the site, all plan views include an arrow indicating which direction on the drawing is north.

1.3.3 Line Types

Many different types of lines are used to draw and describe a structure. Lines are drawn wide or narrow, dark or light, dashed or connected, with each type of line conveying a specific meaning. *Figure 21* shows the most common lines used on construction drawings. The following are descriptions of each type of line:

- *Object line* — A thick line used to show the main outline of a structure, including exterior walls, interior partitions, porches, patios, sidewalks, parking lots, and driveways.
- *Dimension and extension lines* — Thin lines used to show the dimensions of an object. Extension lines extend from both ends of the object being measured. A dimension line is drawn parallel to the object being measured, pointing to the extension lines on both sides. To make them easier to see, extension lines and object lines do not touch. Sometimes a gap is provided in the dimension line and the dimension is shown in the gap.

- *Center line* — Used to designate the center of an area or object and provide a reference point for dimensioning. Center lines are typically used to indicate the centers of objects such as columns, posts, footings, and doorways.
- *Cutting plane line* — Used to indicate a section of a drawing that is cut away and shown in a section view. The arrows at the ends of the cutting plane line indicate the direction in which the section is viewed. Letters identify the section view of that specific part of the structure. More elaborate methods of labeling cutting plane lines are used in larger sets of plans (*Figure 22*) where many sections are being used. The section view may be on the same page as the cutting plane line or another page.
- *Break line* — Used to indicate that an object or area is not being shown in its entirety.
- *Leader line* — Used to connect a note or dimension to a related part of the drawing. Leader lines are usually curved or at an angle from the object to avoid confusion with other lines.
- *Hidden line* — Used to indicate an outline that is invisible to an observer because it is covered by another surface or object that is closer to the observer. Hidden lines are usually dashed.
- *Phantom line* — Used to indicate alternative positions of moving parts, such as a damper's swing or adjacent positions of related parts. Phantom lines may also be used to represent repeated details. Phantom lines are typically dashed, but with longer dashes than hidden lines.

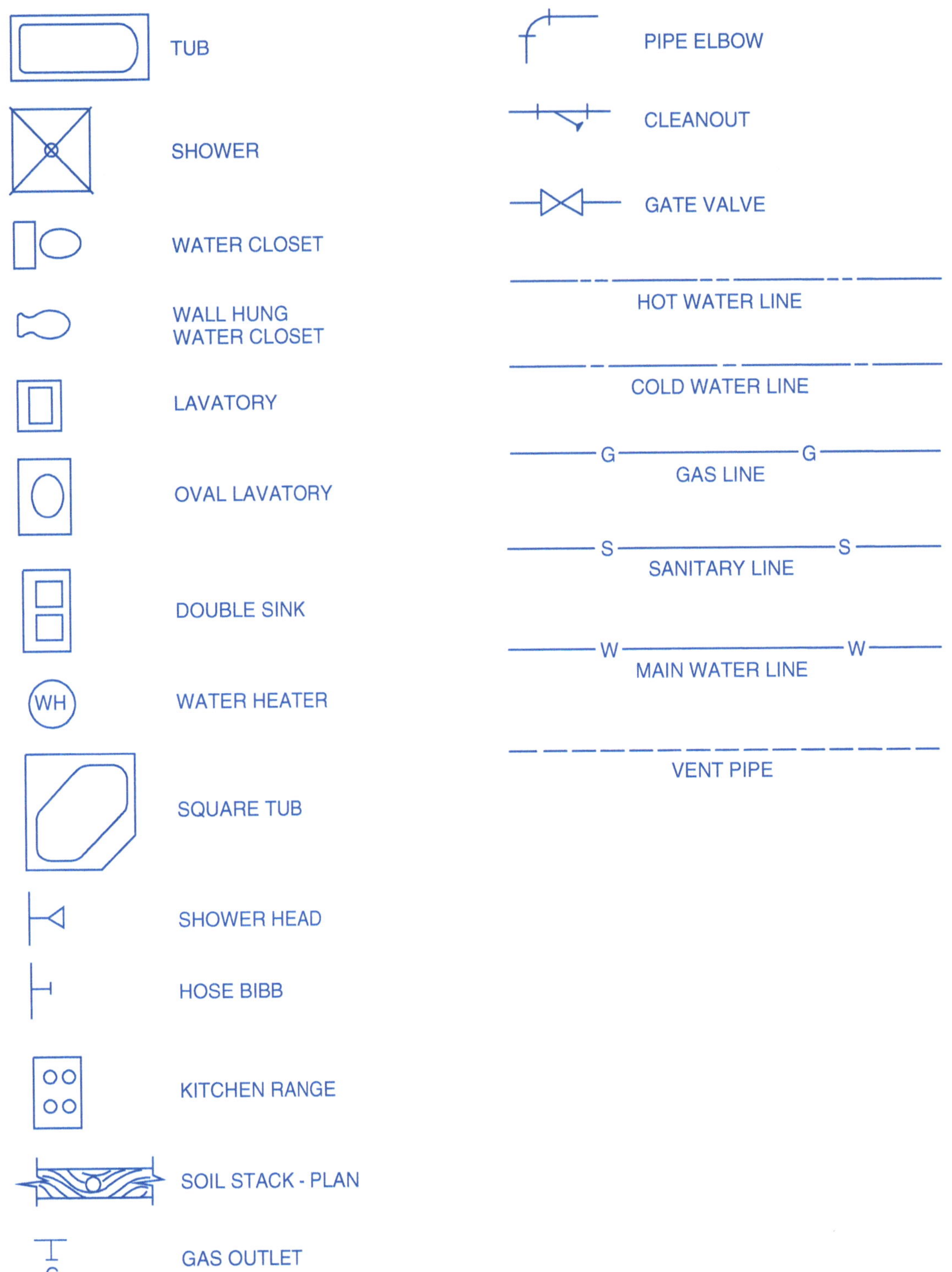

Figure 17 Plumbing symbols.

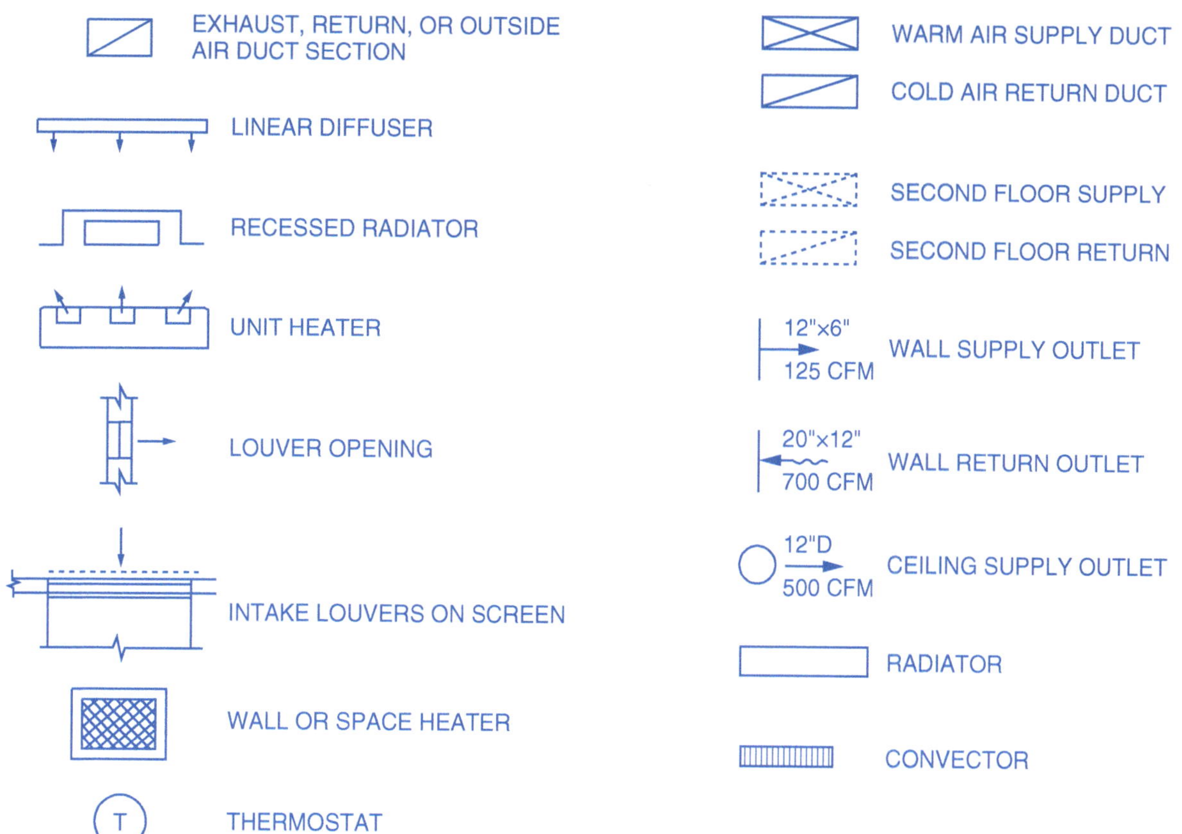

Figure 18 HVAC symbols.

- *Stair indicator line* — A short line with an arrowhead that shows the ascent or descent of a stairway on a floor plan.
- *Contour line* — Used on civil plans to show changes in elevation or contour of the land. Dashed lines are used to show the natural grade, while solid lines show the finish grade to be achieved during construction. Each line across the plot of land represents a different elevation relative to a reference point, such as sea level. Each line is drawn at a uniform change in elevation, such as every 10'. The closer together the contour lines are, the steeper the terrain is.

1.4.0 Roofing Project Documents

Before a roofing project begins, roofing craft professionals must use construction documents to determine basic information about the project and what resources will be needed to complete it (*Figure 23*).

Construction documents for a new construction project include the drawing set and project manual. The project manual contains the project specifications and contracts. For a reroofing project, workers should consult a roof replacement plan.

1.4.1 New Construction

Before installing a roof system on a new building, workers can use construction documents to find out what kind of system will be installed, its size and location, and what kind of deck the building has (*Figure 24*). By starting from the big picture and working your way to more detailed information, you can gain a better understanding of the project.

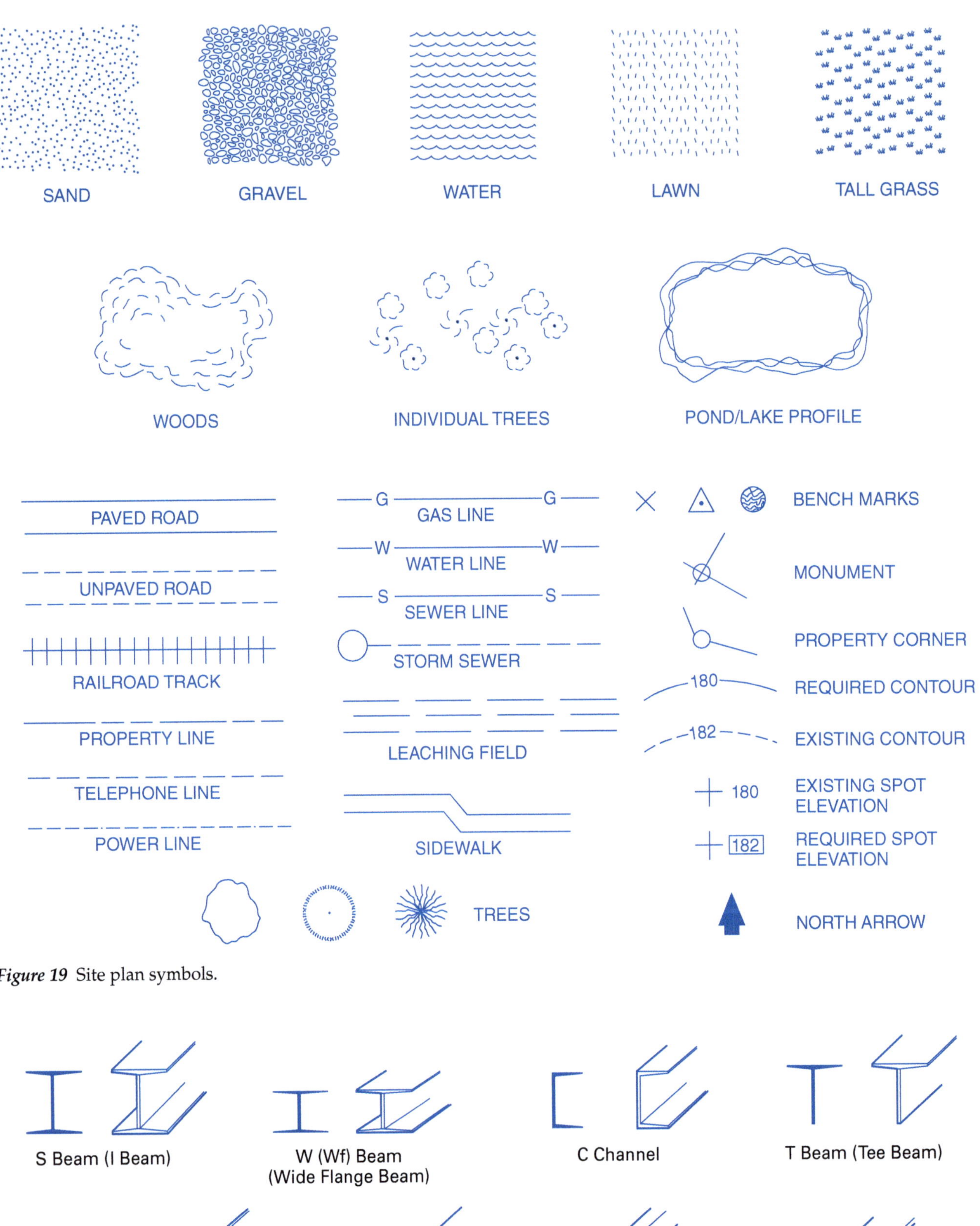

Figure 19 Site plan symbols.

Figure 20 Structural member symbols.

NCCER – *Roofing*

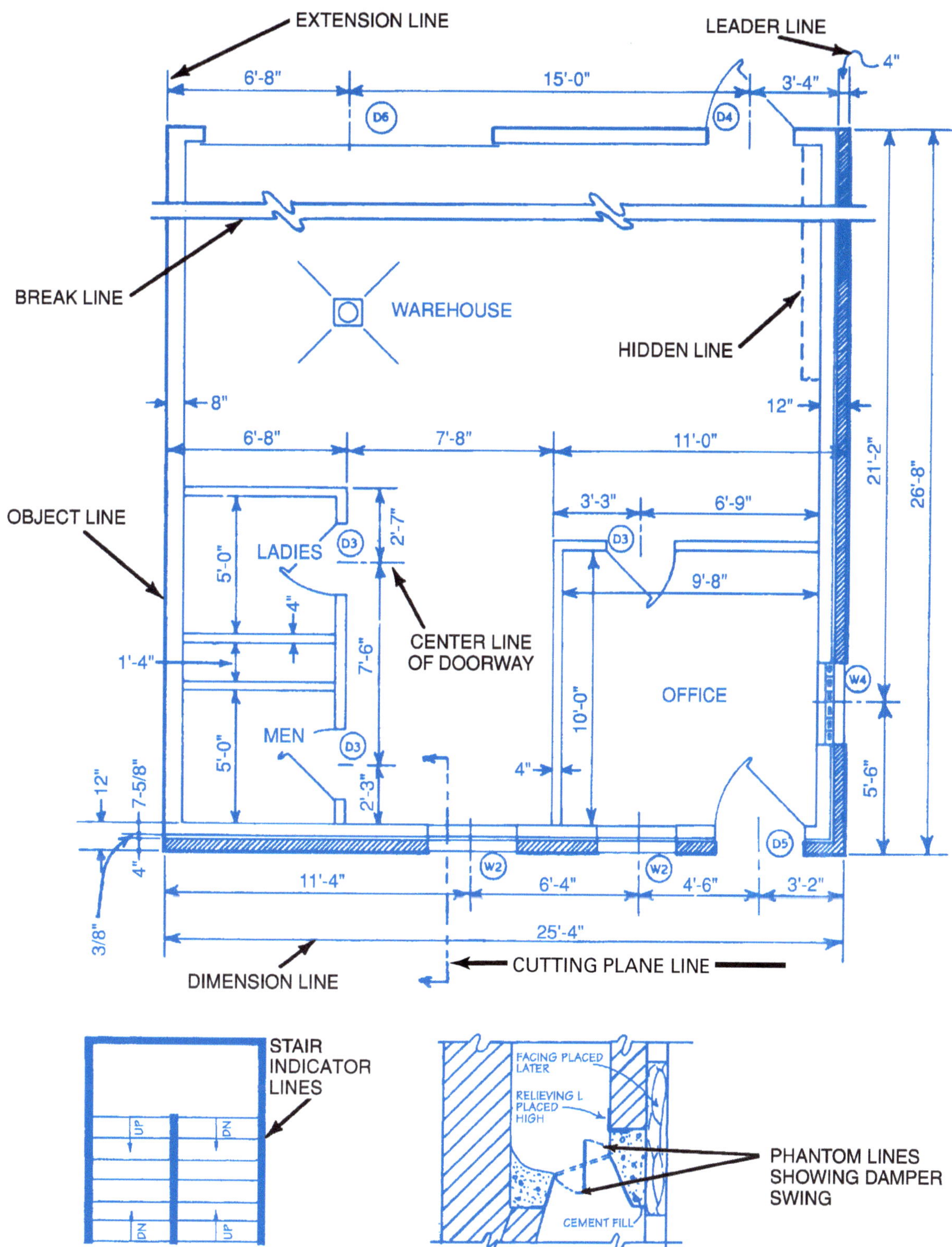

Figure 21 Lines used on construction drawings.

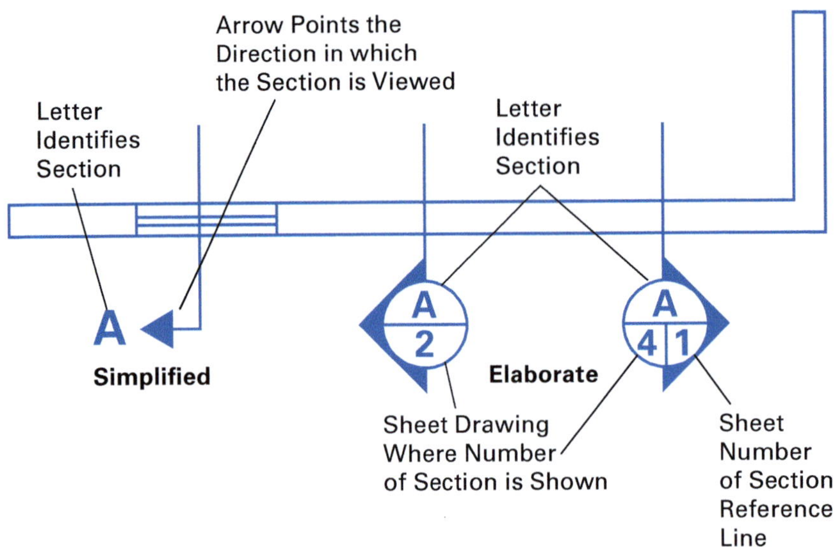

Figure 22 Methods of labeling cutting plane lines.

Figure 23 Craft professionals interpreting construction drawings.

Figure 24 Construction documents on a jobsite.

The following general steps describe how to identify roofing project information in a drawing set:

Step 1 Find the roof plan in the drawing set. Since the roof plan is an architectural drawing, it is typically found in the architectural (A) section.

Step 2 Look through mechanical, electrical, and plumbing (MEP) drawings for penetrations such as the following:
- Plumbing stacks
- Heat stacks
- Vents
- Gas and electrical lines
- AC units
- Refrigerant lines (especially in commercial/industrial buildings, such as factories)
- Flashings, gutters, and valleys

Step 3 Find relevant section and detail views for components such as copings and flashings.
- Drainage details
- Perimeter details
- Penetration details

Project Specifications

Since there is not enough space on the drawings to include all the required technical information about the project, the designer develops a set of specifications (*Figure 25*) to provide important technical information about the work and quality of materials. The drawing set shows what to install, and the specifications explain what materials to use and how to do it.

While drawings show exact dimensions and where objects are located, specifications describe the materials to be used, hardware requirements,

Figure 25 Project specifications.

and other performance guidelines for the project. Contractors and their employees are responsible for adhering to project specifications and meeting standards of quality.

Specifications serve three general purposes:

- Define the responsibilities of the designer, owner, contractor, and subcontractors.
- Supplement the working drawings with detailed technical information regarding the quality of work and materials to be used.
- Support the drawings as part of the contract documents.

For convenience, speed, and ease of use, specifications are formatted into a series of sections dealing with the construction requirements for the various areas of work on a construction project. People who use the specifications must be able to find all the information they need without spending too much time looking for it.

Construction materials and methods have expanded greatly over the years. The written specifications for construction work have similarly expanded from a few pages to documents that can easily exceed 500 pages. After World War II, specifications were standardized and separated into divisions. After 2004, this format expanded into the most common specification writing format used in North America, the MasterFormat®, with five major groupings and 49 divisions. The 2016 version of MasterFormat® is shown in *Figure 26*.

Under the 2016 standard, the Thermal and Moisture Protection division (07) is most important to roofers.

Roofing Foreman

Roofing crews provide a critical service to their communities by installing and maintaining roof systems that protect buildings from the elements. Each roofing crew member plays an important role in a roofing project. A professional roofing foreman provides leadership for the roofing crew. The roofing foreman must be able to interpret construction documents and provide guidance and insight to the other members of the crew. When crew members are unsure about something, they must ask their foreman for help.

Figure Credit: Courtesy of the National Roofing Contractors Assoc.

MasterFormat Groups, Subgroups, and Divisions

PROCUREMENT AND CONTRACTING REQUIREMENTS GROUP

Division 00 – Procurement and Contracting
         Requirements
         Introductory Information
         Procurement Requirements
         Contracting Requirements

SPECIFICATIONS GROUP

GENERAL REQUIREMENTS

Division 01 – General Requirements

FACILITY CONSTRUCTION SUBGROUP

Division 02 – Existing Conditions
Division 03 – Concrete
Division 04 – Masonry
Division 05 – Metals
Division 06 – Wood, Plastics, and Composites
Division 07 – Thermal and Moisture Protection
Division 08 – Openings
Division 09 – Finishes
Division 10 – Specialties
Division 11 – Equipment
Division 12 – Furnishings
Division 13 – Special Construction
Division 14 – Conveying Equipment
Division 15 – Reserved for Future Expansion
Division 16 – Reserved for Future Expansion
Division 17 – Reserved for Future Expansion
Division 18 – Reserved for Future Expansion
Division 19 – Reserved for Future Expansion

FACILITY SERVICES SUBGROUP

Division 20 – Reserved for Future Expansion
Division 21 – Fire Suppression

Division 22 – Plumbing
Division 23 – Heatin, Ventilating, and
         Air-Conditioning (HVAC)

Division 24 – Reserved for Future Expansion
Division 25 – Integrated Automation
Division 26 – Electrical
Division 27 – Communications
Division 28 – Electronic Safety and Security
Division 29 – Reserved for Future Expansion

SITE AND INFRASTRUCTURE SUBGROUP

Division 30 – Reserved for Future Expansion
Division 31 – Earthwork
Division 32 – Exterior Improvements
Division 33 – Utilities
Division 34 – Transportation
Division 35 – Waterway and Marine Construction
Division 36 – Reserved for Future Expansion
Division 37 – Reserved for Future Expansion
Division 38 – Reserved for Future Expansion
Division 39 – Reserved for Future Expansion

PROCESS EQUIPMENT SUBGROUP

Division 40 – Process Interconnections
Division 41 – Material Processing and Handling
         Equipment
Division 42 – Process Heating, Cooling, and
         Drying Equipment
Division 43 – Process Gas and Liquid Handling,
         Purification, and Storage Equipment
Division 44 – Pollution and Waste Control
         Equipment
Division 45 – Industry-Specific Manufacturing
         Equipment
Division 46 – Water and Wastewater Equipment
Division 47 – Reserved for Future Expansion
Division 48 – Electrical Power Generation
Division 49 – Reserved for Future Expansion

19

Introduction and Applications Guide

Figure 26 2016 MasterFormat®.

Augmented Reality

Augmented reality (AR) is quickly becoming a useful tool for the construction industry. BIM can be integrated into AR platforms, enabling users to view 3D models in real world settings. This provides the ability to visualize and interact with BIM data from any location. These tools are often compatible with mobile devices, such as smartphones and tablets, for easy access and collaboration.

1.4.2 Reroofing

For reroofing projects, the original drawing set for the existing roof may not be available or applicable. Instead, a roof replacement plan is created by the building owner or a designer. Although roofers are not involved in the creation of the plan, it is helpful for them to understand how it was created.

A roof replacement plan is created by conducting an on-site investigation of the existing roof system and using as-built measurements. This process involves the following general steps:

Step 1 Perform roofing takeoff and measure physical dimensions of roof.

Step 2 Note positions of components, including the distance of purlins.

Step 3 Note direction and amount of deck slope.

Step 4 Perform pull test and record results.

Step 5 A design professional (such as a roofing contractor, engineer, etc.) uses the collected data to generate a roof replacement plan, including insulation and membrane layout and fastening patterns.

> **NOTE**
>
> *Step 5* is not performed by roof installers.

1.4.3 Building Information Modeling

Building information modeling (BIM) is an intelligent process that uses computer databases and CAD modeling to create a dynamic, interactive model of an entire structure (*Figure 27*). BIM is used to provide insight for the design, construction, and management of buildings. It allows contractors and building owners to visualize the structure and collaborate throughout the construction process. Workers can use mobile devices to access 3D models while on the job.

BIM allows people to virtually walk through a building before it has been built to see how it will look when it is completed. This allows designers to identify problems and conflicts with various components in time to correct them before construction is complete. BIM also allows designers to

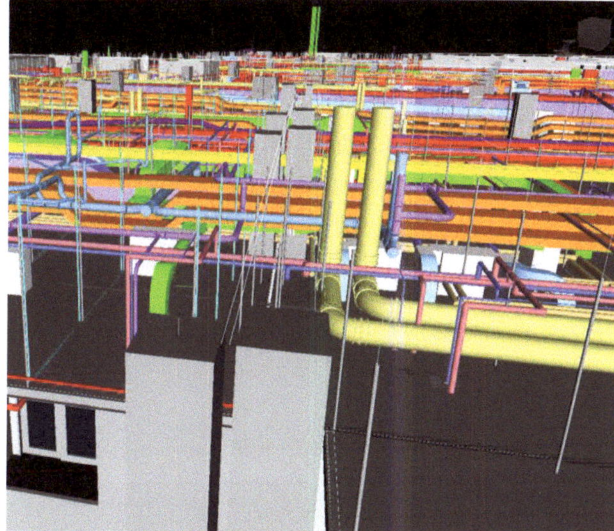

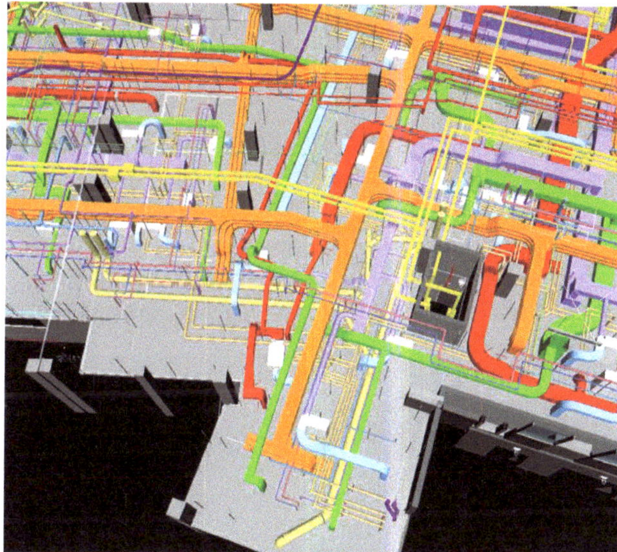

Figure 27 Examples of BIM drawings.

estimate the costs of constructing the building and even the cost of facilities maintenance over time.

BIM can be used to prepare coordination drawings and as-built drawings. BIM has not taken the place of construction drawings, but it does offer a new and improved way to look at the systems within a building. Standards for BIM are being developed in the United States, Canada, and Europe. Other terms for BIM include *virtual building environment (VBE)* and *virtual design and construction (VDC)*.

---

1. *Most* construction drawings are created by _____.

    a. a roofing contractor
    b. a foreman
    c. building designers
    d. building owners

2. When a change is made to a drawing, an entry is added in the _____.

    a. legend
    b. title sheet
    c. title block
    d. revision block

3. Which type of line is usually curved and connects a note or dimension to a related part of the drawing?

    a. Leader line
    b. Object line
    c. Center line
    d. Cutting plane line

4. Which of the following is *true* about project specifications?

    a. They are not included in the project documents.
    b. They provide technical information not included in the drawing set.
    c. They are created by the roofing contractor.
    d. Roofing crews do not need them.

1. Sheets in a drawing set marked with the letter A are _____.

   a. architectural drawings
   b. ancillary drawings
   c. as-built drawings
   d. abbreviations

2. Which type of drawing shows the location of the building on the site from an overhead view?

   a. Roof framing plan
   b. MEP plan
   c. Floor plan
   d. Civil plan

3. Which type of drawing shows the layout (including rooms, walls, and doors) in a structure?

   a. Mechanical plan
   b. HVAC plan
   c. Floor plan
   d. Site plan

4. Fire protection plans show _____.

   a. locations of sprinklers
   b. evacuation routes
   c. smoke detectors
   d. electrical equipment

5. Which of the following might be found on a title sheet?

   a. A door schedule
   b. An index to other drawings
   c. Section drawings
   d. Detail drawings

6. An approximate size by which supplies are known and sold is called _____.

   a. scale size
   b. actual size
   c. proportional size
   d. nominal size

7. Which of the following is used to indicate different kinds of materials, fixtures, and structural members on a drawing?

   a. Symbols
   b. Hidden lines
   c. Specifications
   d. The title block

8. Which type of line shows the main outline of a structure?

   a. Center line
   b. Contour line
   c. Object line
   d. Dimension line

9. Where in the drawing set should you look to find penetrations such as plumbing stacks, vents, gas and electrical lines, and AC units?

   a. MEP plans
   b. Architectural plans
   c. General plans
   d. Site plans

10. Which of the following is one of the purposes of project specifications?

    a. To show topographical information about the site.
    b. To list abbreviations and symbols used on the drawings.
    c. To provide information about materials and quality of work.
    d. To show dimensions and locations of building components.

Fill in the blank with the correct term that you learned from your study of this module.

1. The process of making a set of construction drawings using computer software is called _______________.

2. Components used to seal the edges of a roof system at perimeters and penetrations are called _______________.

3. The detailed plans drawn by an architect or engineer showing all information and dimensions necessary to build a structure are in a(n) _______________.

4. A schematic drawing that depicts the layout, components, and connections of a piping system is called a(n) _______________.

5. Depictions of a completed installation that reflect the structure as it was constructed and show any unexpected changes during the project are called _______________.

6. The approximate or rough dimensions by which a material is known and sold is called its _______________.

7. The process of re-covering or tearing off and replacing an existing roof system is called _______________.

8. The process of surveying, measuring, itemizing, and accounting for all materials and equipment needed for a construction project is called _______________.

9. Drawings shown at a larger scale in order to show specific features or connections are called _______________.

10. Written documents prepared by the architect/engineer for the owner in order to execute a project are in the _______________.

11. Covering pieces on top of a wall that is exposed to the weather are called _______________.

12. Tables that describe and specify various types and sizes of construction materials used in a building are called _______________.

13. Horizontal secondary structural members that transfer loads to the primary structural framing are called _______________.

14. An explanatory table defining all symbolic information contained in a drawing set is called a(n) _______________.

15. An intelligent process that uses computer databases and CAD modeling to create a dynamic, interactive model of an entire structure is called _______________.

16. Drawings that show a record of any unexpected changes made during construction are called _______________.

Trade Terms

As-built drawings	Copings	Nominal size	Riser diagram
Building information modeling (BIM)	Detail drawings	Project manual	Schedules
	Drawing set	Purlins	Takeoff
Computer-aided drafting (CAD)	Flashings	Record drawings	
	Legend	Reroofing	

Clay Thomas

Project Manager
Advanced Roofing, Inc.

How did you choose a career in the industry?

Like many in the industry, roofing chose me. I'm a third-generation roofer and grew up in the family business. Along the way I have worked in other construction fields, but roofing is home to me. (Ah, the smell of asphalt in the morning….)

Who inspired you to enter the industry?

My dad. He was a successful businessman with an eighth-grade education. He believed that a strong work ethic and integrity are the path to success. He was an excellent craftsman. By the time I was out of high school, I knew the job requirements as well as the business.

What types of training have you been through?

As a trainer for the National Roofing Contractors Association (NRCA), NCCER, ABCI, and Advanced Roofing, I get all certifications that my trainees achieve. I strive for continuous improvement, so if it adds value to my skill set and improves my game, I'll be in the class. However, life has taught me the toughest and best lessons. You just have to be willing to accept and learn from them.

How important is education and training in construction?

It is always important to practice your craft to master it. People take notice when you are trying to improve. Stand out above the crowd. For an employer, education and training of their workforce improves the company's knowledge base and provides an internal pipeline of talent. As they say, "knowledge is power."

What kinds of work have you done in your career?

I started sorting nails and making "China caps" in my dad's sheet metal shop. By the age of 15, I was running my own tear-off, dry-in and mop crews. I also spent a few summers laying tile. I graduated engineering school and did design work, but my real passion was in the field. I spent 10 years building luxury high rises on Miami Beach. But ultimately, I returned to my roots in roofing.

Tell us about your present job.

I work in HR supporting the hiring process, creating training content and conducting training internally for Advanced Roofing. I work to help employees develop a career path and achieve their goals. I serve on various committees and boards of directors (BODs) that promote worker training and education. I am also an instructor for the ABCI Roofing Apprentice Program.

What do you enjoy most about your job?

The opportunity to work with and mentor the next generation of roofing professionals. I have also made some great relationships with some fantastic people. People are the lifeblood of our business. Without a team, no roofing gets done.

What factors have contributed most to your success?

Soft skills. Good communication, treating people with respect, integrity… your reputation can make or break you, so make decisions wisely. Always follow through and do what you say you are going to do. Others believing in me and giving me the opportunity to be successful has also been key.

Would you suggest construction as a career to others? Why?

Certainly, especially if you are good with your hands and like the outdoors. Shelter is one of the basic needs. It is also a very diverse industry. There are many different opportunities to pursue. It's an industry where good common sense, treating people fairly, and a strong work ethic can take you far.

What advice would you give to those new to the field?

Respect goes a long way in this business. It's a fairly close-knit industry. Develop a mindset of constant learning and be willing to work hard to learn the trade. To be successful at the top it is important to have worked your way up from the bottom.

Interesting career-related fact or accomplishment:

I was running my own crew during the summers at 15 years old. My dad started me young. Some of the poured "Bermuda" roofs my dad put on in the 1950s and 1960s are still in service today here in south Florida.

How do you define craftsmanship?

To use your mind and hands to deliver a vision—artistry.

Trade Terms Introduced in This Module

As-built drawings: Drawings of a completed installation that reflect the structure as it was constructed and show any unexpected changes during construction.

Building information modeling (BIM): An intelligent process that uses computer databases and computer-aided drafting (CAD) modeling to create a dynamic, interactive model of an entire structure and track its lifecycle.

Computer-aided drafting (CAD): The making of a set of construction drawings with the aid of a computer.

Copings: Coverings on top of a wall exposed to the weather, usually made of metal, brick, or stone. Copings are usually sloped to shed water.

Detail drawings: Drawings shown at a larger scale in order to show specific features or connections.

Drawing set: The set of detailed drawings or plans drawn to scale by an architect and/or engineer, showing all information and dimensions necessary to build or remodel a structure.

Flashings: Components used to seal the edges of a roof system at perimeters, penetrations, walls, expansion joints, valleys, drains, and other places where the roof covering or membrane is interrupted or terminated.

Legend: In maps, plans, and diagrams, an explanatory table defining all symbolic information contained in the document.

Nominal size: Approximate or rough size by which materials are known and sold. Nominal size is usually slightly larger than the actual size.

Project manual: Documents prepared by the architect/engineer for the owner in order to execute a project. Documents in a project manual include bidding documents, contracts, project details, and a list of construction drawings. Also called a *specification book* or *spec book*.

Purlins: Horizontal secondary structural members that transfer loads to the primary structural framing.

Record drawings: Drawings that show a record of any unexpected changes made during the construction of a structure.

Reroofing: The process of re-covering or tearing off and replacing an existing roof system.

Riser diagram: A schematic drawing that depicts the layout, components, and connections of a piping system.

Schedules: Tables that describe and specify the various types and sizes of construction materials used in a building. Door schedules, window schedules, and finish schedules are the most common types. Other types include equipment schedules, beam schedules, column schedules, and footing schedules.

Takeoff: The process of surveying, measuring, itemizing, and accounting for all materials and equipment needed for a construction project.

Additional Resources

This module presents thorough resources for task training. The following reference material is suggested for further study.

National Roofing Contractors Association (NRCA), **www.nrca.net**.
NCCER Module 00105, *Introduction to Construction Drawings*.

Figure Credits

iStock@yangwenshuang, Module Opener

Courtesy of the National Roofing Contractors Assoc., Figure 12

Patrick D Murphy Co., Inc., Architects, Figure 13

iStock@Cineberg, Figure 23

iStock@tawanlubfah, Figure 24

iStock@piyaphun, Figure 25

© 2018 The Construction Specifications Institute, Inc. (CSI). MasterFormat® excerpt and trademarks used under permission from CSI., Figure 26

Courtesy of Lake Mechanical Contractors Inc., Figure 27

Section Review Answer Key

Answer	Section Reference	Objective
1. c	1.0.0	1
2. d	1.2.1	1b
3. a	1.3.3	1c
4. b	1.4.1	1d

This page is intentionally left blank.

NCCER CURRICULA — USER UPDATE

NCCER makes every effort to keep its textbooks up-to-date and free of technical errors. We appreciate your help in this process. If you find an error, a typographical mistake, or an inaccuracy in NCCER's curricula, please fill out this form (or a photocopy), or complete the online form at **www.nccer.org/olf**. Be sure to include the exact module ID number, page number, a detailed description, and your recommended correction. Your input will be brought to the attention of the Authoring Team. Thank you for your assistance.

Instructors – If you have an idea for improving this textbook, or have found that additional materials were necessary to teach this module effectively, please let us know so that we may present your suggestions to the Authoring Team.

NCCER Product Development and Revision
13614 Progress Blvd., Alachua, FL 32615

Email: curriculum@nccer.org
Online: www.nccer.org/olf

❏ Trainee Guide   ❏ Lesson Plans   ❏ Exam   ❏ PowerPoints   Other _______________

Craft / Level: ____ Copyright Date: ___________

Module ID Number / Title: ____

Section Number(s): ____

Description: ____

____

____

____

____

Recommended Correction: ____

____

____

____

____

Your Name: ____

Address: ____

____

Email: ____ Phone: _________________________

This page is intentionally left blank.

Introduction to Steep-Slope Roofing

OVERVIEW

Steep-slope roofs can be seen on residential and industrial buildings across the world. They contain several basic components, including a deck, underlayment, water-shedding covering, and metal flashings. It is important for roofing professionals to be familiar with features of steep-slope roofs so that they are prepared to safely install steep-slope systems.

Module 16105

Trainees with successful module completions may be eligible for credentialing through the NCCER Registry. To learn more, go to **www.nccer.org** or contact us at 1.888.622.3720. Our website, **www.nccer.org**, has information on the latest product releases and training.

Your feedback is welcome. You may email your comments to **curriculum@nccer.org**, send general comments and inquiries to **info@nccer.org**, or fill in the User Update form at the back of this module.

This information is general in nature and intended for training purposes only. Actual performance of activities described in this manual requires compliance with all applicable operating, service, maintenance, and safety procedures under the direction of qualified personnel. References in this manual to patented or proprietary devices do not constitute a recommendation of their use.

Introduction to Steep-Slope Roofing

Objectives

Successful completion of this module prepares you to do the following:

1. Identify types of roof decks and underlayments used in steep-slope roof systems.
   a. Describe types of popular steep-slope roof decks.
   b. Examine common types of underlayments used in steep-slope applications.
2. Recognize the types of steep-slope roof systems.
   a. Categorize common steep-slope roof styles.
   b. Describe basics of steep-slope roof systems.
3. Explain the purpose and location of flashing in steep-slope systems.
   a. Understand the basics of steep-slope flashings.

Performance Tasks

This is a knowledge-based module. There are no Performance Tasks.

Trade Terms

Asphalt shingles
Battens
Clay tiles
Closed-cut valleys
Concrete tiles
Counter-battens
Counterflashing
Deck
Downspouts
Eaves
Elbows
Felt
Gable roof
Gambrel roof

Gutters
Hip roof
Hydrokinetic
Low-slope
Mansard roof
Open valleys
Shed roof
Slate
Steep-slope
Step flashing
Underlayment
Wood shakes
Wood shingles

Industry Recognized Credentials

If you are training through an NCCER-accredited sponsor, you may be eligible for credentials from NCCER's Registry. The ID number for this module is 16105. Note that this module may have been used in other NCCER curricula and may apply to other level completions. Contact NCCER's Registry at 1.888.622.3720 or go to **www.nccer.org** for more information.

Contents

This page is intentionally left blank.

1.0.0 STEEP-SLOPE ROOF DECKS AND UNDERLAYMENTS

Objective

Identify types of roof decks and underlayments used in steep-slope roof systems.

a. Describe types of popular steep-slope roof decks.
b. Examine common types of underlayments used in steep-slope applications.

Trade Terms

Battens: Horizontal strips of solid material, typically wood, that provide a fixing point for connecting a roof system.

Counter-battens: Grids of vertical wood or metal strips installed under horizontal battens that help with ventilation and system installation.

Deck: A structural component of the roof of a building, capable of safely supporting the weight of the roof or weatherproofing system, as well as the additional live loads required by the governing building codes. The deck provides the substrate to which the roof or weatherproofing system is applied.

Felt: A flexible sheet made by the interlocking of fibers with a binder or through a combination of mechanical work, moisture, and heat.

Hydrokinetic: Shedding liquid or snow off of a roof.

Low-slope: A category of roofs that generally includes weatherproof membrane types of roof systems installed on slopes of 3:12 or less.

Steep-slope: A category of roofing that generally includes water-shedding types of roof coverings installed on slopes greater than 3:12.

Underlayment: An asphalt-saturated felt or other composite or synthetic sheet material (sometimes self-adhering) installed between a roof deck and roof covering, usually used in a steep-slope roof construction. Underlayment is primarily used to separate the roof covering from the roof deck, shed water, and provide fire protection and secondary weather protection.

A roof contains many interacting components including the roof deck, air or vapor retarder (if needed), insulation, and primary covering designed to weatherproof the structure (such as a membrane or shingles). In general, a roof assembly consists of the structural deck and roof system.

Every component above the roof deck makes up the roof system. Roof systems are categorized as steep-slope or low-slope based on the roof's degree of incline, known as its *slope*. Steep-slope roofs are common on residential homes (*Figure 1*). Low-slope roofs (sometimes informally called *flat roofs*) are often seen on larger buildings (*Figure 2*).

The slope of a roof is one of the primary differences among roof assemblies. The steepness of a slope allows gravity to move water off a roof, promoting drainage. Without slope, a roof could hold water, causing significant water damage or even building collapse.

The most common method used to express the slope of a roof is a ratio between the vertical rise of the roof and the horizontal run, as illustrated in *Figure 3*. This is usually expressed in inches of rise per foot (12") of run, or in feet of rise per 12' of run. In this example, the rise is 8' and the run is 12', so the slope of the roof is 8:12.

Roof slope can be expressed in a variety of ways. Slope is often informally called *pitch*. For example, you may hear workers refer to a roof with a slope of 4:12 as a *4:12-pitch roof* or a *third-pitch roof*. (Note that the term *pitch* can also refer to a roofing material derived from coal tar.) A 4:12 slope can also be referred to as a *four-in-twelve slope*. Examples of ways a 4:12 slope may be written include 4-12, 4/12, and 4 in 12. It can also be expressed in degrees. *Table 1* shows roof slope expressed in both degrees and ratios.

The NCCER Roofing program defines low- and steep-slope roofs as follows:

- *Low-slope* — 3:12 slope or less
- *Steep-slope* — Above 3:12 slope

Table 1 Roof Slope Conversion Chart

Roof Slope	Angle
1:12	4.76°
2:12	9.46°
3:12	14.04°
4:12	18.43°
5:12	22.62°
6:12	26.57°
7:12	30.26°
8:12	33.69°
9:12	36.37°
10:12	39.81°
11:12	42.51°
12:12	45°

Figure 1 Steep-slope roof.

Steep-slope roof systems are designed to shed water. While a low-slope roof is hydrostatic, steep-slope roof is hydrokinetic. A hydrostatic roof, such as one with a structural metal-panel system, is watertight and typically installed on roofs with lower slopes. A hydrokinetic roof, such as one with an asphalt-shingle system, relies on gravity to shed rainwater and is only installed on steeply sloped roofs.

Steep-slope roofs typically include an underlayment, a covering, and metal flashings. Insulation is usually located beneath the roof deck (on the floor of the attic) in steep-slope roofs, so insulation in steep-slope systems is typically not installed by the roofing professional. Steep-slope and low-slope systems have different basic components, as illustrated in *Figure 4*.

1.1.0 Steep-Slope Decks

Although roofing professionals are not responsible for installing a roof deck, they must be able to identify damage to the deck and ensure that it provides a suitable substrate for attaching the roof system. If a defect is discovered on the roof deck, corrections must be made before system installation begins. Ensure that a deck is clean, dry, and smooth before installation begins.

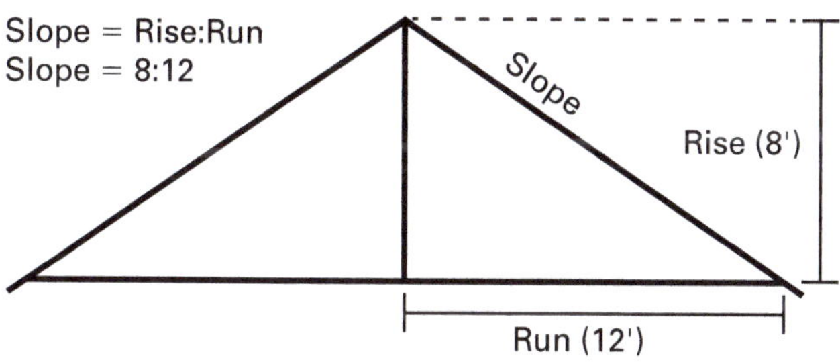

Figure 2 Low-slope roof.

Slope = Rise:Run
Slope = 8:12

Figure 3 Roof slope.

1.1.1 Wood Decks

Decks are referred to by their material type. Wood decks are the most common in steep-slope systems due to their affordability and availability. Wood decks also conveniently serve as a nailable substrate onto which the underlayment and roof system may be directly applied using roofing nails or screws. As is the case with all deck types, wood decks should be fastened in accordance with local building code requirements. Roofing professionals should also inspect the deck before applying any roof-system features.

Wood decks can be divided into two categories: wood panels, typically made from plywood or oriented strand board (OSB); and wood planks or boards.

Wood-panel decks are constructed from plywood or OSB. Both types of panels should be spaced with $\frac{1}{8}$" gaps at panel edges to account for panel expansion. Plywood panels (*Figure 5*) are made with veneers, or thin strips of wood peeled from logs and glued together. The National Roofing Contractors Association (NRCA) states a preference for plywood decks over OSB decks since OSB decks can lose structural stability when exposed to moisture.

OSB panels (*Figure 6*) are made from compressed wood strands and are more commonly used than plywood panels due to lower costs. As is the case with all decks, a roofing professional should feel confident that the deck onto which the roofing system is to be installed will be able to hold the weight of the system over a long period of time.

Both wood-plank and wood-board (*Figure 7*) roof decks are made from solid pieces of lumber. However, wood planks are longer and thicker (2" to 5" thick [5 cm to 12 cm]) than wood boards (less than 2" thick [less than 5 cm]).

1.1.2 Concrete Decks

When using concrete for a roofing deck, the foundation and framing of the building must be strong enough to support a heavier deck. There are various ways to use concrete as a decking material. Structural concrete can be cast on-site and has the capacity to handle very heavy loads. Decks may be made from lightweight structural concrete, which creates a lower-density deck. Decks can integrate concrete and steel by using steel as a base on top of which concrete is applied. A concrete deck may have batten or counter-battens installed to assist with system installation. Bat-

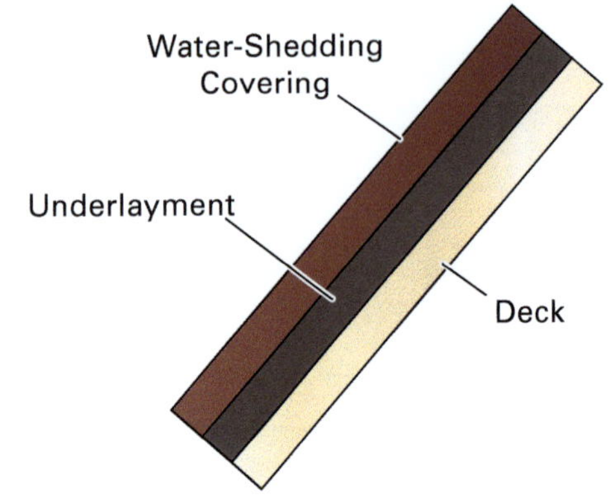

(A) Steep-Slope Roof System Components (Water-Shedding)

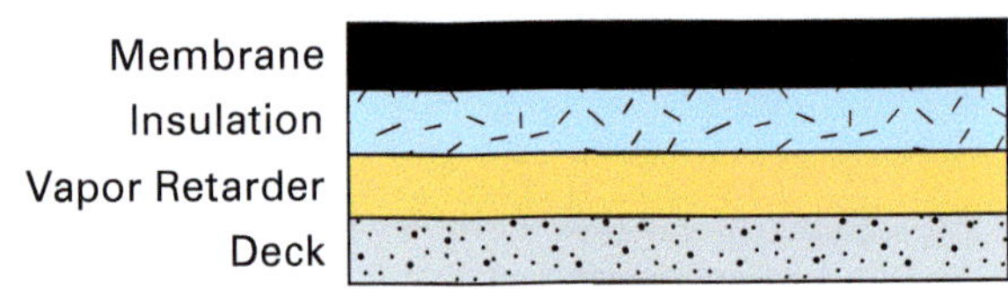

(B) Low-Slope Roof System Components (Weatherproof)

Figure 4 Basic components of roof systems.

Figure 5 Plywood roof deck.

NCCER – *Roofing*

Figure 6 Oriented strand board (OSB) roof deck.

tens are strips of solid material, typically wood, that provide a fixing point for connecting the roof system. Counter-battens are grids of vertical battens usually installed under horizontal battens that help with ventilation and system installation.

1.1.3 Steel Decks

Some steel decks are considered energy efficient, and they can be made sustainably from recycled materials. Steel decks offer a longer deck lifespan than alternate deck material options. Steel decks typically are not used with residential structures. Steep-slope steel decks are more often seen on commercial and institutional roofing.

Steel decks often have wood panels installed over the deck to provide a substrate that makes roof system attachment easier. Installing battens and counter-battens allows for better ventilation by raising and separating the nailable substrate from the deck surface.

1.2.0 Common Underlayments

Underlayment (*Figure 8*) is used to separate the roof covering and deck, providing an additional weatherproofing barrier for any moisture that seeps past the roof covering and flashings. Underlayments offer temporary protection for the roof deck before the rest of the roof system is installed.

Underlayments must comply with national and local building codes. They must also have a Class A, B, or C fire rating.

> **NOTE**
>
> Before installing underlayment, you should be confident that the building's attic space is well ventilated. Proper attic ventilation and condensation control are covered in more detail in NCCER Module 16104, *Substrates, Decks, and Roof Insulation*.

Figure 7 Wood-board roof deck.

There are common underlayment types with specific features with which a roofing professional should be familiar. Before underlayment installation, be sure the roof deck is clean, dry, and smooth. Roof decks must also be suitable for the roof covering that will be installed with the system; for example, a $\frac{1}{2}$" (12 mm) plywood deck should not have a heavy, concrete-tile covering installed over it. Always consult the roof system's manufacturer instructions to check for more specific underlayment material requirements and warnings before beginning installation. Roofing professionals should be familiar with the following common underlayment types:

- Asphalt felts
- Synthetic
- Polymer-modified bitumen

Figure 8 Installing an underlayment.

Load Types

A roof assembly must be capable of supporting both dead loads and live loads. Dead loads, or weight from objects on the roof that will remain permanent, include the weight of the roof system itself. Live loads, or temporary, variable weight added to the structure, account for any momentary added weight. Live loads can include anything from the weight of roofing professionals during installation to snow collecting after a storm before it melts. Your local building code should be reviewed and followed to ensure the roof deck and system can account for the combined weight of dead loads and live loads.

Live Load
Temporary weight that the roof must support but does not have to withstand permanently.

Example: snow collecting on roof after snowfall.

Dead Load
Permanent weight that the structure must be able to withstand at all times.

Example: the roof system itself.

OSB Installation

OSB panels have a grade stamp displayed on one side containing important information about the boards' features and installation conditions. Since successful OSB installation is dependent on the panels' direction, be sure to consult the grade stamp to make sure you are properly placing the panels.

More information on OSB can be found at the website for APA – The Engineered Wood Association, **www.apawood.org**.

Figure Credit: "osb texture" by Rick is licensed under CC BY 2.0

When applying base underlayment, do the following:

- Use the appropriate safety equipment (including fall protection and gloves).
- Unroll the base underlayment parallel to the roof's eaves before fastening it in place.
- Continue installing underlayment in shingle-style application, overlapping each sheet with the minimum overlapping requirements according to manufacturer guidelines.

1.2.1 Asphalt Felts

Asphalt felts have been commonly used in roofing for years. This underlayment is divided into two categories: organic and inorganic.

Felt underlayment was once typically fastened using staples. However, using roofing nails, nail caps, and plastic wind strips is recommended instead to prevent water from leaking through staple punctures in the underlayment material.

While organic felts may contain components such as cotton rag, wood fiber, and other cellulose fiber, inorganic felts are composed of glass fibers or polyester fibers.

1.2.2 Synthetic Underlayments

Since synthetic underlayment entered the market in the early 2000s, it has become a preferred product. Synthetic underlayment rolls are larger than roofing felt and weigh much less, making them easier and quicker to install. This underlayment also works better than felt in a variety of weather conditions because it more effectively fastens to the deck, captures less heat, and is more flexible in cold weather.

ASTM Approved

ASTM International, a world-wide organization that analyzes and develops technical standards for products, publishes Roofing Standards about the proper fabrication, installation, and potential hazards associated with Roofing products.

If a specific underlayment product receives ASTM approval, it is held in higher regard.

For more information, visit the ASTM website at **www.astm.org/Standards/roofing-standards**.

Most synthetic underlayments are made by spinning or weaving polyethylene or polypropylene. The manufacturing of synthetic underlayment allows for flexibility and variation—some rolls might be made more thinly than others, or they might be made with specific attention paid to UV protection. Be sure to check with the manufacturer to understand the specific features of the underlayment you are applying. It is important to confirm that the underlayment you use is compatible with the rest of the roof's system.

1.2.3 Polymer-Modified Bitumen

Polymer-modified bitumen sheets, typically reinforced with a fiberglass or polyester mat, may be used as a steep-slope underlayment. This underlayment should be installed using a method similar to felt-underlayment installation.

A subset of polymer-modified bitumen is self-adhering underlayment, which is usually made from polymer-based material. Self-adhering underlayment typically provides a water and ice-dam protection membrane, and it is sometimes installed in just a single layer. As its name suggests, this underlayment binds to a roof deck without requiring the use of an additional attachment method.

The following steps are frequently taken for self-adhering underlayment installation:

Step 1 Carefully unroll the underlayment across the surface of the roof deck parallel to the roof's eaves.

Step 2 Set the underlayment in place. Remove the protective film from beneath the material and begin applying pressure to attach the self-adhering side of the membrane directly to the deck.

Step 3 Use a weighted roller to bond the underlayment material firmly onto the deck.

Step 4 When overlapping sheets of underlayment to cover the roof deck, consult the manufacturer guidelines.

Self-adhering underlayment seals around fasteners, providing additional protection against wind-driven rain and ice dams. This underlayment is sometimes only applied to sections of a deck, as opposed to the entire deck, depending on a given region's weather patterns. Self-adhering underlayment may restrict air flow and lead to issues such as deck swelling or mold growth. It is important to make sure the roof system and attic are properly ventilated when installing underlayment.

1.0.0 Section Review

1. Which deck material is the *most* common in steep-slope roof systems?

    a. Steel
    b. Wood
    c. Concrete
    d. Battens

2. Which of the following separates the roof covering and deck, providing an additional weatherproofing barrier?

    a. FSI
    b. Flashings
    c. Underlayments
    d. Decks

2.0.0 STEEP-SLOPE ROOF TYPES

Objective

Recognize the types of steep-slope roof systems.
  a. Categorize common steep-slope roof styles.
  b. Describe basics of steep-slope roof systems.

Trade Terms

Asphalt shingles: Shingles manufactured by coating a reinforcing material (paper felt or fiberglass mat) with an asphalt-based coating and having mineral granules on the side exposed to the weather.

Clay tiles: A roof covering made by molding clay into a tile shape and baking it.

Concrete tiles: A roof covering made from shaping a mixture of cement, sand, and water into tiles.

Eaves: The horizontal edges at a roof's perimeter.

Gable roof: A roof style named after its prominent gables, or the triangular shapes formed between edges on the opposing sides of an intersecting slope.

Gambrel roof: Also called a *Dutch roof*; this roof style is usually symmetrical with two slopes on each side—its lower slope steeper than its upper slope.

Hip roof: A roof style formed by slopes on all four sides of a building coming together at the top of the roof to form a ridge.

Mansard roof: Also called a *French roof*; a decorative steep-slope roof on the perimeter of a building.

Shed roof: A roof style that slopes only once in one direction; frequently featured on small buildings or integrated with other roof styles.

Slate: A naturally occurring roofing material made from rocks.

Wood shakes: A roof covering split from lumber logs; they typically have rough and textured surfaces.

Wood shingles: A roof covering cut from wood; they typically have flat and smooth surfaces.

There are many different styles and roof coverings used in steep-slope systems. Each type of steep-slope roof system has some of the same basic features. *Figure 9* illustrates basic terminology that roofing professionals will use when installing steep-slope systems.

All roofs have different areas with which you should be familiar:

- *Attic* — The area between the roof and the living area of a home.
- *Dormer* — Projection through the sloping plane of a steep-slope roof, usually housing a window.
- *Eaves* — The horizontal edges at the roof's perimeter.
- *Fascia* — A vertical-faced trim board installed to the joists, rafters, or trusses at the perimeter edge of the roof.
- *Field* — The main, open part of the roof over which a roof covering is installed.
- *Flashing* (noun) — Components used to fortify, weatherproof, and/or seal roofing systems at edges, penetrations, and transitions.
- *Flashing* (verb) — Fortifying roof edges, penetrations, and transitions to achieve a weatherproof system.
- *Hips* — The meeting of two roof slopes that form an external angle. Water runs away from hips.
- *Rake* — The angled edge at a roof's perimeter.
- *Ridge* — The meeting of two roof slopes that form an external angle horizontal to both slopes. Water runs away from ridges.
- *Soffit* — Underside of an overhanging eave, often enclosed, that is frequently used in the ventilation of attic spaces.
- *Valley* — The meeting of two roof slopes that form an internal angle between both slopes. Water runs into valleys.
- *Vent* — Added to a roof to allow for the movement of air. Different types of vents include attic, gable, plumbing, ridge, and soffit vents.

> **NOTE**
>
> This section covers the most common steep-slope roof styles and coverings. You may encounter additional styles not addressed in this section. If you work on a roof style or with a roof covering that you have questions about, be sure to talk with your supervisor.

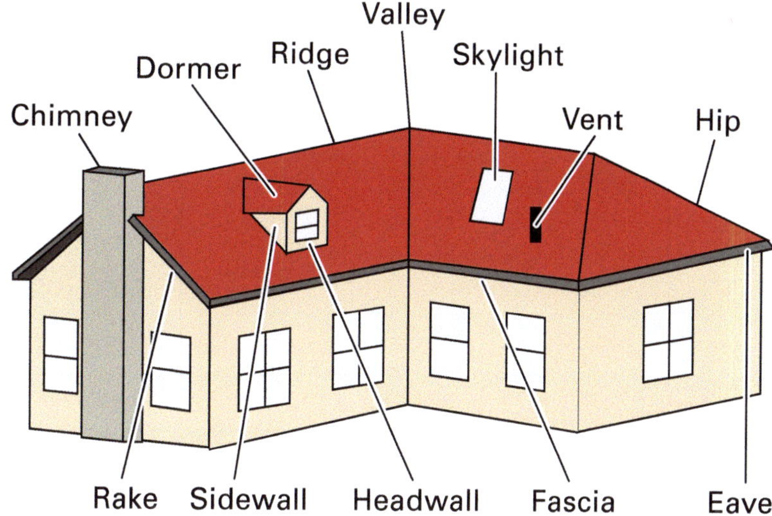

Figure 9 Anatomy of a steep-slope roof.

2.1.0 Steep-Slope Styles

While on the job, roofing professionals will encounter variations of steep-slope roof styles, some of which are overviewed in *Figure 10*. It is important to familiarize yourself with these common styles.

> **NOTE**
>
> Roofs often combine the various styles discussed in this section. For example, a mansard roof might feature a shed roof style on a lower plane. Roofs may also combine low- and steep-slope styles. It is possible for a jobsite to have different contractors working on the low- and steep-slope sections of the roof.

2.1.1 Shed Roof

A shed roof (*Figure 11*) slopes in only one direction. Although this roof style contains just one plane, it qualifies as a steep-slope rather than low-slope style because it is hydrokinetic, or water-shedding.

The shed roof gets its name from the single-slope water "shedding" roof. It is a simple roof style that is well-suited for small structures, but it can also be found on larger roofs. It is also commonly seen integrated with other styles of roofs, such as the mansard- or hip-roof styles that will be discussed later in this section.

Shed roofs have a simple design that makes them easier to construct than other popular roof styles.

2.1.2 Gable Roof

A gable roof, also called a *gable-end roof*, is named after its prominent gables, or the triangular shapes formed between edges on the opposing sides of an intersecting slope (*Figure 12*).

The gable design can vary greatly in terms of slope and height of the roof. Gable roofs are most popular in regions with colder climates. Areas with heavy snow may require gables with a shallow slope that is better able to take on the extra weight of snowfall. By contrast, regions with heavy rainfall benefit more from a steeper pitch to increase water-shedding.

Gable roofs (*Figure 13*) may be more at risk of wind damage than some other roof styles. However, this style remains widely popular due to its simplicity, affordability, and flexibility of design.

There are several popular variations of the gable roof style:

- *Side gable* — Two sides pitched at an angle meet at the ridge in the center of the building.
- *Cross gable* — Two or more gables intersect at an angle to create a dynamic roof shape.

Did You Know?

Regional Differences

A shed roof may be referred to by different names depending on the roof's region. It is sometimes called a *mono-pitched roof* or a *pent roof*. In Australia and New Zealand, shed roofs are called *skillion roofs*.

Commonly Found Steep-Slope Roof Styles

Roof Style	Description	Example
Shed Roof	Roof contains only one sloping plane.	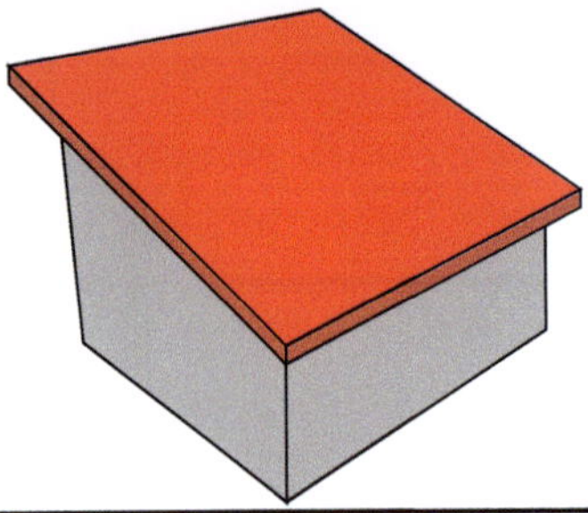
Gable Roof	Roof is single-ridge and terminates at gable ends.	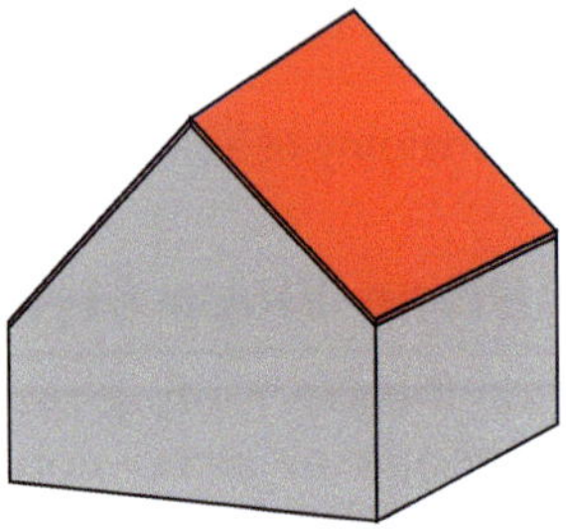
Hip Roof	Two intersecting roof planes form an inclined external angle.	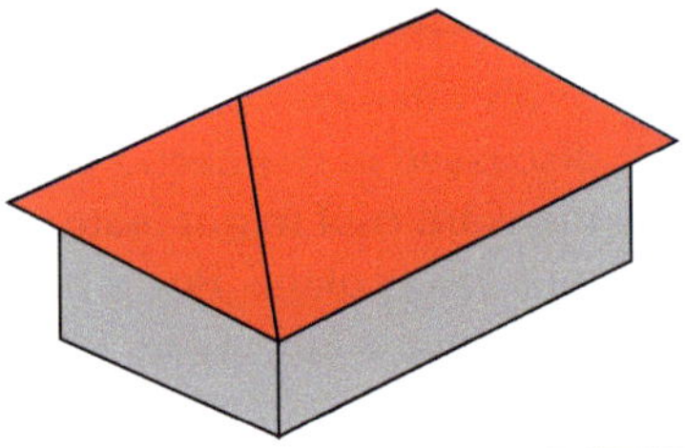
Gambrel Roof	Roof has two slopes on each side; the top roof area has less slope than the low roof area.	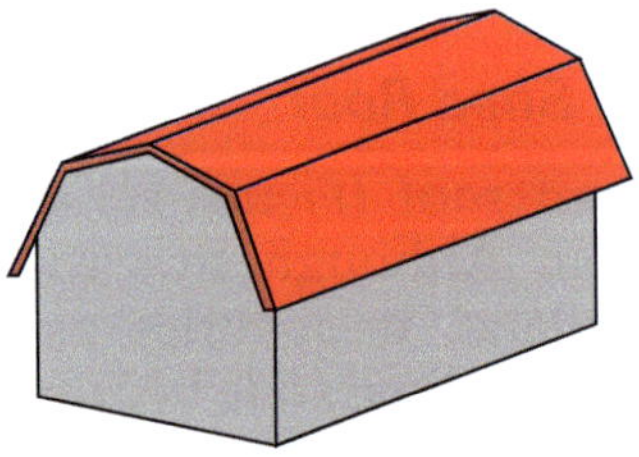
Mansard Roof	Roof has two slopes on each side; the top roof area has less slope than the low roof area.	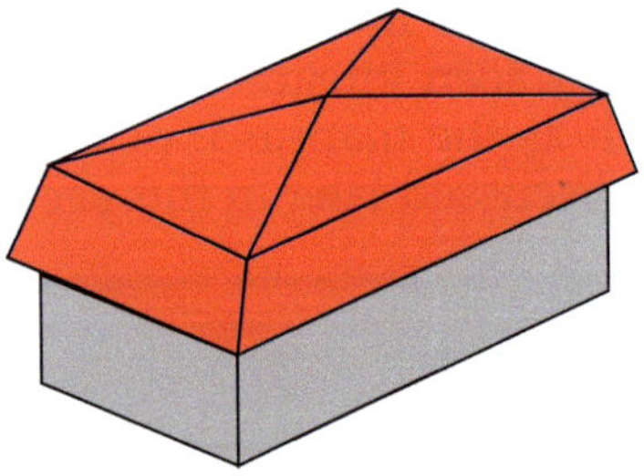

Figure 10 Common steep-slope styles.

NCCER – *Roofing*

Figure 11 Small building with a shed roof.

Figure 12 Gable shape outlined by a red triangle.

Figure 13 Gable roof with solar energy panels.

Figure Credit: iStock@Florelena

- *Front gable* — Also known as a *gablefront house*, has the gable facing the building's entrance.
- *Dutch gable* — A hybrid of the gable- and hip-roof styles.

2.1.3 Hip Roof

A hip roof is formed by slopes on all four sides of a building coming together at the top of the roof to form a ridge (*Figure 14*).

Figure 14 Hip roof.

Hip roofs offer more stability than shed- or gable-style roofs. This added stability makes hip roofs a popular choice for windy and snowy regions. The slanted sides of the roof promote water-shedding and the solid framework of this style offers support for live loads such as snowfall. Due to a more complex design that requires more time and materials, hip roofs are typically more expensive than gable roofs.

Like gable roofs, hip roofs take a variety of forms:

- *Simple Hip* — Gently slopes down on all four sides.
- *Cross Hipped* — Consists of two "L-shaped" hip roofs installed perpendicular to one another.
- *Half Hipped* — Integrates the gable- and hip-styles, replacing the upper points of a gable roof with a small hip.

2.1.4 Gambrel Roof

A gambrel roof (*Figure 15*), also called a *Dutch roof*, usually has a symmetrical style with two slopes on each side—its lower slope steeper than its upper slope. This roof style is featured on barns as well as Dutch- and Colonial-style buildings. The gambrel roof's shape allows for

Figure 15 Barn with a gambrel roof.

more space inside the building than taller roof styles like the gable roof.

2.1.5 Mansard Roof

A **mansard roof**, or *French roof*, shares some similarities with a gambrel roof. With mansard roofs, a double slope on each side meets to form a low slope on the top layer and a steeper slope on the bottom layer. *Figure 16* shows a mansard roof; the low-slope area of this roof, installed above the dormers, is not visible from ground level.

2.2.0 Steep-Slope Systems

Steep-slope systems you will encounter on the job include asphalt shingles, wood shakes and shingles, metal roof coverings, clay and concrete tiles, slate, and synthetic coverings. This section introduces important features of these systems.

Each roof system calls for a specific underlayment that is designed to meet the installation and performance demands of that particular system. For example, a metal-roof system needs a more heat-resistant underlayment, whereas a tile-roof system needs a longer-lasting underlayment. Be sure to consult manufacturer instruction and building codes so that the underlayment installed pairs well with the system installed.

2.2.1 Asphalt Shingles

Asphalt shingles (*Figure 17*) are the most popular steep-slope roof covering. Popular types of asphalt shingles include 3-tab, dimensional, and architectural shingles. Asphalt shingles are hydrokinetic, installed to successively lap upper courses over lower courses to help shed the water to the eave. This roof covering is made of a base material (such as felt or fiberglass), asphalt, and surfacing material (granules). Asphalt shingles are usually sold in strips that prevent wind uplift and make installation easier and quicker. The granules on the material's surface provide extra fire resistance.

Asphalt shingles should be installed in overlapping patterns to help the roof shed water. Start installing the strips from the eaves of the roof so that you can properly overlap the shingles. Add flashing to vulnerable areas that require extra

Figure 16 Mansard roof from a ground-level view.

Figure 17 Asphalt-shingle installation.

protection from moisture. To attach this system, use roofing nails specified by the manufacturer instructions.

2.2.2 Wood Shakes and Shingles

Wood shakes (*Figure 18*) and wood shingles are split from logs and reshaped for commercial use. They vary in thickness and are usually thicker than asphalt shingles near their bottom ends. Wood shakes and shingles are sold with fire retardants and preservatives to prevent fires and help the covering last longer. Natural preservatives make wood shakes and shingles resistant to moisture.

At least two roofing nails should be used to fasten each wood shake or shingle to the deck. As with asphalt shingles, wood shakes and shingles should be installed starting from the eaves to aid with water-shedding.

Wood shakes and shingles are typically made from a variety of trees, such as red cedar, pine, cypress, and redwood. The manufacturer treats the wood with paint, stain, or varnish to guard against termites and other wood-degrading insects.

2.2.3 Metal Roof Coverings

Steep-slope metal roof coverings (*Figure 19*) come in the following categories:

- Architectural metal panels
- Structural metal panels
- Metal shingles

Architectural metal panels and metal shingles are intended for steep-slope roofs. Structural metal panels can be used in both steep- and low-slope applications. Metal systems are made by press-forming a metal covering to make it into a specified shape. Metal panels are usually attached onto the deck with fasteners, ridge caps, and closure strips. Metal shingles are installed in overlapping courses similar to asphalt-shingle installation.

Steep-slope metal roof coverings are more durable and require less maintenance than coverings made from wood or asphalt. Steel, a popular metal-covering material, withstands heavy

Figure 18 Wood-shake roof.

winds and provides the roof with fire resistance. Other metals used for these systems include aluminum, copper, and zinc. Since metal roof coverings are prone to collecting condensation on the undersides of the system, this system's vapor retarder, insulation, and ventilation need to be carefully considered.

Metal-roof systems are very heavy and must be installed on a building and deck that are capable of supporting the weight of a metal system.

Metal roofs are generally more slippery than other roof systems, so it is important to take extra safety precautions that account for this difference.

2.2.4 Clay and Concrete Tiles

Clay tiles are made by molding clay into a tile shape and firing the molds in a kiln at 2,000°F (1,093°C). As a result, these tiles are available in many different shapes, sizes, and colors. Clay tile systems may come with accessory tiles to be used on ridges, hips, and gable ends. Manufacturers sell clay tiles in a variety of styles, including one-piece, two-piece, interlocking, and flat tiles.

Concrete tiles (*Figure 20*) are made from a mixture of cement, sand, and water. These tiles sometimes have lugs on their bottom surface so that they can be hung onto a batten strip during installation. Concrete tile systems may also include accessory pieces to help seal ridges, hips, and gable ends.

Figure 19 Metal roof.

Clay and concrete tiles may be secured in a variety of methods. Roofing nails can attach the tiles to a nailable substrate. Tiles may also be secured by wiring or hanging their lugs over a batten system. Clay and concrete tile systems are durable and provide reliable weather resistance.

Tile systems are heavy. A building and roof deck must be able to support the heavy weight of tile-system installation.

2.2.5 Slate

Slate (*Figure 21*) is a naturally occurring roofing material made from rocks. Slate systems can come in a variety of textures and colors depending on their source materials. Roofs may be constructed with "rough" or "smooth" slate, designed with varying lengths, thicknesses, and colors.

Copper slate nails are compatible with this roof system. All roofing slate must have a minimum of two nails that both penetrate all layers of the roofing assembly to securely fasten the system to the deck.

Slate is a non-combustible material, and its high density makes it weatherproof. However, due to the heaviness of slate systems, a building should be evaluated for structural integrity to make sure that its deck can support the dead weight of a slate covering.

CAUTION

Cutting tiles can expose a roofing professional to safety hazards such as silica dust. When cutting tiles, be sure to follow OSHA's requirements for respiratory protection as well as any additional safety guidelines.

Figure 20 Concrete-tile roof.

2.2.6 Synthetic Covering

Synthetic coverings are manufactured products designed to look like shingles or tiles but made with plastic or rubber and other materials. They are much lighter than most of the other systems, and they are naturally fire retardant and algae resistant.

Since this system did not enter the market until the early 1990s, it is important to check local building codes and exercise caution when installing a synthetic covering.

Figure 21 Slate roof.

2.0.0 Section Review

1. Vertical-facing trim boards installed to block moisture are called _____.

    a. dormers
    b. fascia
    c. sidewalls
    d. hips

2. Which type of roof material is made from a mixture of cement, sand, and water?

    a. Concrete tiles
    b. Metal tiles
    c. Clay tiles
    d. Asphalt shingles

3.0.0 STEEP-SLOPE FLASHING

Objective

Explain the purpose and location of flashing in steep-slope systems.

a. Understand the basics of steep-slope flashings.

Trade Terms

Closed-cut valleys: Valleys with shingles that extend across the valley from one side while the shingles on the other side are trimmed back from the centerline.

Counterflashing: Secondary flashing installed along the top edge of primary flashing to protect the surface and its fasteners.

Downspouts: Pipes that carry water from the gutter to the ground or a drain.

Elbows: Help direct the flow of a downspout.

Gutters: Help drain water away from a building and are usually attached with straps and support brackets.

Open valleys: Valleys that trim the roof-covering material on both sides to expose metal valley flashing that helps shed the water down a roof.

Step flashing: Weatherproofing installed from under roof shingles or tiles and up to sidewall.

Every roof has two functional areas: field and flashing. The field of a roof is its main, open section. Flashing occurs at any area of a roof system where the field is interrupted or terminated. An interruption occurs where an object, such as a vent pipe or dormer, penetrates the roof. A roof terminates at a wall or edge. Since steep-slope roofs are hydrokinetic, or water-shedding (*Figure 22*), make sure that all vulnerable areas of the roof are flashed to help water shed off a roof.

3.1.0 Steep-Slope Flashing and Weatherproofing

Steep-slope roofs require flashing and weatherproofing to make sure they correctly shed water. Flashing is thin material crafted from tin, aluminum, copper, or other flexible membrane materials. This material is used to seal areas that may be vulnerable to leakage. Flashing can be installed using adhesives, roofing nails, or roofing screws. Flashing must be installed whenever the roof plane is interrupted, including on valleys, hips, pipes, headwalls, and sidewalls.

3.1.1 Eave and Rake Flashing

Eaves are the horizontal edges at the roof's perimeter. They require flashing where the roof terminates. Roof rakes, the slanting edges of a gable roof where the roof terminates, also require flashing.

Drip edge flashings help water successfully shed off a roof edge or rake by directing rainfall off the roof. Drip edge metals commonly come in T- and L-type shapes (*Figure 23*) and should be installed flush with the roof's decking and fascia (the vertical-facing trim board that runs below the roof system). Drip metal is commonly fastened to the deck in overlapping succession with roofing nails. This flashing must cover the perimeter of a roof (*Figure 24*).

3.1.2 Vertical Surface Flashings

Vertical interruptions of the roof's field require flashing between the roof system and protruding surface to prevent water leakage. Types of steep-slope penetrations include chimneys, skylights, and dormers.

Apron flashing weatherproofs the part of a roof that intersects a headwall, or the space where the roof terminates at a wall parallel to the roof's eave. Headwalls are found on dormers, chimneys, skylights, and other vertical roof interruptions.

Sidewall flashings should be installed at the junction between a sidewall and the sloped portion of a roof. Step flashing is typically used to weatherproof sidewalls for systems that have roof coverings with more than one layer (such as overlapping shingles); channel flashings, or *pan flashings*, are used on sidewalls with single-layer roof systems (such as metal panels).

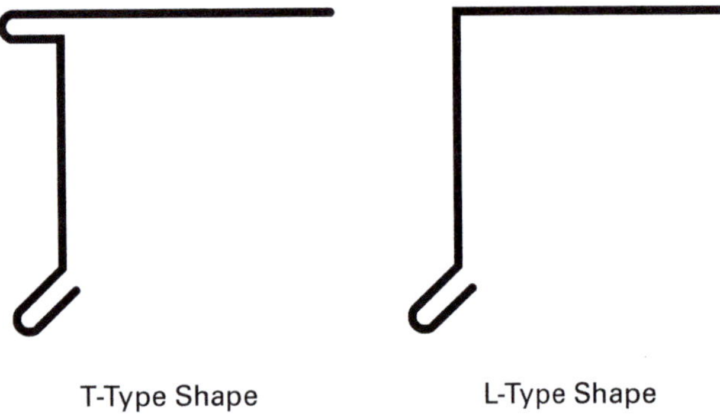

Figure 22 Water shedding off a steep-slope roof.

T-Type Shape   L-Type Shape

Figure 23 Common types of drip-edge flashings.

The backside or upside slope of a chimney or curbed roof penetration requires a backer flashing or cricket flashing. Backer flashings are used to flash penetrations less than 24" (61 cm) wide, and they are often used with skylights. Cricket flashings are used at the upslope side of a chimney or with penetrations 24" (61 cm) or greater.

Kickout flashings are used to direct water away from where an eave intersects a continuous vertical surface. This type of flashing is most often used on a lower-level roof edge that intersects one of building's walls.

Apron, step, cricket, and backer flashings also require counterflashing to prevent water from seeping through their top edges. Counterflashing is installed along the top edge of the base flashing to protect that base flashing and its fasteners. *Figure 25* shows an example of apron flashing, step flashing, and counterflashing.

NOTE

Skylights may not require counterflashing because skylights often come with flashing and counterflashing systems built into their frames.

Figure 24 Drip-edge installation.

3.1.3 Penetration Flashings

Penetrations, such as pipes, fans, and furnaces, interrupt the roof field and require flashing. Flat flanges are sometimes used around a penetration to weatherproof the area. Flanges are installed under the roof covering on the upslope section of the penetration and over the roof covering on the downslope section. A cylinder or box is attached to the flange and seals around the penetration to provide more weatherproofing. Plumbing boots and vent boots (*Figure 26*) are usually pre-formed flanges used for stack pipes or vents that penetrate the roof's covering. Boots are sometimes called *collars*.

3.1.4 Ridge and Hip Accessories

Ridge and hip accessories help weatherproof the external angle of a roof where two adjacent slopes intersect. A ridge cap or hip cap should extend over both sides of the intersection and the roof cover to help shed water down and off the roof (*Figure 27*). Ridge and hip caps should be installed in shingle fashion with each new cap lapping over the previous cap. Begin installation at the lower end of a hip or ridge, and add additional weatherproofing whenever required by building codes and manufacturer instructions.

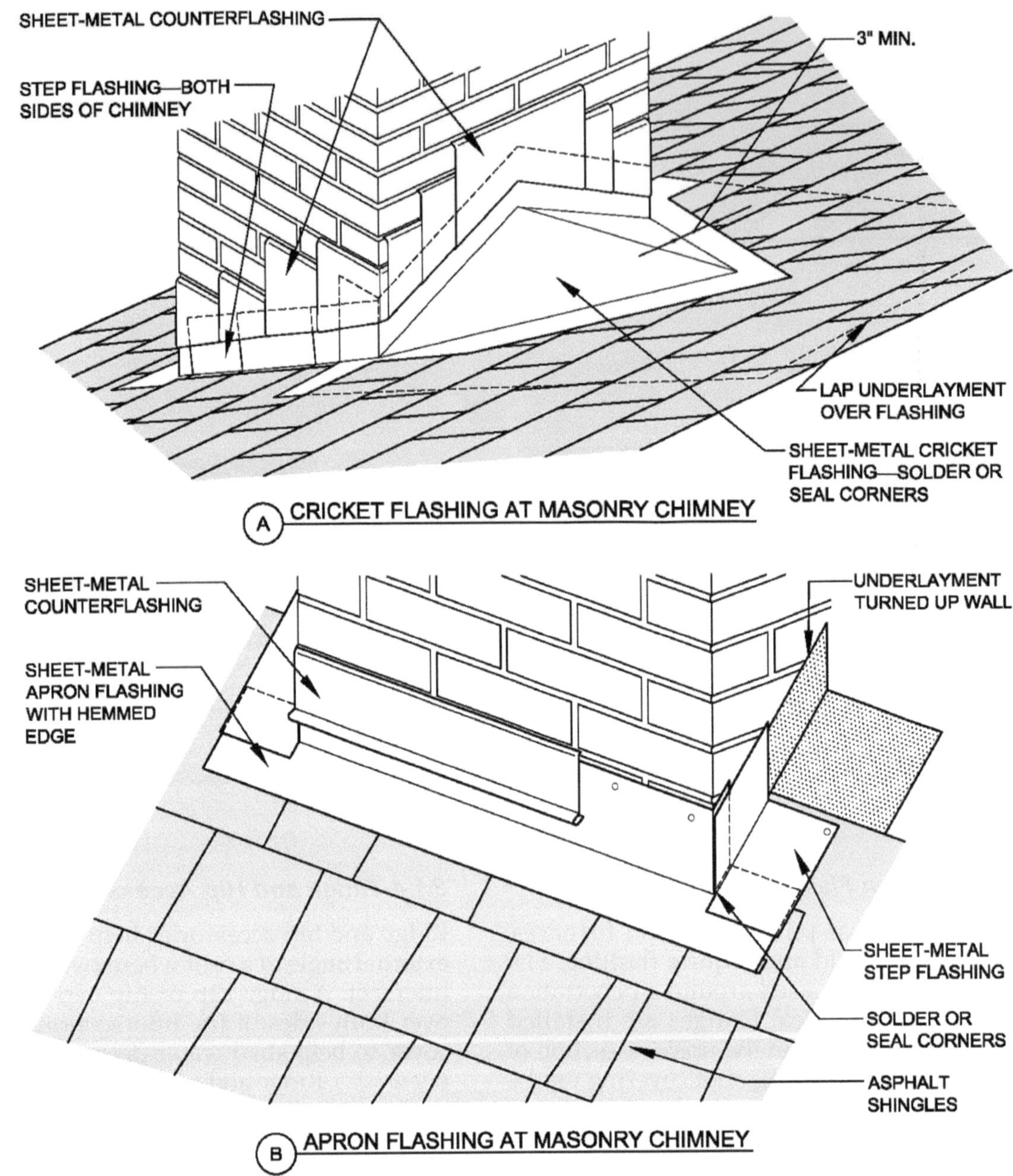

Figure 25 Flashing at a vertical-surface penetration.

Figure 26 Boot applied to a penetration.

Figure 27 Ridge-cap installation.

3.1.5 Valley Flashings

Valleys are the internal angles formed by the intersection of two sloping planes. Valleys require special attention during system installation because they are vulnerable to leakage due to the greater volume of water collected in these areas.

Steep-slope valleys come in two categories: open and closed. Closed-cut valleys (*Figure 28*) have shingles or tiles that extend across the valley from one side while the shingles on the other side are trimmed back from the centerline. Open valleys (*Figure 29*) trim the roof-covering material on both sides to expose metal valley flashing that helps shed water down and off a roof.

Since some shingle and tile systems only allow open or closed valleys in their installations, be sure to consult manufacturer instructions before valley installation.

<table>
<tr><td>CAUTION</td><td>Penetrations should not be installed within a valley because they may interrupt the downflow of water.</td></tr>
</table>

3.1.6 Gutters and Downspouts

Gutters (*Figure 30*) drain water safely away from a building and come in a variety of shapes and materials. Gutters are usually attached with straps and support brackets. Straps may be attached to the front and back of a gutter to keep the front rim a measured distance away from the back of the gutter. Support brackets hold up the weight of the gutter and must be able to withstand harsh weather conditions. Gutter sections can be joined through soldering or overlapped sealing. The front face of the gutter should be installed 1" (25 mm) lower than the back face to help the gutter drain water.

Downspouts are pipes that carry water from the gutter to the ground or a drain. Downspouts are secured to walls with hangers and straps that attach to the building with noncorrosive fasteners. Downspouts are connected to gutters with a downspout adapter. The flange of the outlet tube is fastened and then sealed or soldered to the gutter to ensure that any collected water streams away from the building. Elbows help direct the flow of a downspout. Style A elbows angle the downspout upwards or downwards while Style B elbows angle the downspout to the left or right.

Figure 28 Closed-cut valley.

Figure 29 Open valley.

Figure 30 Gutter installation.

3.0.0 Section Review

1. Which of the following is used to divert water away from areas that may be vulnerable to leakage?
   a. Valleys
   b. Rakes
   c. Penetrations
   d. Flashings

2. Which of the following is a pipe that carries water from the gutter to the ground?
   a. Parapet
   b. Eave
   c. Downspout
   d. Valley

1. Which of the following components is *not* part of a roof system?

    a. Deck
    b. Shingles
    c. Underlayment
    d. Flashings

2. What aspect of a steep-slope roof allows gravity to move water off a roof, promoting drainage?

    a. The deck material
    b. Insulation
    c. Slope
    d. Penetrations

3. Which of the following types of steep-slope decks is considered the *most* energy efficient?

    a. OSB
    b. Plywood
    c. Batten and counter-batten
    d. Steel

4. Before installing an underlayment, you *must* _____.

    a. install gutters
    b. ensure insulation is present
    c. install flashings
    d. ensure the deck is clean, dry, and smooth

5. The first row of asphalt shingles should be installed _____.

    a. starting at the eave
    b. starting at the rake
    c. starting at a sidewall
    d. before the underlayment

6. Which type of steep-slope roof-covering material has *both* panel systems and shingle systems popularly available?

    a. Asphalt
    b. Metal
    c. Wood
    d. Clay

7. Synthetic roof coverings are made of _____.

    a. asphalt
    b. slate
    c. plastic or rubber
    d. copper

8. Drip edge flashings should be installed _____.

    a. on dormers
    b. at the roof's perimeter decking and fascia
    c. at the junction between a sidewall and the sloped portion of a roof
    d. where the roof intersects a wall

9. Which of the following is an example of a penetration?

    a. A hip
    b. A ridge
    c. A drip edge
    d. A dormer

10. Which type of flashing helps weatherproof the area where two adjacent slopes intersect?

    a. Ridge flashing
    b. Flat flange
    c. Kickout flashing
    d. Step flashing

Fill in the blank with the correct term learned from your study of this module.

1. ________________ are strips of solid material that provide a fixing point for connecting a roof system.

2. ________________ help drain water away from a building and are usually attached with straps and support brackets.

3. ________________ help direct the flow of a downspout.

4. ________________ is secondary flashing installed at the top edges of primary flashings.

5. A roof with four roof planes coming together at a peak with four separate roof legs is a(n) ________________.

6. ________________ are valleys that trim the roof-covering material on both sides to expose metal valley flashing.

7. A(n) ________________ is a roof style that slopes only once in one direction.

8. ________________ is a category of roofing that generally includes water-shedding types of roof coverings installed on slopes exceeding 3:12.

9. ________________ is weatherproofing installed from under roof shingles or tiles and up to sidewall.

10. A roof covering called ________________ is made by molding clay into a tile shape and baking it.

11. ________________ are manufactured by coating a reinforcing material with an asphalt-based coating and having mineral granules on the side exposed to the weather.

12. A(n) roof covering called ________________ is made from shaping a mixture of cement, sand, and water into tiles.

13. ________________ means water-shedding.

14. ________________ are a roof covering split from lumber logs and typically have a rough and textured surface.

15. ________________ are pipes that carry water from the gutter to the ground or a drain.

16. ________________ is an asphalt-saturated felt or other composite or synthetic sheet material installed between a roof deck and roof covering, usually used in a steep-slope roof construction.

17. A category of roofs that generally includes weatherproof membrane types of roof systems installed on slopes at or less than 3:12 is ________________.

18. A(n) ________________ is also called a French roof and is a decorative steep-slope roof on the perimeter of a building.

19. ________________ are the horizontal edges at a roof's perimeter.

20. A(n)_______________ is a roof style named after its prominent gables, or the triangular shapes formed between edges on the opposing sides of an intersecting slope.

21. _______________ are a roof covering cut from wood and typically have a flat and smooth surface.

22. A naturally occurring roofing material made from rocks is _______________.

23. A(n) _______________ is a roof style that is usually symmetrical with two slopes on each side—its lower slope steeper than its upper slope.

24. A structural component of the roof of a building, capable of safely supporting the weight of the roof or weatherproofing system, as well as the additional live loads required by the governing building codes is a(n) _______________.

25. _______________ are grids of horizontal and vertical battens that help with ventilation and system installation.

26. _______________ is a flexible sheet made by the interlocking of fibers with a binder or through a combination of mechanical work, moisture, and heat.

27. Valleys with shingles that extend across the valley from one side while the shingles on the other side are trimmed back from the centerline are _______________.

Trade Terms

Asphalt shingles	Deck	Gutters	Slate
Battens	Downspouts	Hip roof	Steep-slope
Clay tiles	Eaves	Hydrokinetic	Step flashing
Closed-cut valleys	Elbows	Low-slope	Underlayment
Concrete tiles	Felt	Mansard roof	Wood shakes
Counter-battens	Gable roof	Open valleys	Wood shingles
Counterflashing	Gambrel roof	Shed roof	

Cornerstone of Craftsmanship

Amy Staska
Subject Matter Expert
NRCA

How did you choose a career in the industry?

The way many do, at first—I needed a job and knew someone who knew someone. I did leave NRCA for a time to train in another industry, but then I chose roofing when I returned to the association in 2011.

How important is education and training in construction?

How much space can I take up here? I think training is not just about changing behavior—though that is the top-line reason and super important—but it is also a recruitment and retention tool. It creates loyalty and helps make people feel good about their developing skills and contributions to their companies. It leads to camaraderie, belonging, and esteem.

What kinds of work have you done in your career?

I have delivered and developed adult training in various settings; ironically, all non-profit: Universities, ministries, and the National Roofing Contractors Association (NRCA).

Tell us about your present job.

I have the pleasure of heading the team that develops and delivers training for the roofing industry, on a national level. From webinars to in-person classes to online training, we love to create opportunities for contractors and other roofing professionals to pass on training opportunities to their employees—to help them be better at their work, but also to understand their value within an amazing craft.

What do you enjoy most about your job?

On a macro level, knowing we (NRCA) provide resources most small companies cannot produce on their own. We empower small businesses to empower their people.

Day to day, I love working with my team to create new and creative resources to expand peoples' understanding of themselves and their careers.

Would you suggest construction as a career to others? Why?

Absolutely. I am not a college naysayer, like many trying to recruit into the industry today; however, it's not everyone's jam to continue in their schooling and end up in a job where they work with words, mathematical formulas, or diplomatic negotiations among countries. Some people are wired to work with their hands and thrill at seeing skylines full of homes, businesses, and other spaces they helped create with their own hands. It is a visceral sense of creativity, accomplishment, and pride.

What advice would you give to those new to the field?

Work hard; show up on time; always have an attitude to learn; bring respect to the work and others within it.

Interesting career-related fact or accomplishment:

In my 20s, I made major changes in my life to pursue medical school, including becoming an EMT so I could get over my paralyzing squeamishness … only to realize it was the educational aspect of being a physician I really wanted. Thank goodness I avoided the student loan debt of medical school before realizing I actually wanted to be in the training/education field.

Trade Terms Introduced in This Module

Asphalt shingles: Shingles manufactured by coating a reinforcing material (paper felt or fiberglass mat) with an asphalt-based coating and having mineral granules on the side exposed to the weather.

Battens: Horizontal strips of solid material, typically wood, that provide a fixing point for connecting a roof system.

Clay tiles: A roof covering made by molding clay into a tile shape and baking it.

Closed-cut valleys: Valleys with shingles that extend across the valley from one side while the shingles on the other side are trimmed back from the centerline.

Concrete tiles: A roof covering made from shaping a mixture of cement, sand, and water into tiles.

Counter-battens: Grids of vertical wood or metal strips installed under horizontal battens that help with ventilation and system installation.

Counterflashing: Secondary flashing installed along the top edge of primary flashing to protect the surface and its fasteners.

Deck: A structural component of the roof of a building, capable of safely supporting the weight of the roof or weatherproofing system, as well as the additional live loads required by the governing building codes. The deck provides the substrate to which the roof or weatherproofing system is applied.

Downspouts: Pipes that carry water from the gutter to the ground or a drain.

Eaves: The horizontal edges at a roof's perimeter.

Elbows: Help direct the flow of a downspout.

Felt: A flexible sheet made by the interlocking of fibers with a binder or through a combination of mechanical work, moisture, and heat.

Gable roof: A roof style named after its prominent gables, or the triangular shapes formed between edges on the opposing sides of an intersecting slope.

Gambrel roof: Also called a *Dutch roof*; this roof style is usually symmetrical with two slopes on each side—its lower slope steeper than its upper slope.

Gutters: Help drain water away from a building and are usually attached with straps and support brackets.

Hip roof: A roof style formed by slopes on all four sides of a building coming together at the top of the roof to form a ridge.

Hydrokinetic: Shedding liquid or snow off of a roof.

Low-slope: A category of roofs that generally includes weatherproof membrane types of roof systems installed on slopes of 3:12 or less.

Mansard roof: Also called a *French roof*; a decorative steep-slope roof on the perimeter of a building.

Open valleys: Valleys that trim the roof-covering material on both sides to expose metal valley flashing that helps shed the water down a roof.

Shed roof: A roof style that slopes only once in one direction; frequently featured on small buildings or integrated with other roof styles.

Slate: A naturally occurring roofing material made from rocks.

Steep-slope: A category of roofing that generally includes water-shedding types of roof coverings installed on slopes greater than 3:12.

Step flashing: Weatherproofing installed from under roof shingles or tiles and up to sidewall.

Underlayment: An asphalt-saturated felt or other composite or synthetic sheet material (sometimes self-adhering) installed between a roof deck and roof covering, usually used in a steep-slope roof construction. Underlayment is primarily used to separate the roof covering from the roof deck, shed water, and provide fire protection and secondary weather protection.

Wood shakes: A roof covering split from lumber logs; they typically have rough and textured surfaces.

Wood shingles: A roof covering cut from wood; they typically have flat and smooth surfaces.

Additional Resources

This module presents thorough resources for task training. The following reference material is suggested for further study.

APA – The Engineered Wood Association, **www.apawood.org**.

National Roofing Contractors Association (NRCA), **www.nrca.net**.

NCCER Module 16201, *Asphalt Shingle Roof Systems.*

NCCER Module 16202, *Clay and Concrete Tile Roof Systems.*

NCCER Module 16203, *Wood Roof Systems.*

NCCER Module 16204, *Slate Roof Systems.*

NCCER Module 16205, *Metal Roof Systems.*

NCCER Module 75901, *Fall Protection Orientation.*

Occupational Outlook Handbook, Roofers, Bureau of Labor Statistics, US Department of Labor, **www.bls.gov/ooh/construction-and-extraction/roofers.htm**.

Occupational Safety and Health Administration (OSHA), US Department of Labor, **www.osha.gov**.

OSHA 3755-05, *Protecting Roofing Workers.* US Department of Labor, **www.osha.gov/Publications/OSHA3755.pdf**.

Figure Credits

iStock@Olga_Gavrilova, Module Opener

Courtesy of the National Roofing Contractors Assoc., Figures 2, 24–26, 29

iStock@JenDen2005, Figure 5

iStock@brizmaker, Figure 6

Courtesy of the National Roofing Contractors Assoc., Figure 7

Courtesy of GAF, Figures 8, 17, 27–28

iStock@HildaWeges, Figure 11

iStock@filmfoto, Figure 13

iStock@Westhoff, Figure 15

iStock@jimplumb, Figure 16

iStock@cbglp2, Figure 18

iStock@titine974, Figure 19

Eagle Roofing Products, Figure 20

"slate roof copper gutter" by Brock Builders is licensed under CC BY 2.0, Figure 21

iStock@Ivan Elizondo, Figure 22

iStock@ronstik, Figure 30

Section Review Answer Key

SECTION 1.0.0

Answer	Section Reference	Objective
1. b	1.1.1	1a
2. c	1.2.0	1b

SECTION 2.0.0

Answer	Section Reference	Objective
1. b	2.0.0	2
2. a	2.2.4	2b

SECTION 3.0.0

Answer	Section Reference	Objective
1. d	3.1.0	3a
2. c	3.1.6	3a

NCCER CURRICULA — USER UPDATE

NCCER makes every effort to keep its textbooks up-to-date and free of technical errors. We appreciate your help in this process. If you find an error, a typographical mistake, or an inaccuracy in NCCER's curricula, please fill out this form (or a photocopy), or complete the online form at **www.nccer.org/olf**. Be sure to include the exact module ID number, page number, a detailed description, and your recommended correction. Your input will be brought to the attention of the Authoring Team. Thank you for your assistance.

Instructors – If you have an idea for improving this textbook, or have found that additional materials were necessary to teach this module effectively, please let us know so that we may present your suggestions to the Authoring Team.

NCCER Product Development and Revision

13614 Progress Blvd., Alachua, FL 32615

Email: curriculum@nccer.org
Online: www.nccer.org/olf

❏ Trainee Guide ❏ Lesson Plans ❏ Exam ❏ PowerPoints Other ___________________

Craft / Level: _______________________________________ Copyright Date: ______________

Module ID Number / Title: ____

Section Number(s): ____

Description: ____

____

____

____

Recommended Correction: ____

____

____

____

Your Name: ____

Address: ____

____

Email: ____ Phone: _________________________

This page is intentionally left blank.

Introduction to Low-Slope Roofing

OVERVIEW

Buildings of all shapes and sizes have low-slope roofs—supermarkets, stadiums, office buildings, schools, and other commercial and industrial facilities are just some examples. Low-slope roofs have many essential components to keep water out and insulate a building. These components include insulation, a waterproof covering (or membrane), and flashings. Roofers must be familiar with roof system components and how they function to keep buildings warm and dry.

Module 16106

Trainees with successful module completions may be eligible for credentialing through the NCCER Registry. To learn more, go to **www.nccer.org** or contact us at 1.888.622.3720. Our website, **www.nccer.org**, has information on the latest product releases and training.

Your feedback is welcome. You may email your comments to **curriculum@nccer.org**, send general comments and inquiries to **info@nccer.org**, or fill in the User Update form at the back of this module.

This information is general in nature and intended for training purposes only. Actual performance of activities described in this manual requires compliance with all applicable operating, service, maintenance, and safety procedures under the direction of qualified personnel. References in this manual to patented or proprietary devices do not constitute a recommendation of their use.

16106 V1.0

From *Roofing, Trainee Guide*. NCCER.
Copyright © 2021 by NCCER. Published by Pearson. All rights reserved.

16106
INTRODUCTION TO LOW-SLOPE ROOFING

Objectives

Successful completion of this module prepares you to do the following:

1. Define low-slope roof systems and identify their basic components.
   a. Identify low-slope deck types.
   b. Identify common insulation and cover board types.
   c. Describe types of low-slope roof systems and membranes.
2. Explain how low-slope roof systems are installed and how they function.
   a. Describe positive drainage and explain how tapered insulation is used to direct water.
   b. Identify methods used to attach low-slope roof systems.
   c. List methods for seaming a roof membrane.
   d. Explain the purpose of flashings and where they are needed.

Performance Tasks

This is a knowledge-based module. There are no Performance Tasks.

Trade Terms

Base flashing
Bitumen
Building envelope
Closed-cell
Combustible
Counterflashing
Crickets
Deck
Field
Flash off

Flashings
Gutters
Low-slope
Pavers
Positive drainage
R-value
Saddles
Scupper
Steep-slope
Substrate

Industry Recognized Credentials

If you are training through an NCCER-accredited sponsor, you may be eligible for credentials from NCCER's Registry. The ID number for this module is 16106. Note that this module may have been used in other NCCER curricula and may apply to other level completions. Contact NCCER's Registry at 1.888.622.3720 or go to **www.nccer.org** for more information.

Contents

This page is intentionally left blank.

1.0.0 LOW-SLOPE ROOFS

Objective

Define low-slope roof systems and identify their basic components.

 a. Identify low-slope deck types.
 b. Identify common insulation and cover board types.
 c. Describe types of low-slope roof systems and membranes.

Trade Terms

Bitumen: A dark, sticky substance found in asphalts, tars, and pitches. May also refer to any material composed mainly of bitumen, such as asphalt or coal tar.

Building envelope: The systems of a building that separate the interior of the building from the outside, including the roof, walls, windows, and doors.

Closed-cell: The structure of cellular insulation in which the tiny cellular structures are packed closely together but are not connected to each other. The cells of the material themselves may be solid or hollow.

Combustible: Capable of burning.

Deck: A structural component of the roof of a building, capable of safely supporting the weight of the roof or waterproofing system, as well as the additional live loads required by the governing building codes. The deck provides the substrate to which the roof or waterproofing system is applied.

Low-slope: A category of roofs that generally includes weatherproof membrane types of roof systems installed on slopes at or less than 3:12.

R-value: A measurement of how well a material resists the transfer of heat through itself. Also called the *thermal resistance*.

Steep-slope: A category of roofing that generally includes water-shedding types of roof coverings installed on slopes exceeding 3:12.

Substrate: The surface upon which a roofing or waterproofing membrane is applied (such as the structural deck or rigid board insulation).

A roof assembly contains many interacting components, including the roof deck, air or vapor retarder (if needed), insulation, and primary covering designed to weatherproof the structure (such as a membrane or shingles). In general, a roof assembly consists of the structural deck and roof system.

Every component above the roof deck makes up the roof system. Roof systems are categorized as low-slope or steep-slope based on the roof's degree of incline, known as its *slope*. Low-slope roofs (sometimes informally called *flat roofs*) are often seen on larger buildings (*Figure 1*). Steep-slope roofs are common on residential homes (*Figure 2*).

Figure 1 Low-slope roof.

Figure 2 Steep-slope roof.

The slope of a roof is one of the primary differences among roof assemblies. The steepness of a slope allows gravity to move water off a roof, promoting drainage. Without slope, a roof could hold water, causing significant water damage or even building collapse.

The most common method used to express the slope of a roof is a ratio between the vertical rise of the roof and the horizontal run, as illustrated in *Figure 3*. This is usually expressed in inches of rise per foot (12") of run, or in feet of rise per 12' of run. In this example, the rise is 8' and the run is 12', so the slope of the roof is 8:12.

Roof slope can be expressed in a variety of ways. Slope is often informally called *pitch*. For example, you may hear workers refer to a roof with a slope of 4:12 as a *4:12-pitch roof* or a *third-pitch roof*. (Note that the term *pitch* can also refer to a roofing material derived from coal tar.) A 4:12 slope can also be referred to as a *four-in-twelve slope*. Examples of ways a 4:12 slope may be written include 4-12, 4/12, and 4 in 12. It can also be expressed in degrees. *Table 1* shows roof slope expressed in both degrees and ratios.

The NCCER Roofing program defines low- and steep-slope roofs as follows:

- *Low-slope* — 3:12 slope or less
- *Steep-slope* — Above 3:12 slope

Table 1 Roof Slope Conversion Chart

Roof Slope	Angle
1:12	4.76°
2:12	9.46°
3:12	14.04°
4:12	18.43°
5:12	22.62°
6:12	26.57°
7:12	30.26°
8:12	33.69°
9:12	36.37°
10:12	39.81°
11:12	42.51°
12:12	45°

Low-slope roof systems are assembled in layers on the deck surface. Basic components of a low-slope roof system include insulation boards and a weatherproof membrane. When required, they may also include a vapor retarder. Low-slope and steep-slope systems have different basic components, as illustrated in *Figure 4*.

> **NOTE**
>
> Throughout NCCER's Roofing program, the term *steep-slope* refers to a roof slope above 3:12. This differs from OSHA's definition of steep-slope roofs (slope above 4:12). When installing or repairing roof systems, always check company, local, state, and federal codes. If any of the applicable codes contradict each other, follow the strictest one.

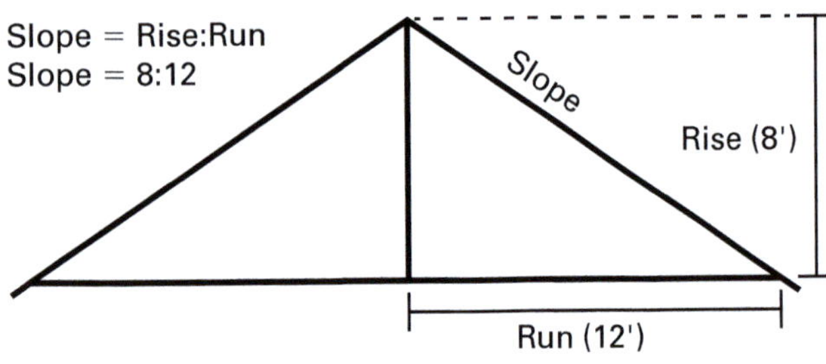

Figure 3 Roof slope.

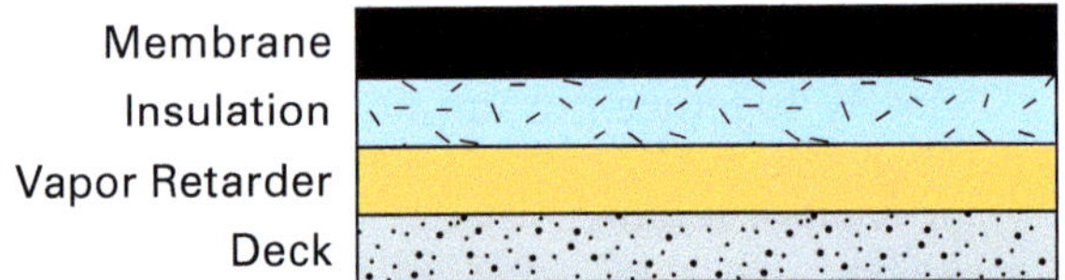

(A) Low-Slope Roof System Components (Weatherproof)

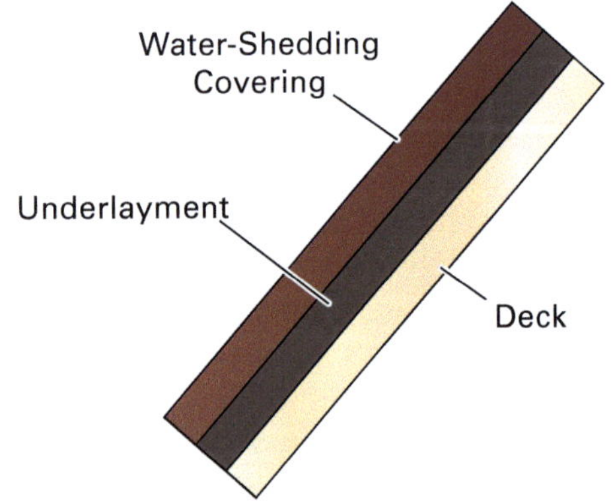

(B) Steep-Slope Roof System Components (Water-Shedding)

Figure 4 Basic components of roof systems.

NCCER – *Roofing*

1.1.0 Low-Slope Decks

The deck is the structure that supports the roof system. It forms the top of the building before any weatherproofing components are installed. Location and climate can influence the material, slope, rafter spacing, and other structural characteristics of a roof deck. Decks are categorized based on whether or not they can burn (combustible or non-combustible), as well as if they can readily accept mechanical fasteners (nailable or non-nailable). *Table 2* shows how roof decks are categorized.

Decks are installed by the general or deck contractor. Roofers are not responsible for installing the deck, but they must understand the different types of decks and how to prepare them safely.

Figure 5 Steel deck.

1.1.1 Steel Decks

Steel decks (*Figure 5*) are common in low-slope roof systems. They are made of steel sheets or panels. Each panel is formed with ribs and flutes for strength (*Figure 6*). They are considered nailable because they can readily accept certain types of mechanical fasteners.

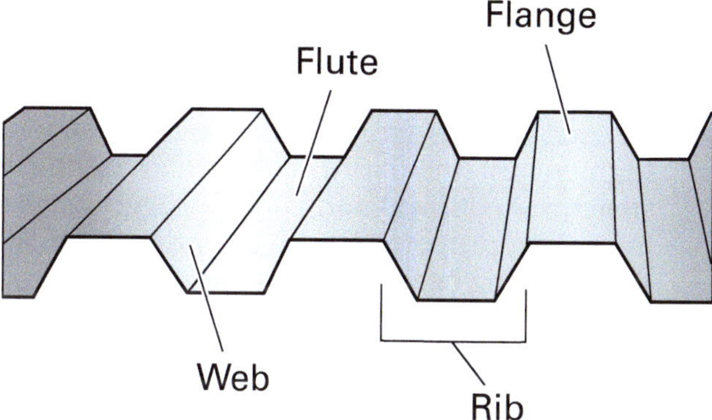

Figure 6 Anatomy of a steel panel.

1.1.2 Concrete Decks

Concrete is a common deck material in low-slope roof systems (*Figure 7*). Concrete decks may be poured in place at the jobsite, or they may be precast into panels and set into place with a crane.

Concrete is made by combining aggregates (such as crushed stone, slate, or other materials), Portland cement, and water. Structural concrete contains crushed stone and gravel, making it very heavy and strong. Lightweight structural concrete contains shale, clay, or slate to make it lighter. Lightweight concrete may be used on top of structural concrete to create slope and promote drainage.

Table 2 Categorizing Types of Roof Decks

	Nailable	Non-nailable
Combustible	Wood panels	
	Wood planks	
Non-combustible	Cementitious wood fiber panels	Structural concrete (normal and lightweight)
	Steel	
	Lightweight insulating concrete	

Figure 7 Concrete deck.

Another type of concrete that may be used in roof assemblies is lightweight insulating concrete. It is often used on top of an existing structural steel deck for its insulating properties. It may also be used over structural concrete to provide slope. Lightweight insulating concrete is lighter than structural concrete and contains aggregates with insulating properties, such as perlite, vermiculite, or insulating beads.

1.1.3 Wood Decks

Wood decks (*Figure 8*) are made of planks, boards, or panels. Planks and boards are narrow and set close together to form the deck's surface. Both are sawn lumber, but planks are larger than boards. Panels are made of plywood or oriented strand board (OSB). They may be referred to as *structural wood panels* or *sheathing*.

1.1.4 Cementitious Wood Fiber Decks

Cementitious wood fiber panels are composed of a binder and wood fiber composite, creating a surface that resembles shredded wheat. You may hear them referred to as *Tectum*®, which is a brand often associated with these panels. They provide a finished interior ceiling and promote soundproofing. This makes them a popular choice for auditoriums, gymnasiums, theaters, schools, and libraries. A cementitious wood fiber deck is shown in *Figure 9*.

<table>
<tr><td>CAUTION</td><td>Cementitious wood fiber panels are porous and absorb more water than other deck types. Moisture can make them weak and brittle. They must be protected from moisture and replaced if they get wet.</td></tr>
</table>

Perlite

Perlite, a common insulating material, is made from volcanic rock. The rock is heated until it explodes, transforming the rock into small white pieces. In addition to being used in some types of concrete, perlite is often added to soil and potting mixes to aerate plants and strengthen their roots.

Figure Credit: iStock@praisaeng

Figure 8 Low-slope wood deck during reroofing project.

NCCER – *Roofing*

Figure 9 Cementitious wood fiber panel deck.

1.2.0 Insulation and Cover Boards

Insulation and cover boards provide thermal resistance for the roof system. They are usually installed on top of the deck to provide a substrate for applying the roof membrane (*Figure 10* and *Figure 11*). A roof is one of the largest surface areas of a building envelope through which interior heat can escape. Insulation helps to maintain the inside temperature of a building at a more constant, comfortable level.

Heat always flows from hotter areas to colder areas. This is called *heat transfer*. In the winter, heat inside the building will want to move toward the cold outside. In the summer, heat outside the building will want to move toward the cooler air inside. Since heat naturally rises, buildings can lose the most heat through the roof.

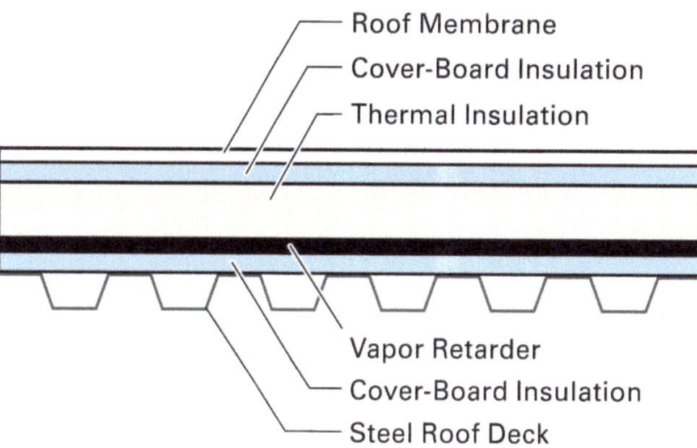

Figure 10 Example of insulation in a low-slope roof assembly.

Insulation resists heat transfer. The R-value of a given insulation material is a measurement of its resistance against heat transfer. The larger the R-value, the better the insulation.

1.2.1 Polyisocyanurate (Iso)

Polyisocyanurate insulation is a common closed-cell insulation material used in roofing (*Figure 12*). It is manufactured from rigid polyisocyanurate foam sandwiched between two facers. Facers include aluminum foils, fiberglass-reinforced cellulosic mats, coated or uncoated polymer-bonded fiberglass mats, or other rigid board materials. Polyisocyanurate is often called *polyiso* or *iso* for short.

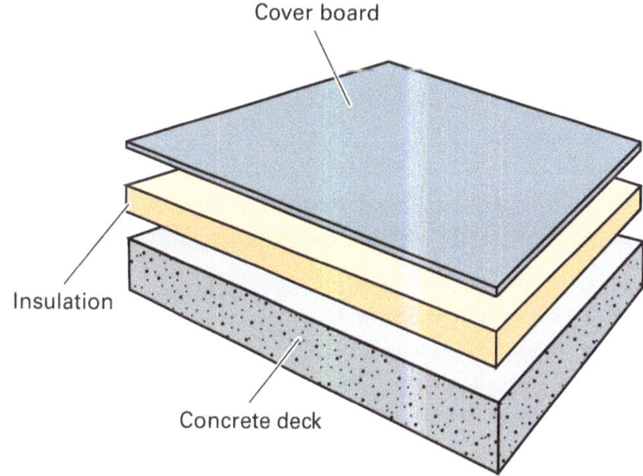

Figure 11 Cover board is a thin layer that serves as the membrane substrate.

Figure 12 Polyisocyanurate insulation.

The following properties of polyisocyanurate make it an effective insulation material:

- **Bitumen** and adhesive compatibility
- Impact resistance
- Fire resistance
- Durability
- Moisture resistance
- Thermal resistance
- Attachment capability

1.2.2 Gypsum

Gypsum, also known as *drywall*, is a material primarily used to finish walls and ceilings. Gypsum used for roofing consists of a gypsum-based core sandwiched between fiberglass-mat facings (*Figure 13*). Gypsum is not typically classified as an insulator. It is used as a thermal barrier to provide fire resistance, substrates for air and vapor retarders, and cover boards below roof membranes. Gypsum is often called *gyp* for short.

<table>
<tr><td>CAUTION</td><td>Gypsum should not be used in hot process applications.</td></tr>
</table>

The following properties of glass-faced gypsum make it well suited for use in roof assemblies:

- Adhesive compatibility
- Fire resistance
- Moisture resistance
- Stable R-value
- Attachment capability
- Dimensional stability
- Compressive strength

1.2.3 Perlite

Perlite board insulation used for roofing is a rigid insulating material manufactured by combining expanded volcanic minerals with organic fibers and binders (*Figure 14*). Generally, the top surface of perlite board roof insulation is treated

Figure 13 Glass-faced gypsum.

with an asphalt emulsion to minimize bitumen absorption.

The following properties of perlite board insulation make it an effective insulation material:

- Bitumen and adhesive compatibility
- Impact resistance
- Fire resistance
- Thermal resistance
- Stable R-value
- Attachment capability

1.2.4 Wood Fiber

Wood fiber insulation (*Figure 15*) is a rigid material manufactured from wood or cane fibers and various binders. It is a natural and renewable

Figure 14 Perlite board insulation.

Figure 15 Wood fiber board insulation.

material that has insulating and soundproofing qualities.

The following properties of wood fiber board insulation make it an effective insulating material:

- Bitumen and adhesive compatibility
- Impact resistance
- Durability
- Thermal resistance
- Stable R-value
- Attachment capability
- Dimensional stability

1.2.5 Polystyrene (EPS and XPS)

The two most common types of polystyrene insulation used in roofing are expanded polystyrene (EPS) and extruded polystyrene (XPS). The primary difference between them is the manufacturing process.

EPS is manufactured by expanding beads in a mold and fusing them together with heat and pressure. It is often referred to as *beadboard* because of the visible beads. EPS roof insulation is shown in *Figure 16*.

XPS is manufactured in a continuous extrusion process that produces a closed-cell foam insulation. XPS roof insulation is shown in *Figure 17*.

Another difference between EPS and XPS is how it is used in a roof system. EPS is traditionally incorporated within the roof system. In contrast, XPS is often installed on top, external to the roof system.

The following properties of polystyrene insulation make it an effective insulating material:

- Compatibility with asphalt at low asphalt temperatures
- Impact resistance
- Durability
- Moisture resistance
- Thermal resistance
- Stable R-value

1.2.6 Stone Wool

Stone wool is manufactured by combining and heating natural minerals until they are molten, and then spinning them into a fibrous material that is often referred to as *stone wool*. The stone wool fibers are bound together with a binding agent to form a rigid insulation board.

The following properties of stone wool insulation make it an effective insulating material:

- Compatibility with bitumen and adhesive
- Fire resistance
- Durability
- Thermal resistance
- Stable R-value
- Dimensional stability

1.3.0 Low-Slope Roof Systems

There are a variety of different membranes used to weatherproof a low-slope roof. The following are some of the most common types:

- Single-ply systems
- Bituminous systems
- Liquid-applied systems

Structural metal panels (*Figure 18*) may also be used as a low-slope roof system.

Figure 16 EPS roof insulation.

Figure 17 XPS roof insulation.

Figure 18 Structural metal roof system.

1.3.1 Single-Ply Systems

A single-ply roof membrane consists of one layer of weatherproofing. There are two categories of single-ply membranes: thermoset and thermoplastic (*Figure 19*). EPDM is by far the most common thermoset, and PVC and TPO are thermoplastics. Many single-ply membranes are manufactured with polyester or glass fiber reinforcement that strengthens each sheet to help it resist tears and punctures. Some membranes are not reinforced and are highly elastic.

The main difference between thermosets and thermoplastics is how they react to heat. Thermoset materials are stable and can be adhered for strong, reliable service. Laps are sealed with adhesives or tape. Thermoplastic materials can be repeatedly reshaped when heated. Laps are seamed with hot-air welding.

Thermoset materials have been used for low-slope roofs for decades (*Figure 20*). They can be very cost effective, reliable, and resistant to high temperatures. Once installed, thermoset materials cannot be reshaped.

Thermoplastic roof systems (*Figure 21*) may be chosen because their light color reflects heat instead of absorbing it, making them energy efficient. They are also recyclable, durable, and puncture resistant. They have a clean and pleasing finished look and are easy to maintain and repair.

1.3.2 Bituminous Systems

Bituminous roof systems contain a substance called *bitumen*. Bitumen is the liquid binder used in asphalt. There are two common types of bituminous roof systems: built-up roof (BUR) systems and polymer-modified systems.

BUR systems are made of multiple layers of coated sheets, saturated felts, fabrics, or mats assembled in an overlapping pattern with alternate layers of bitumen. They are called *built up* because the layers are assembled (built up) on the roof. A bitumen adhesive, such as cold-applied bitumen, hot-mopped asphalt, or polymer-modified asphalt, is applied onto the substrate and sheets or felts are rolled into it (*Figure 22*). The system is then surfaced with mineral aggregate, bituminous materials, liquid-applied coating, or granule-surfaced cap sheets. *Figure 23* shows an example of general components found in a BUR system.

Polymer-modified systems (often referred to as *mod-bit* or *modified*) have been used in the United States since the 1970s. Polymer-modified bitumen is bitumen that has been modified by adding a polymer. Polymers used are synthetic compounds derived from petroleum.

Polymer-modified systems are comprised of ply sheets and cap sheets, similar to a BUR system. However, these ply sheets are pre-manufactured by embedding reinforcing fabrics in polymer-modified bitumen. The sheets are then installed in layers to create a multi-ply roof membrane. The sheets may be installed in the following ways:

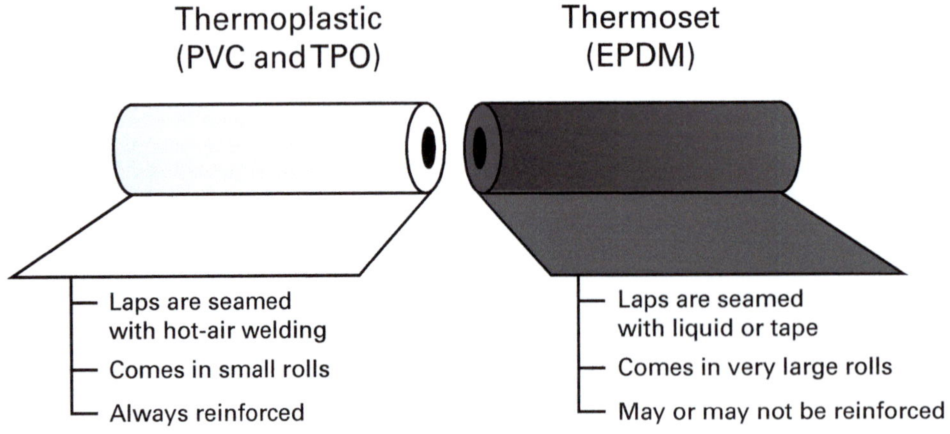

Figure 19 Thermoplastic versus thermoset roof systems.

NCCER – *Roofing*

Figure 20 Thermoset (EPDM) roof system.

Figure 21 Thermoplastic roof system.

- Adhering in cold-applied adhesives
- Adhering in hot-mopped or polymer-modified asphalt
- Heating with a roofing torch until the bitumen flows and infuses the sheet to the substrate (*Figure 24*)
- Using self-adhering sheets

1.3.3 Liquid-Applied Systems

There are two categories of liquid-applied systems: liquid membrane roof systems (*Figure 25* [A]) and roof coating systems (*Figure 25* [B]). They are generally made of polymer-based resin or polymer-modified compounds.

Liquid membrane roof systems are generally applied in two coats that include reinforcements, such as polyester fabric or fleece. The resin or compound cures to form a weatherproof membrane. Liquid-applied systems may include additional surfacing, such as aggregate or a coating.

These roof systems have gained popularity, especially in reroofing projects.

Roof coating systems are applied over an existing roof system. This may be done to restore the roof system and extend its warranty.

Figure 22 BUR membrane installation.

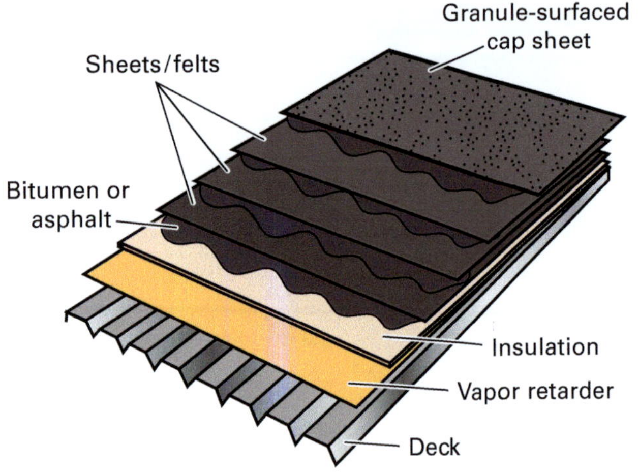

Figure 23 Components of a BUR system.

(A) Liquid Membrane Roof System

Figure 24 Torching equipment for bituminous systems.

(B) Roof Coating System

Figure 25 Liquid-applied systems.

1.0.0 Section Review

1. The structure that supports the entire roof system is the _____.

    a. insulation
    b. cover boards
    c. deck
    d. underlayment

2. Which of the following is a purpose of insulation in a roof system?

    a. Prevent heat from escaping
    b. Weatherproof the structure
    c. Provide structural stability
    d. Protect against water vapor

3. What kind of roof membrane is TPO?

    a. BUR
    b. EPDM
    c. Thermoset
    d. Thermoplastic

2.0.0 LOW-SLOPE SYSTEM INSTALLATION AND FUNCTION

Objective

Explain how low-slope roof systems are installed and how they function.

a. Describe positive drainage and explain how tapered insulation is used to direct water.
b. Identify methods used to attach low-slope roof systems.
c. List methods for seaming a roof membrane.
d. Explain the purpose of flashings and where they are needed.

Trade Terms

Base flashing: Plies or strips of roof membrane material used to close off and/or seal a roof at the horizontal-to-vertical intersections, such as at a roof-to-wall juncture. Base flashing covers the edge of the field membrane and extends up the vertical surface.

Counterflashing: A roof system component, usually composed of metal, used to cover or shield the upper edges of the membrane base flashing or wall flashing.

Crickets: Components used to divert water away from a chimney, wall, expansion joint, or other interruption in the field of a roof.

Field: The main, uninterrupted surface area of a roof.

Flash off: When an adhesive cures or dries just enough for two pieces to stick together. It is tacky to the touch but will not leave any residue on your glove.

Flashings: Components used to weatherproof or seal edges of a roof system at perimeters, penetrations, walls, expansion joints, valleys, drains, and other places where the roof covering is interrupted or terminated. For example, membrane base flashing covers the edge of the field membrane, and cap flashings or counterflashings shield the upper edges of the base flashing.

Gutters: Channeled components installed along the perimeter of a roof to carry water to drains or downspouts.

Pavers: Precast slabs, usually made of concrete, used to weigh down (or ballast) a roof system and provide a functional finish and walking surface.

Positive drainage: A condition in which a roof is designed and shaped to ensure proper drainage of the roof area within 48 hours after rainfall.

Saddles: Small, sloped structures that help to channel surface water to drains, frequently located in a valley. Saddles are often constructed like a small hip roof or pyramid with a diamond-shaped base. Also called *diamond crickets*.

Scupper: A drainage outlet through a wall, parapet wall, or raised roof edge, typically lined with a sheet-metal sleeve.

All roof systems are designed to prevent water from entering the building through the roof. If water pools or gets into seams on the roof, it can cause damage and corrosion to the roof system, leading to leaks and even structural damage. When installing a roof system, every step in the process must prevent water from entering the building or ponding on the roof (*Figure 26*).

2.1.0 Positive Drainage

Low-slope roofs often look completely flat, but they always have some slope to promote positive drainage and prevent water from pooling. Any water on the roof must run off and/or completely dry within 48 hours. Positive drainage means the shape of the roof moves water off the roof or toward drains.

> **NOTE**
>
> Check local building codes for the precise definition of positive drainage in your area.

Figure 26 Ponding on a roof.

There are several ways to create positive drainage in a low-slope roof system, including the use of tapered insulation and **crickets** to direct water to drains. Low-slope roofs may have drains in the **field** of the roof, or an outlet called a **scupper** (*Figure 27*). Drains are located at low points on the roof. Scuppers are commonly used as secondary drainage for low-slope roofs with parapet walls. **Gutters** are installed along the perimeter of a roof to carry water from the roof to the drains or downspouts.

2.1.1 Tapered Insulation

Tapered insulation is prefabricated to be thicker on one end to create a slope when placed on a flat surface. It can be used to move water toward drains.

Tapered insulation comes in pieces, each labeled according to where they should be placed in sequence. They are shaped to fit together and create a smooth slope, as shown in *Figure 28*. In this example, the set includes letters AA, A, B, and C. Pieces labeled AA are the thinnest, and pieces labeled C are the thickest.

2.1.2 Crickets and Saddles

Sometimes, one-directional slope is not enough. Water may pool because there is no slope down the middle of the field to guide it to the drains. In these cases, water may need extra help finding the drain.

Crickets and **saddles** are pieces of insulation specially cut or manufactured to guide water toward drainage (*Figure 29*). A cricket is used to divert water away from a horizontal intersection of the roof, such as a skylight or wall. A saddle is a sloped structure that channels surface water to drains. Saddles are also called *diamond crickets*.

2.2.0 Attachment Methods

Roof systems may be attached to a building in three general ways: mechanically attached, adhered, and ballasted. Another increasingly popular method is induction welding.

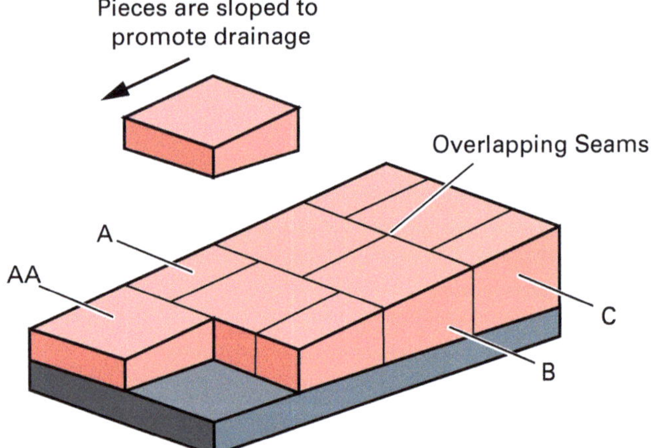

Figure 28 Example of tapered insulation.

Figure 27 A drain and overflow scupper on a thermoplastic roof.

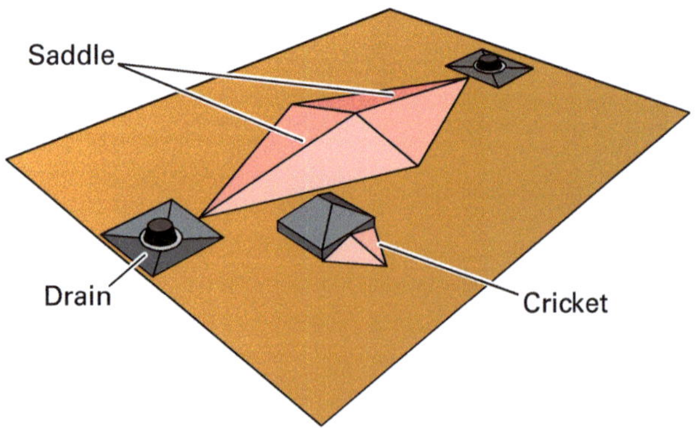

Figure 29 Crickets and saddles.

Battling Water

When it comes to roofing, water is the enemy. Everything roofing crews do is an effort to keep water out of buildings. To accomplish this task, roofers must understand the following basic principles of water movement:

- **Water runs downhill.** It may run sideways or even uphill when its natural flow is interrupted or when it finds seams or another surface to run along.
- **Roofs are most effective when water drains off as quickly as possible.** Water looks for another path when anything gets in its way. A good roof keeps water moving toward drains or roof edges to prevent it from sneaking into other places.
- **With time, water destroys everything in its path.** Water can corrode materials, carve new paths, and weaken roof systems.

Figure Credit: iStock@Beeldbewerking

2.2.1 Mechanical Attachment

Mechanically attached roofs are common because they can be faster and less expensive to install than other types. They are also easy for the manufacturer to inspect and verify proper installation, making them a popular choice for building owners. Mechanically attached roofs are attached using mechanical fasteners (*Figure 30*).

The following steps outline a general procedure for mechanically attaching a single-ply roof system:

> **NOTE**
>
> The following procedure is an example and assumes insulation has already been installed over the deck. Always comply with company procedures and manufacturer instructions when installing a roof membrane.

Step 1 Lay down the membrane roll.

Step 2 Drill the fasteners directly through the insulation boards and into the deck below at the edge of the membrane roll (*Figure 31*).

Step 3 Lay down the next membrane roll, positioning it to cover the fasteners with the edge of the sheet.

Step 4 Repeat Steps 1–3, laying down each roll successively and overlapping each lap.

Step 5 Seal the laps to create a watertight seam. For an EPDM system, adhere them with tape. For a thermoplastic system, heat weld the membranes together with a hot-air welder.

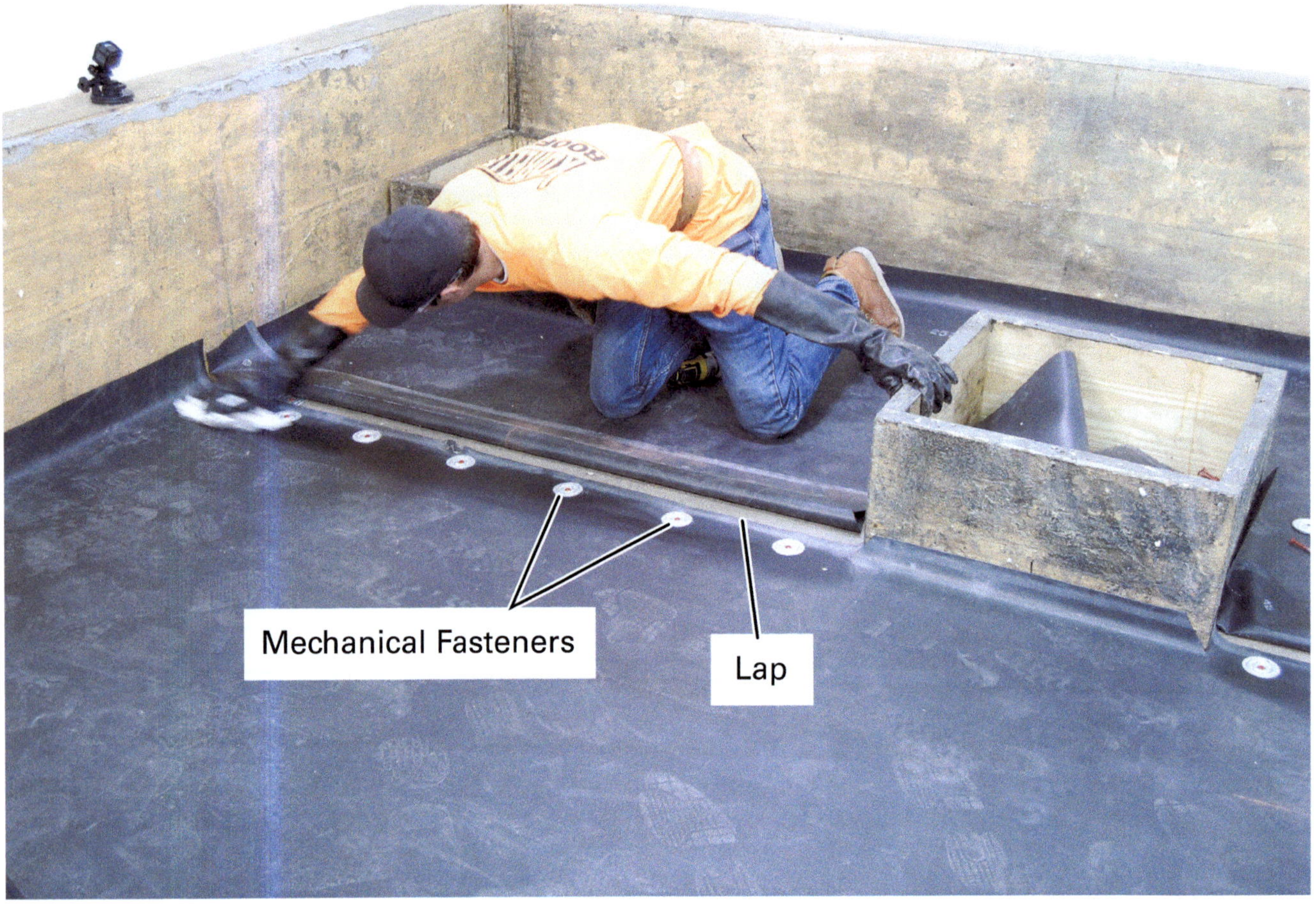

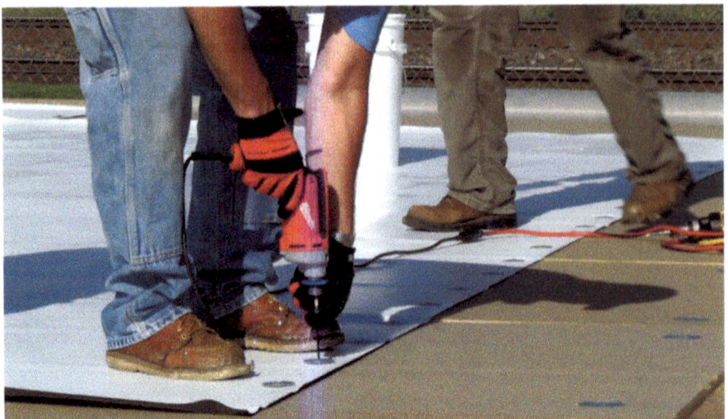

Figure 30 Mechanically attached EPDM roof membrane.

2.2.2 Adhesives

Adhered roof systems are glued to the substrate using an adhesive. Building owners or designers may choose this method for better wind uplift prevention. However, this method may be slower and more expensive. Large quantities of adhesive must be sprayed or rolled on at an appropriate temperature. Then, the adhesive must flash off, or cure, to just the right tackiness before the membrane is pressed onto it with a weighted roller (*Figure 32*).

Figure 31 Fastening a thermoplastic membrane.

2.2.3 Ballast

Ballasted or loose-laid roof systems are often used on buildings that have walkways or high-traffic areas on the roof. They are attached using a ballast material on top of the membrane to weigh it down (*Figure 33*). Ballast materials may include rounded aggregate, pavers, or vegetation.

2.2.4 Thermoplastic Induction Welding

Induction welding is increasing in popularity as an attachment method. Many roofing contractors believe that this method reduces labor on larger projects. It also eliminates some cold weather restrictions for adhered systems and wind concerns for mechanically attached systems.

This method uses specially coated metal plates placed under a thermoplastic membrane. After the membrane is laid on top of the plates, a thermomagnetic induction welder is used to heat the plates. The heat melts the coating and the membrane, fusing them together. At the end of the process, magnets are placed on top of the membrane (where the plates are) for a few minutes, keeping it secure until it cools and bonds.

Figure 32 Weighted roller.

Figure 33 Ballasted roof system.

2.3.0 Seaming and Welding

Seams are the areas where membrane sheets overlap. They must be made watertight to prevent water from entering between the membrane sheets. Roofers use the following methods to accomplish this:

- *Adhesives* — Adhesives are used between the sheets to bind them together at the seam (*Figure 34*).

Electromagnetic Induction

The process of induction is when a conductor becomes electrically charged from being near another electrically charged object or becomes magnetized by being inside a magnetic field. This action is called *induction* because the objects don't have to physically touch to become electrically charged (the charge is induced).

- *Seam tape* — Seam tape is used between the sheets to bind them together at the seam (*Figure 35*).
- *Hot-air welding* — Extremely hot air, typically over 500°F (260°C), is blown into the seam, melting and bonding the sheets together (*Figure 36*).

2.4.0 Field and Flashings

There are two main areas of any roof: the field and the flashings. The field is the mostly unobstructed expanse or main area of the roof. The roof field can be interrupted by penetrations. Penetrations are any areas where something goes through or gets in the way of the roof membrane. Water has the highest chance of getting into a building at penetrations, so they must be treated carefully by installing flashings.

Flashings protect vulnerable areas from water intrusion. They weatherproof or seal the roof at its perimeter, penetrations, walls, expansion joints,

Figure 34 Adhesive seam on an EPDM roof system.

and drains—any place where the roof membrane is interrupted. Building codes refer to places where flashings should be installed. Areas where flashings must be installed include skylights, drains, scuppers, and parapet walls. Examples of different types of flashings are shown in *Figure 37*.

The term *flashing* is often used as both a noun and a verb. For example, a drain on a low-slope roof system may require a flashing (noun), and you may also see directions for flashing (verb) a drain. In this case, flashing a drain means the act of installing materials that seal this penetration.

Every flashing is constructed of two basic components: **base flashing** and **counterflashing**. Another building component may be used as a counterflashing.

In low-slope systems, flashings are often made of the same material as the roof membrane. Sheets of membrane material are cut to fit the flashing area. They may be either adhered or attached using mechanical fasteners.

Roof system specifications may use different kinds of flashings with different components depending on the area of the roof to be flashed. The following are some basic flashing components:

- *Cant strips* — Cant strips help to provide a transition where the horizontal roof deck and vertical projection of a flashing intersect. They are made of various materials, including perlite, wood fiber, glass fiber, mineral wool, and wood. Sometimes, they are poured in place on a cement deck.
- *Adhesives* — A variety of adhesives are used in roof systems, such as hot asphalt and cold-applied asphalt roof cement. Some modified bitumen flashing sheets are adhered by heating the bitumen on the back of the sheet. Generally, flashing membranes are adhered and then mechanically attached along their top edge.

Figure 35 Seam tape.

(A) Heat Welding a Seam with Hand Seamer

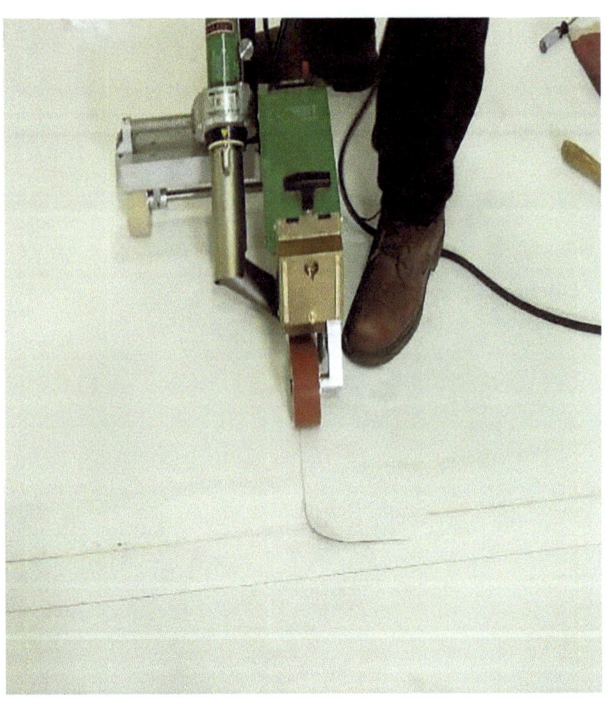

(B) Heat Welding a Seam with Hot-Air Welder

Figure 36 Heat welding a seam.

- *Flashing ply sheets* — Flashing base plies, sometimes called *backer sheets,* are the preliminary plies of a flashing membrane and may consist of the same ply sheets used in the field of the roof. The surface ply flashing sheets are made of reinforced sheet material. It is common to find them made from modified bitumen field membrane sheets. They also may have a factory surfacing already applied.

(A) Curb Flashing

(C) Wall Flashing

(D) Wall Flashing Splice

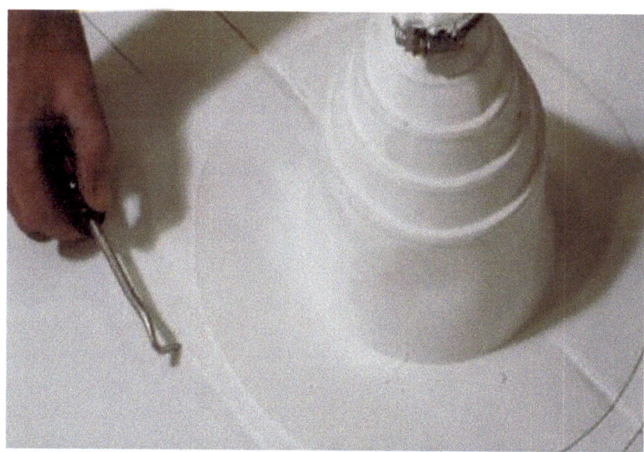

(E) Pipe Flashing

Figure 37 Types of flashings.

Most flashings are secured to their vertical or horizontal substrate using fasteners as described in the specifications. Flashing ply laps are applied so that water will run over the seams and not buck against them. The smooth, exposed parts of all flashing materials require surfacing to protect the flashing from damage as it's exposed to the elements.

Metal flashing components are made of many kinds of metal. Roofing professionals must be aware of all flashing components and how each is to be installed before beginning work. The specifications and detail drawings will indicate the type, quantity, and measurements of all required components to install the complete flashing system.

Though they may use the same materials, all roof flashing details need to address the need for base flashing and counterflashing components. Some common roof flashing detail examples include the following:

- Roof-to-wall transitions
- Eaves
- Skylight and hatch curbs
- Mechanical equipment curbs
- Expansion joints
- Pipe penetrations
- Roof drains
- Scuppers
- Penetration pockets

1. Any water on a roof must dry within ______.

    a. 8 hours
    b. 24 hours
    c. 48 hours
    d. 72 hours

2. Which of the following attachment methods uses fasteners such as nails or screws?

    a. Mechanically attached
    b. Ballasted
    c. Adhered
    d. Induction welded

3. Areas where membrane sheets overlap are called ______.

    a. penetrations
    b. seams
    c. pavers
    d. saddles

4. What indicates the type, quantity, and measurements of all required flashing components for a specific roof system?

    a. Penetrations
    b. Specifications and detail drawings
    c. Building codes
    d. Attachment methods

1. Which of the following components is *not* part of a roof system?

   a. Deck
   b. Insulation
   c. Membrane
   d. Flashings

2. What aspect of a roof allows gravity to move water off the roof or toward drains?

   a. The deck material
   b. Insulation
   c. Its slope
   d. Penetrations

3. Which of the following deck types is combustible?

   a. Structural concrete
   b. Lightweight concrete
   c. Steel
   d. Wood

4. Which of the following statements about heat is *true*?

   a. Heat always moves from hotter areas to colder areas.
   b. Heat always moves from colder areas to hotter areas.
   c. Heat naturally sinks.
   d. Insulation increases heat transfer.

5. Which of the following is *not* typically classified as an insulator?

   a. Polyisocyanurate
   b. EPS
   c. XPS
   d. Gypsum

6. Thermoset roof systems ______.

   a. can be reshaped after they are installed
   b. are recyclable
   c. are seamed with adhesives or tape
   d. are seamed with hot-air welding

7. Which type of roof is made of multiple layers of sheets (such as felts or mats) assembled on the roof with alternate layers of bitumen?

   a. BUR
   b. Thermoplastic
   c. Thermoset
   d. Liquid applied

8. Which of the following components is made of specially cut insulation and used to divert water away from a horizontal intersection of the roof?

   a. Scupper
   b. Drain
   c. Cricket
   d. Induction plate

9. Which attachment method uses coated metal plates placed under the membrane?

   a. Induction welded
   b. Mechanically attached
   c. Ballasted
   d. Adhered

10. What is installed to weatherproof or seal roof penetrations?

    a. Scuppers
    b. Flashings
    c. Adhesive
    d. Saddles

Fill in the blank with the correct term learned from your study of this module.

1. A material that is capable of burning is referred to as _______________.

2. The main, uninterrupted surface area of a roof is called the _______________.

3. The systems of a building that separate the interior of the building from the outside make up the _______________.

4. Plies or strips of membrane material used to seal a roof at a horizontal-to-vertical intersection are called _______________.

5. The category of roofs that generally includes water-shedding roof systems is called _______________.

6. Precast slabs of concrete used to ballast a roof system are called _______________.

7. A roof system component used to cover or shield the upper edges of the membrane base flashing or wall flashing is called a(n) _______________.

8. Components used to weatherproof or seal edges of a roof system at perimeters, penetrations, walls, expansion joints, valleys, etc. are called _______________.

9. The dark, cement-like substance found in asphalts, tars, pitches, and asphaltites is called _______________.

10. The category of roofs that generally includes weatherproof membrane roof systems is called _______________.

11. A condition in which a roof is designed and shaped to ensure drainage of the area within 48 hours after rainfall is called _______________.

12. A drainage outlet through a wall, parapet wall, or raised roof edge, typically lined with a sheet-metal sleeve, is called a(n) _______________.

13. The structural component of a roof that supports the weight of the roof system is called the _______________.

14. Components used to divert water away from a chimney, wall, expansion joint, or other interruption in the field of a roof are called _______________.

15. Small, sloped structures that help channel water to drains are called _______________.

16. Before an adhered membrane is pressed onto the adhesive with a weighted roller, the adhesive must _______________.

17. Channeled components installed along the perimeter of a roof to carry water to drains or downspouts are called _______________.

18. Insulation in which tiny cellular structures are packed closely together but not connected is referred to as being _______________.

19. A measurement of how well a material resists the transfer of heat through itself is called its _______________.

20. The surface upon which a roofing or waterproofing membrane is applied is called the _______________.

Trade Terms

Base flashing	Counterflashing	Flashings	R-value
Bitumen	Crickets	Gutters	Saddles
Building envelope	Deck	Low-slope	Scupper
Closed-cell	Field	Pavers	Steep-slope
Combustible	Flash off	Positive drainage	Substrate

Cornerstone of Craftsmanship

Brian Davis
Director of Technical Services
GAF

How did you choose a career in the industry?

I've always had an interest in buildings, structures, and design. After completing my degree in architecture, I decided to put my resume on the market and see what types of jobs came available. I worked with a contractor over a few summers installing asphalt shingle roofs, so I had some roofing experience under my belt. GAF was one of the first companies to contact me. After doing some research, I found out they were one of the largest roofing manufacturers looking for a computer-aided drafting (CAD) operator for their roofing details. They liked the fact that I had some prior roofing knowledge and offered me a position with GAF right out of college. I have been in the industry ever since.

Who inspired you to enter the industry?

Many people have inspired me to enter and stay in this industry over the years. I've had a few great mentors within the roofing organization, such as Helene Hardy-Pierce and William Woodring. Combined, they have over 80 years of roofing experience in the industry and a willingness to share that knowledge with me and others.

What types of training have you been through?

Over the years, I've invested a lot of time in training. I believe training is very important. The word *training* to me varies. It can be as simple as a conversation, call, or video. Or it can be more complex, such as hands-on or in-person training. Either way, I believe you can never have enough training. Products, technologies, and building codes are always evolving. Staying trained and educated will only help you succeed.

How important is education and training in construction?

Education and training in construction are very important. Receiving the proper training not only teaches you the correct way of doing things, it also teaches you to do them safely.

What kinds of work have you done in your career?

GAF has been a great company to work for, and my 15-year career has solely been with them. I've mostly been on the technical end of roofing. I started as a contractor services analyst drafting roofing details. I then managed a tapered polyiso insulation department for about 10 years. I currently manage the publication of our application and specification manuals for both low- and steep-slope roofs. I also work cross-functionally with other departments as well as contractors, architects, and specifiers on various roofing projects.

Tell us about your present job.

In my current role, I oversee a team of five. My main responsibilities include maintaining all of our technical manuals and literature, roofing detail drawings, and product packaging. I also get to attend and participate in roofing industry organizations such as the National Roofing Contractors Association (NRCA), ARMA, and SPRI on a regular basis.

What do you enjoy most about your job?

I enjoy getting to travel and meet new people. It is also rewarding to know that what we are doing has a huge impact on people's lives and the environment.

What factors have contributed most to your success?

Asking questions and learning as much as possible. Training, thinking outside the box, and hard work have also contributed to my success.

Would you suggest construction as a career to others? Why?

Construction in general has a lot of pathways. Roofing is one that has a lot of potential for growth, as there is a high demand for skilled workers right now. NRCA's ProCertification® program helps these workers get certified in specific roof systems installations. You can then take those skills and hit the ground running.

What advice would you give to those new to the field?

Getting started with a solid foundation is key. Receiving the proper training early is important. The ability to demonstrate proper knowledge, safety, and skill is essential to someone starting out in roofing.

Interesting career-related fact or accomplishment?

I'm an accredited LEED AP (Leadership in Energy & Environmental Design Accredited Professional) and GRP (Green Roof Professional). Accreditation was during my time at GAF.

How do you define craftsmanship?

To me, craftsmanship is when someone takes pride in the quality of work and the attention to detail when making something.

Trade Terms Introduced in This Module

Base flashing: Plies or strips of roof membrane material used to close off and/or seal a roof at the horizontal-to-vertical intersections, such as at a roof-to-wall juncture. Base flashing covers the edge of the field membrane and extends up the vertical surface.

Bitumen: A dark, sticky substance found in asphalts, tars, and pitches. May also refer to any material composed mainly of bitumen, such as asphalt or coal tar.

Building envelope: The systems of a building that separate the interior of the building from the outside, including the roof, walls, windows, and doors.

Closed-cell: The structure of cellular insulation in which the tiny cellular structures are packed closely together but are not connected to each other. The cells of the material themselves may be solid or hollow.

Combustible: Capable of burning.

Counterflashing: A roof system component, usually composed of metal, used to cover or shield the upper edges of the membrane base flashing or wall flashing.

Crickets: Components used to divert water away from a chimney, wall, expansion joint, or other interruption in the field of a roof.

Deck: A structural component of the roof of a building, capable of safely supporting the weight of the roof or waterproofing system, as well as the additional live loads required by the governing building codes. The deck provides the substrate to which the roof or waterproofing system is applied.

Field: The main, uninterrupted surface area of a roof.

Flash off: When an adhesive cures or dries just enough for two pieces to stick together. It is tacky to the touch but will not leave any residue on your glove.

Flashings: Components used to weatherproof or seal edges of a roof system at perimeters, penetrations, walls, expansion joints, valleys, drains, and other places where the roof covering is interrupted or terminated. For example, membrane base flashing covers the edge of the field membrane, and cap flashings or counterflashings shield the upper edges of the base flashing.

Gutters: Channeled components installed along the perimeter of a roof to carry water to drains or downspouts.

Low-slope: A category of roofs that generally includes weatherproof membrane types of roof systems installed on slopes at or less than 3:12.

Pavers: Precast slabs, usually made of concrete, used to weigh down (or ballast) a roof system and provide a functional finish and walking surface.

Positive drainage: A condition in which a roof is designed and shaped to ensure proper drainage of the roof area within 48 hours after rainfall.

R-value: A measurement of how well a material resists the transfer of heat through itself. Also called the *thermal resistance*.

Saddles: Small, sloped structures that help to channel surface water to drains, frequently located in a valley. Saddles are often constructed like a small hip roof or pyramid with a diamond-shaped base. Also called *diamond crickets*.

Scupper: A drainage outlet through a wall, parapet wall, or raised roof edge, typically lined with a sheet-metal sleeve.

Steep-slope: A category of roofing that generally includes water-shedding types of roof coverings installed on slopes exceeding 3:12.

Substrate: The surface upon which a roofing or waterproofing membrane is applied (such as the structural deck or rigid board insulation).

Additional Resources

This module presents thorough resources for task training. The following reference material is suggested for further study.

National Roofing Contractors Association (NRCA), **www.nrca.net**.
NCCER Module 16205, *Metal Roof Systems*.
NCCER Module 16206, *Thermoplastic Roof Systems*.
NCCER Module 16207, *EPDM Roof Systems*.
NCCER Module 16208, *Built-Up Roof Systems*.
NCCER Module 16209, *Modified Bitumen Roof Systems*.
NCCER Module 16210, *Liquid-Applied Roofing*.

Figure Credits

Courtesy of the National Roofing Contractors Assoc., Figures 1, 5, 7, 22, 27, 30, 34, 35, 36 (A), 37 (A–C)
Patrick D Murphy Co., Inc., Architects, Figures 8, 9, 14, 18, 33
Courtesy of GAF, Figures 12, 13, 15, 21, 25 (both images), 31, 36 (B), 37 (D–E)
iStock@Svetlana123, Figure 20
iStock@Michael Vi, Figure 26

Section Review Answer Key

SECTION 1.0.0

Answer	Section Reference	Objective
1. c	1.1.0	1a
2. a	1.2.0	1b
3. d	1.3.1	1c

SECTION 2.0.0

Answer	Section Reference	Objective
1. c	2.1.0	2a
2. a	2.2.1	2b
3. b	2.3.0	2c
4. b	2.4.0	2d

NCCER CURRICULA — USER UPDATE

NCCER makes every effort to keep its textbooks up-to-date and free of technical errors. We appreciate your help in this process. If you find an error, a typographical mistake, or an inaccuracy in NCCER's curricula, please fill out this form (or a photocopy), or complete the online form at **www.nccer.org/olf**. Be sure to include the exact module ID number, page number, a detailed description, and your recommended correction. Your input will be brought to the attention of the Authoring Team. Thank you for your assistance.

Instructors – If you have an idea for improving this textbook, or have found that additional materials were necessary to teach this module effectively, please let us know so that we may present your suggestions to the Authoring Team.

NCCER Product Development and Revision
13614 Progress Blvd., Alachua, FL 32615

Email: curriculum@nccer.org
Online: www.nccer.org/olf

❏ Trainee Guide   ❏ Lesson Plans   ❏ Exam   ❏ PowerPoints   Other _______________

Craft / Level: _______________________________________ Copyright Date: _______________

Module ID Number / Title: ____

Section Number(s): ____

Description: ____

____

____

____

Recommended Correction: ____

____

____

____

Your Name: ____

Address: ____

____

Email: ____ Phone: _______________________

This page is intentionally left blank.

Substrates, Decks, and Roof Insulation

OVERVIEW

The deck is the structural component of a roof assembly. Roofing materials, such as underlayment in steep-slope systems or insulation in low-slope systems, are attached to the deck. Roof systems are designed and installed to protect the building from the elements, as well as to prevent the roof system itself from damage due to any moisture inside the building. This module overviews the types and categories of roof decks and roofing insulation, describes deck deficiencies that must be addressed before roof system installation, and explains how heat and vapor movement are addressed in roof systems.

Module 16104

Trainees with successful module completions may be eligible for credentialing through the NCCER Registry. To learn more, go to **www.nccer.org** or contact us at 1.888.622.3720. Our website, **www.nccer.org**, has information on the latest product releases and training.

Your feedback is welcome. You may email your comments to **curriculum@nccer.org**, send general comments and inquiries to **info@nccer.org**, or fill in the User Update form at the back of this module.

This information is general in nature and intended for training purposes only. Actual performance of activities described in this manual requires compliance with all applicable operating, service, maintenance, and safety procedures under the direction of qualified personnel. References in this manual to patented or proprietary devices do not constitute a recommendation of their use.

16104
SUBSTRATES, DECKS, AND ROOF INSULATION

Objectives

Successful completion of this module prepares you to do the following:

1. Identify types of decks and substrates and explain how deck movement is addressed.
   a. List types of decks and considerations associated with them.
   b. Explain the purpose of expansion joints.
2. Describe how ventilation, vapor retarders, and insulation are used to control the effects of condensation and heat transfer in roof systems.
   a. List the three general ways heat transfer occurs.
   b. Explain basic principles of vapor movement, describe installation procedures for vapor retarders, and describe steep-slope ventilation systems.
   c. Review general procedures for installing, handling, and storing insulation and cover boards.

Performance Tasks

Under supervision, you should be able to do the following:

1. Visually inspect a deck for obvious defects.
2. Lay out two layers of insulation with staggered and offset joints.
3. Fasten insulation with mechanical attachment.

Trade Terms

Carpenter	Evaporation
Competent person	Infrared
Condensation	Water vapor
Dew point	

Industry Recognized Credentials

If you are training through an NCCER-accredited sponsor, you may be eligible for credentials from NCCER's Registry. The ID number for this module is 16104. Note that this module may have been used in other NCCER curricula and may apply to other level completions. Contact NCCER's Registry at 1.888.622.3720 or go to **www.nccer.org** for more information.

Contents

This page is intentionally left blank.

1.0.0 SUBSTRATES AND ROOF DECKS

Objective

Identify types of decks and substrates and explain how deck movement is addressed.

a. List types of decks and considerations associated with them.
b. Explain the purpose of expansion joints.

Performance Task

1. Visually inspect a deck for obvious defects.

Trade Terms

Carpenter: A craftworker who constructs, assembles, installs, and repairs structures made of wood or other materials.

Competent person: As defined by OSHA, an individual who is capable of identifying existing and predictable hazards in the surroundings or working conditions which are unsanitary, hazardous, or dangerous to employees, and who has the authorization to take prompt corrective measures to eliminate such hazards.

A roof deck is the structural component of a roof assembly. In steep-slope roof systems, an underlayment is typically attached directly to the deck before installing shingles or tiles. In low-slope systems, insulation and cover boards are attached to the deck before installing the waterproof membrane.

Before installing a roof system, a roofing contractor should visually inspect the roof deck to determine it can provide a suitable substrate that will accommodate the application of the specified roof system. A substrate is any surface upon which a roofing or waterproofing membrane is applied.

NOTE

It is good practice to have a competent person visually inspect the deck from underneath first to look for areas of deterioration before anyone gets on the roof. The roofing contractor cannot be responsible for the structural integrity of the deck. However, roofers should be able to recognize obvious signs of deck problems, if present, during their initial inspection of the deck. Always notify your crew leader if you notice damage or suspect unsound conditions.

A deck should be dry, broom clean and free of contaminants or debris, reasonably smooth, free of voids or depressions, and securely attached to the underlying structural elements. A deck should remain rigid to avoid excessively deflecting under live loads.

For reroofing projects, removing existing roofing may reveal problems with the roof deck. If the deck is damaged or corroded, it will need to be repaired before roofing materials can be installed. Any damage should be repaired by a decking contractor.

1.1.0 Deck Types

Location and climate can influence the material, rafter spacing, deck slope, and other structural characteristics of a roof deck. Steep-slope roof systems usually have wood decks. A variety of other types of decks are used for low-slope and commercial applications.

Decks are categorized based on whether they can burn (combustible or non-combustible), as well as if they can readily accept mechanical fasteners (nailable or non-nailable). *Table 1* shows how roof decks are categorized.

Table 1 Categorizing Types of Roof Decks

	Nailable	Non-nailable
Combustible	Wood panels	
	Wood planks	
Non-combustible	Cementitious wood fiber panels	Structural concrete (normal and lightweight)
	Steel	
	Lightweight insulating concrete	

1.1.1 Low-Slope Roof Decks

Low-slope roof decks may be made of steel, concrete, cementitious wood fiber panels, or dimensional lumber. Some of the most common low-slope decks are shown in *Figure 1*. Before beginning roofing work on a low-slope roof deck, always ensure it is clean, dry, and free of obvious defects, such as dents, holes, or rust. It is important to use blowers or brooms to remove debris before installing the roof system. If you see something you are not sure about, notify your crew leader.

Steel Decks

Steel roof decks (*Figure 1* [A]) are considered nailable because they are compatible with certain types of mechanical fasteners that attach to the ribs of the metal panels. Steel decks generally require insulation or cover board to be installed above the deck before installing the membrane. Steel panels should be properly aligned and squarely intersect walls and structural framing.

Rusty or corroded steel panels should be replaced (*Figure 2*). Leaves, dirt, and debris can easily collect in the flutes of a steel deck. Use a blower or broom to ensure they are clean and dry before installing roofing materials.

(A) Steel Deck

(B) Concrete Deck

(C) Cementitious Wood
Fiber Panel Deck

(D) Low-Slope Wood Deck

Figure 1 Common low-slope roof decks.

Figure 2 Repairing a rusted steel deck.

Concrete Decks

Concrete decks (*Figure 1* [B]) may be made of various types of concrete. Concrete is made by combining aggregates (such as crushed stone), Portland cement, and water. Structural concrete contains crushed stone and gravel, making it very heavy and strong. Lightweight structural concrete contains shale, clay, or slate to make it lighter. Lightweight concrete may be used on top of structural concrete to create slope and promote drainage.

Lightweight insulating concrete is lighter than structural concrete and contains aggregates with insulating properties, such as perlite, vermiculite, or insulating beads. It may be used on top of an existing structural steel deck. It may also be used over structural concrete to provide slope.

Concrete decks are usually poured at the jobsite and must be allowed time to cure before roof systems are installed. Moisture contained in concrete roof decks that have not adequately cured can damage roof systems.

Concrete decks are non-nailable, so insulation or cover boards must be attached to them using adhesives rather than mechanical fasteners (*Figure 3*).

Cementitious Wood Fiber Panel Decks

Cementitious wood fiber panels, often referred to as *Tectum®*, are composed of a binder and wood fiber composite, creating a surface that resembles shredded wheat (*Figure 1* [C]). They provide a finished interior ceiling and promote soundproofing.

Cementitious wood fiber panels are nailable and non-combustible. The panels must be handled with care, as the edges are easily damaged. They are also subject to damage from weather. If left unprotected, they may become stained or deteriorated and need to be replaced.

Figure 3 Installing insulation to a concrete deck with two-component adhesive.

Wood Decks

Low-slope wood decks are usually made of solid wood boards or planks (*Figure 1* [D]). They can be damaged by moisture and weathering. Wood panels should be placed in storage to protect them from weather while on the jobsite and after installation. Roofing work should begin as soon as possible after the deck is installed.

Some wood is treated with preservatives that may be corrosive to some fasteners. Consult the manufacturer for specific recommendations for applying over preservative-treated wood decks.

1.1.2 Steep-Slope Decks and Underlayment

Wood is the most common construction material in the United States and Canada for steep-slope decks. Wood decks come in two types: wood sheathing and dimensional lumber.

Wood sheathing comes in a variety of thicknesses, between $\frac{1}{2}$" and $\frac{3}{4}$". It most commonly comes in 4' × 8' panels. The two types of wood sheathing typically used for roof decks are plywood and oriented strand board (OSB). Plywood is made of veneers (thin sheets of wood) that are glued together under heat and pressure. OSB is made of wood strands that are oriented at right angles before being formed, compressed, and glued into panels.

Dimensional lumber is solid, sawn wood that may be either boards or planks. Boards and planks are distinguished by thickness and edge profile:

- Boards are less than 2" thick and have square edges.
- Planks are 2" thick or greater and typically have joints on the edges.

Rafters and trusses support the roof deck. They form the structure onto which carpenters attach the deck (*Figure 4*). A **carpenter** is a craftworker who constructs, assembles, installs, and repairs structures made of wood or other materials. Roof rafters and trusses are installed by carpenters instead of roofers. However, since the roof system is attached onto this structural assembly, it is important for roofers to be able to tell a good structural deck apart from a questionable one. Whether panels or boards, the wood roof deck is mechanically fastened to rafters and trusses.

Roof deck fasteners can be nails or screws. Local building codes specify sizes and types of nails to use for rafter, truss, and deck construction.

In the United States, a 2-penny nail (labeled *2d*) is 1" long. Each penny increase is $\frac{1}{4}$" increase in length, up to the 10-penny nail. There is no 11-penny nail, and the 12-penny nail is $3\frac{1}{4}$" long. A 16-penny nail is $\frac{1}{4}$" longer than a 12-penny. *Table 2* shows how nails are sized. Nails are commonly referred to as *sinkers* (for example, a 16-penny sinker).

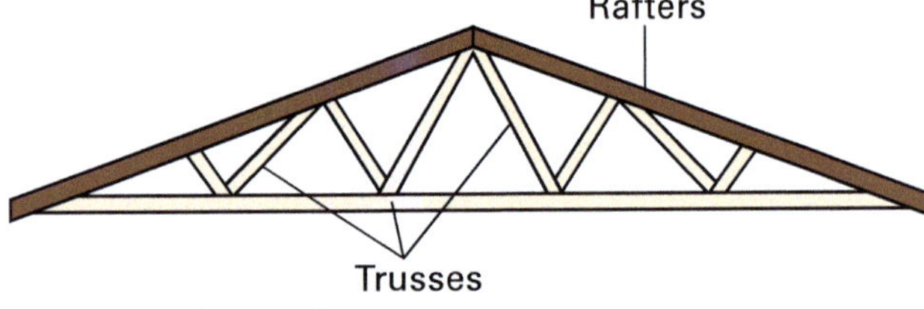

Figure 4 Rafters and trusses.

Table 2 Penny-Inch Nail Equivalents

Nail Size	Length in Inches
2d	1"
3d	$1\frac{1}{4}$"
4d	$1\frac{1}{2}$"
5d	$1\frac{3}{4}$"
6d	2"
7d	$2\frac{1}{4}$"
8d	$2\frac{1}{2}$"
9d	$2\frac{3}{4}$"
10d	3"
12d	$3\frac{1}{4}$"
16d	$3\frac{1}{2}$"

Rink shank nails may also be used on a roofing job. These nails have ridges or grooves to provide increased holding power.

Carpenters use nails that are coated or galvanized for weather protection on roof decks. For rafters and trusses, carpenters use a 16d sinker. Dimensional lumber sheathing requires 16d or greater sinkers. An 8d sinker is used for both wood panels and board decking up to $\frac{1}{4}$" thick.

> **CAUTION**
>
> Nail sizes are not interchangeable. Using a fastener that is too small or too large can cause structural problems.

In new construction, the deck will be installed by someone else before roofing crews arrive. For reroofing, the deck is already in place. In all cases, roofers must pay attention to the condition of the deck and notify their crew leader if they notice any issues.

A deck may have holes due to wear and tear or because a skylight has been removed. Any holes on the deck greater than 2" must be covered and marked so no one steps through them.

Look for any sheathing that is wet or rotted (*Figure 5*). Roofing crews may need to replace some of the boards before installing a new covering.

> **CAUTION**
>
> Plywood and OSB are both specifically rated for roof decks. There should be an APA stamp on any wood sheathing product used in steep roof decking. If there is no stamp on the sheathing, do not use it.

Figure 5 Rotted wood deck.

NCCER – *Roofing*

The substrate onto which the new roof covering will be installed may be the deck itself or an existing layer of a roof covering, such as asphalt shingles. In either case, the substrate must be clean, dry, and in good condition before installing a new roof system on top of it.

Always sweep off debris, replace any damaged sheathing, and ensure the surface is ready for a new covering to be installed.

Underlayment

Underlayment is the first layer of weatherproofing installed on a steep-slope roof deck. Underlayments are also commonly called *felts*. An underlayment is an asphalt-saturated felt installed between a roof deck and a roof covering, usually used in a steep-slope roof construction (*Figure 6*). Underlayment is primarily used to separate the roof covering from the deck, shed water, and provide secondary weather protection for the roof.

The underlayment covers the entire surface of the roof deck. The roof covering, such as asphalt shingles, is applied directly over the underlayment. Underlayment gets its name because it is the first layer of weatherproofing material laid under the shingles.

Underlayments are usually mechanically attached to the deck with fasteners. There are also self-adhering underlayments that seal around fasteners, providing additional protection against wind-driven rain and ice dams (*Figure 7*).

Self-adhering underlayment may be used on sections of the deck or the entire deck, depending on the weather in the region.

Underlayment comes in rolls. Depending on the material, each roll may cover about 400 square feet. Roofing material is often measured in squares. A square is equal to 100 square feet, so 400 square feet is equal to 4 squares.

The following is a general procedure for installing underlayment:

Step 1 Make sure the deck is clean, dry, and in good condition. Use a broom or blower to remove debris and replace any damaged or rotted wood sheathing.

Step 2 Install underlayment in the largest pieces you can safely handle on the roof. Starting at the bottom edge, roll it out along the eave.

Step 3 Work your way up the slope of the deck with each successive row. Be sure to overlap the previous row by a minimum of 2". Overlap ends at least 4".

Check the manufacturer's specifications to be sure about the amount of overlap needed.

Step 4 Attach the underlayment to the deck with roofing nails.

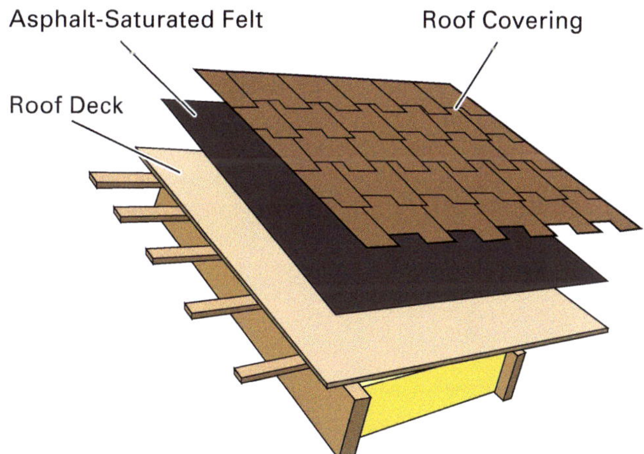

Figure 6 Basic steep-slope roof construction.

Figure 7 Ice dam.

If an installation calls for a two-ply underlayment, it means two layers are needed for extra protection. In a two-ply installation, the overlap must be half of a sheet width plus 1". The end of the piece should be overlapped a minimum of 4". Always check manufacturer's specifications for the correct amount of overlap.

1.2.0 Deck Movement and Expansion Joints

Roof systems are susceptible to damage from expansion, contraction, and movement of the roof deck. This movement occurs due to external factors, such as changes in weather and temperature. Roof expansion joints are used to minimize the effects of stresses and movements of a building's components. The effects of these stresses have the potential to cause damage to the roof system by splitting, buckling, or ridging.

Roof systems must be designed to accommodate deck movement around expansion joints without becoming damaged. When installing a roof system, always follow the specifications for flashing around expansion joints. An example of a steep-slope expansion joint flashing is shown in *Figure 8*.

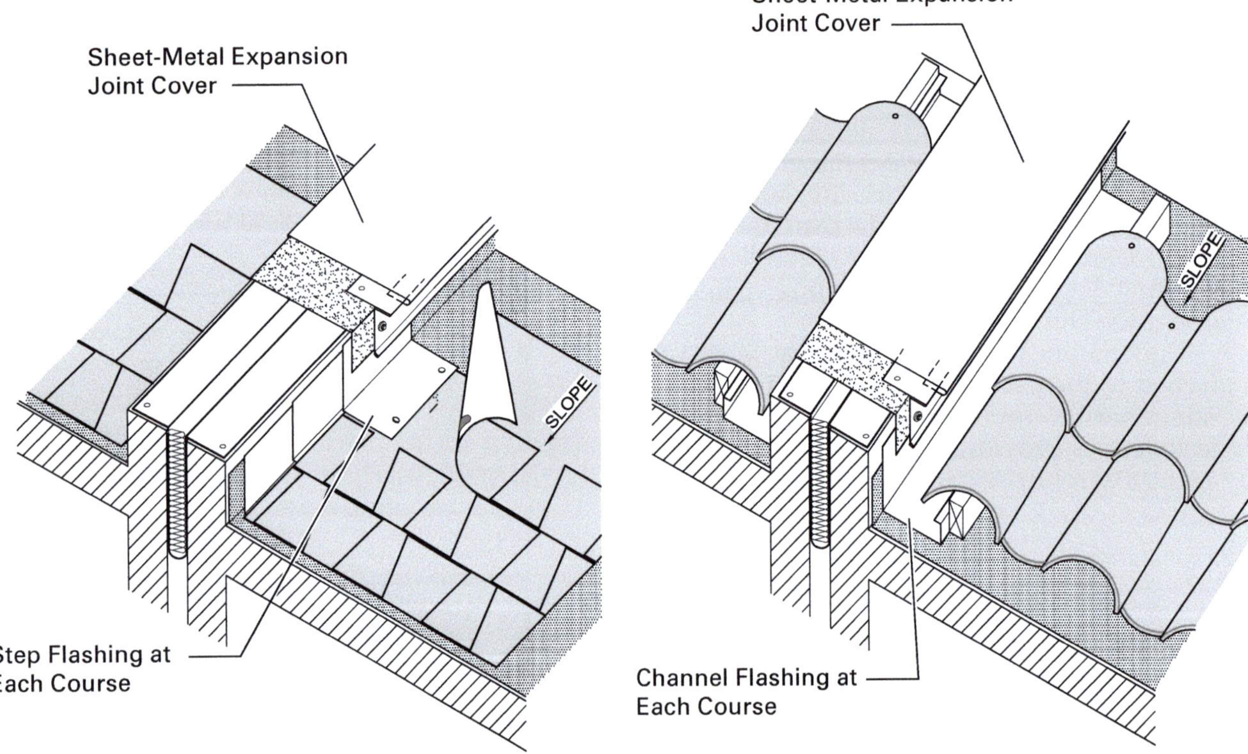

Figure 8 Steep-slope expansion joint flashings.

2.0.0 CONDENSATION CONTROL AND INSULATION

Objective

Describe how ventilation, vapor retarders, and insulation are used to control the effects of condensation and heat transfer in roof systems.

a. List the three general ways heat transfer occurs.
b. Explain basic principles of vapor movement, describe installation procedures for vapor retarders, and describe steep-slope ventilation systems.
c. Review general procedures for installing, handling, and storing insulation and cover boards.

Performance Tasks

2. Lay out two layers of insulation with staggered and offset joints.
3. Fasten insulation with mechanical attachment.

Trade Terms

Condensation: The conversion of water vapor or other gas to liquid phase as the temperature drops or atmospheric pressure rises.

Dew point: The temperature at which air becomes saturated with water vapor and has a relative humidity of 100 percent.

Evaporation: The natural change of liquid to vapor at a temperature below its boiling point.

Infrared: A type of invisible light energy that is emitted by all objects. This type of energy is most effective at producing radiant heating when absorbed.

Water vapor: Water in a vapor (gas) form, especially when below the boiling point and diffused in the atmosphere.

Buildings are designed to maintain a comfortable indoor climate. Because hot air rises, a significant amount of heat can be lost through a building's roof. Insulation prevents this heat loss and maintains a stable temperature inside the building.

Roofers will not usually install insulation in buildings with steep-slope roofs, as the insulation is under the roof deck or below the attic. However, low-slope roof systems include layers of insulation to prevent heat loss through the top of the building.

Temperature differences between the air outside and inside the building can cause water vapor in the air to condense (convert to liquid), damaging roofing materials. In some low-slope roof systems, a vapor retarder may be required to protect the insulation from this moisture. Some steep-slope roofs have ventilation systems to prevent the accumulation of moisture in the roof system.

From the way insulation is laid out to the fasteners selected, every step in the process of installing a roof is designed to account for the effects of heat, vapor, and condensation. It is important for roofers to understand basic principles of heat transfer and vapor movement.

2.1.0 Heat Transfer

Heat transfer occurs in three ways: conduction, convection, and radiation. *Figure 9* shows how heat transfer affects a building envelope. Heat from the sun radiates to the roof (*radiation*). This heat conducts through the roof and into the building (*conduction*). Heat within the building rises through *convection* and enters the attic space.

2.1.1 Conduction

Conduction is the transfer of energy (heat) between two objects in physical contact. For example, if a metal rod is heated over a flame, heat travels by conduction from the hot end to the cooler end. Heat transfer by conduction can occur within an object or substance, as well as between different substances that are in contact with each other.

⌀ Going Green

Saving Energy

A conventionally insulated building can suffer a 55-percent heat loss by heat escaping through its exterior walls and roof and by the infiltration of air from the outdoors. This results in higher heating costs and increased use of fossil fuels at the power generation plants. These are among the many reasons why so much emphasis is placed on the proper insulation of buildings.

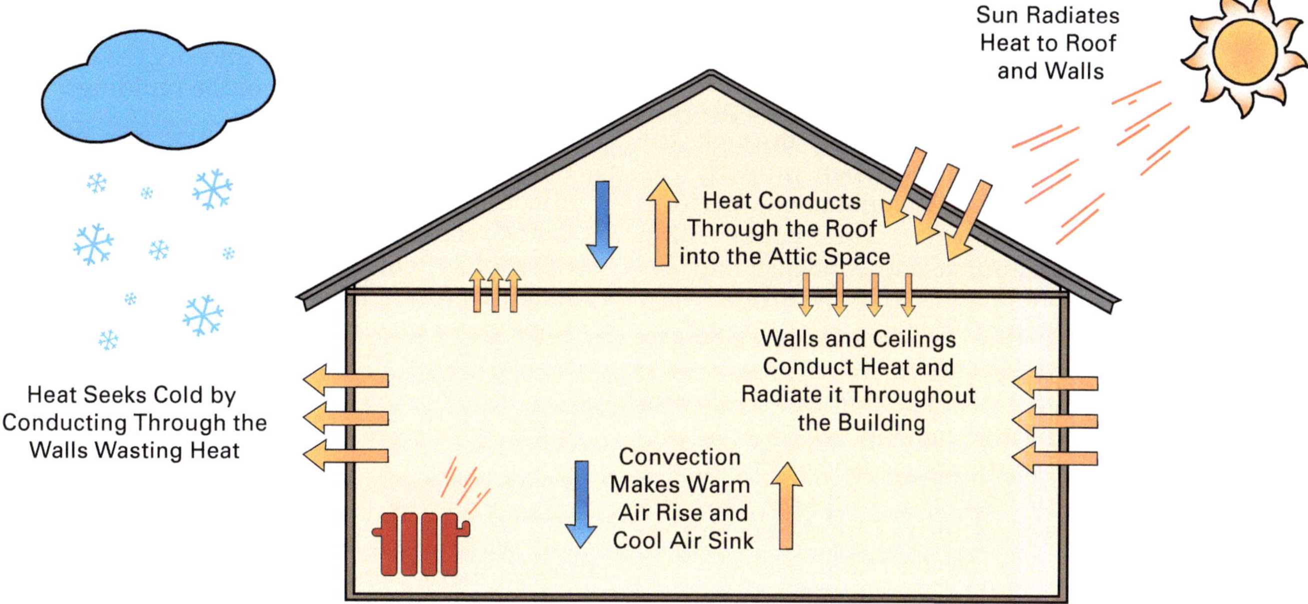

Figure 9 Heat transfer in a building envelope.

Certain building materials conduct heat faster than others. For example, metal conducts heat faster than wood. Poor conductors like wood and fiberglass are good insulators.

2.1.2 Convection

Convection is the transfer of heat through the movement of warm air or liquid to a cooler place. An example of convection is warm, less dense air rising from a radiator as cooler, denser room air falls and forces it upward. This is why temperatures are warmer at the ceiling in a room and cooler at the floor. A good insulating material blocks convection of heat at the surface to which it is applied.

2.1.3 Radiation

Heat transferred by radiation does not need to travel through a material. Radiant heating occurs when light energy is produced by a source, such as the sun, and absorbed by an object. Light energy includes radio waves, infrared, visible light, ultraviolet, X-rays, and gamma rays. Light energy warms the air very slightly as it passes through. For example, a roof with the sun shining on it absorbs the light energy, converting it to heat. The roof might be heated to 180°F (82°C) by the rays of the sun, but the air through which the same light passes travels may only be warmed to 80°F (27°C).

Objects radiate infrared light energy. The higher the temperature of the object, the greater the quantity of heat it radiates. Objects can also absorb infrared energy. Darker and rougher surfaces absorb more radiant heat than light-colored surfaces.

Heat Transfer Experiments

- *Conduction* — Take a piece of copper pipe and attach a thermometer to one end. When the temperature reading has stabilized, heat the other end of the pipe and watch the temperature rise. Then try it with a piece of steel pipe of the same diameter and length and see if the temperature seems to rise at the same rate.
- *Convection* — Obtain a lightweight plastic bag. While holding the bag upside down, fill it with hot air from a hair dryer. Remove the air dryer, then let the bag go and watch it rise because the air inside is less dense than the cooler surrounding air. As the air cools, the bag will begin to descend.
- *Radiation* — Obtain two thermometers. Place one in the shade and one in direct sunlight. Note how much higher the temperature reading is on the thermometer in direct sunlight.

2.2.0 Condensation Control

Moisture can cause significant damage to building materials. The primary purpose of any roof assembly is to keep moisture in its liquid phase—water—from entering a building through the roof. Most roof assemblies, when properly designed, constructed, and maintained, perform this function well. However, water vapor within buildings can accumulate and condense into water, damaging roof systems from the inside.

Condensed moisture can deteriorate roof decks and corrode fasteners. It can also promote the growth of mold and mildew that can weaken or even destroy building components, such as plywood and wood joists, and cause health and aesthetic concerns for building occupants. Moisture that becomes trapped in roof insulation can reduce the insulation's effectiveness.

To prevent moisture accumulation in roof assemblies, building designers must know the source and amount of moisture inside the building. Potential sources of moisture within buildings include the following:

- *Construction moisture* — Significant amounts of water vapor may be generated by construction processes, such as plastering or painting walls and ceilings. Water vapor generated by construction processes is typically temporary, though major concentrations in confined areas may cause permanent damage to some parts of the building. Temporary measures, such as additional ventilation, may be used to prevent moisture damage during these projects.
- *Building occupancy moisture* — A common cause of moisture problems in buildings is excessive interior humidity. Water vapor may be generated from interior sources, such as swimming pools, laundry facilities, or manufacturing processes that require the use of substantial amounts of water.
- *Climate conditions* — The climate in which a building is located will have a significant effect on the type, direction, and degree of moisture that moves in and out of the building.

If inside moisture is a concern, roof systems will be designed to account for it and prevent its effects on roofing materials. In low-slope systems, vapor retarders are used as the bottom layer of the roof system to prevent moisture from getting into the roof insulation. Buildings with steep-slope roofs may have ventilation systems to prevent moisture from collecting.

2.2.1 Principles of Vapor Movement

Water can exist in three phases: solid (ice), liquid (water), and gas (vapor). The phase of water generally depends on its temperature and pressure. At atmospheric pressure conditions (that is, within the earth's atmosphere), water is generally:

- Solid at temperatures below its freezing point (32°F, or 0°C)
- Liquid at temperatures between 32°F (0°C) and 212°F (100°C)
- Gas at temperatures above its boiling point (212°F, or 100°C)

Water commonly turns from liquid to vapor by evaporation, even when the surrounding ambient temperature is less than the boiling point. Heat energy can be transferred to water molecules and cause them to pass from the liquid phase into the gas phase. When the vapor is cooled, it will lose energy and return to its liquid phase (also known as *condensing*).

When moisture-saturated air is cooled, some of the moisture vapor in the air condenses. This process is referred to as *condensation*. The temperature at which air becomes saturated with moisture vapor and condensation begins to form is referred to as the air's dew point.

If buildings are not properly designed to control the passage of water vapor, it can cause condensation. Water vapor moves across or through barriers by means of two mechanisms: air leakage and diffusion. Air leakage generally allows much greater amounts of water vapor movement than diffusion.

Areas that may not be completely airtight, such as roof penetrations, interruptions, and perimeters, can be areas of significant air leakage. Air leakage occurs due to air-pressure differences across the building envelope. The following factors contribute to these differences:

- *Stack effect* — As warm interior air rises to the top of a building, the differing densities between the warm interior air and cool outside air produces a small amount of pressure at the roof, creating air leakage.
- *Wind pressure* — Pressure from wind can cause air leakage in a roof system, especially because wind speeds are faster on the roof than on the ground.
- *Mechanical ventilation* — Operation of building mechanical equipment contributes small pressure according to whether supply and/or exhaust fans are used to create positive pressure (air supply rate greater than exhaust) or negative pressure (air exhaust rate greater than supply) inside a building.

Diffusion is when water vapor passes through a material. Some materials allow diffusion to occur more rapidly than others. A material's ability to allow diffusion of water is measured by its permeance. Permeance is usually expressed in units called *perms*. Materials with lower perm values are more resistant to the diffusion of water vapor than materials with higher perm values. A material with a perm rating of 0.0 allows virtually no moisture vapor to diffuse through it.

2.2.2 Low-Slope Vapor Retarders

A vapor retarder is a component that can be added to a roof system when there is a reason to believe moisture may come from inside a building and get into the insulation of a roof system. Vapor retarders protect the roof system from damage done by condensation. Vapor retarders must have very low perm ratings.

Vapor retarders are typically constructed of bituminous sheets or plastic film sheets. They fall into the following two categories:

- *Bituminous vapor retarders* — One type of bitumen vapor retarder is made of two plies of fiberglass ply sheets, also called *felts*. The sheet layers are installed parallel to one another with the second layer being centered over the first layer's seams. Bituminous vapor retarders may also be self-adhering and installed using a torch, or they may have built-in adhesive on one side.
- *Plastic vapor retarders* — Plastic vapor retarders are made of plastic sheets. They are 5' to 20' rolls of plastic film and can be hundreds of feet long. Plastic vapor retarders are uncommon, but it is good to have a basic understanding in case you need to install one.

The terms *bitumen* and *asphalt* may be used interchangeably when referring to roofing products.

Installing Bituminous Vapor Retarders

Methods of installing bituminous vapor retarders include the following:

- Adhered with hot-mopped bitumen
- Adhered with cold-applied asphalt
- Torched
- Peel-and-stick (often called *sticker sheets* that include a built-in adhesive on the back and are rolled into place)

The following is a general procedure for installing a vapor retarder on a non-nailable deck using hot-mopped bitumen:

Step 1 Make sure the deck is clean, dry, and in good condition.

Step 2 Use a primer to help the bitumen adhere to the deck.

Step 3 Once the primer is dry, apply a layer of hot-mopped bitumen.

Mopping hot bitumen is a very dangerous procedure. Always use extreme care and wear personal protective equipment (PPE), such as work boots, long pants, and a face shield, to protect yourself from burns when working with hot bitumen.

Step 4 Roll out two layers of felt, continuing to apply the hot-mopped bitumen to the felts.

The bitumen film is the primary vapor-resistant component. The felt simply serves to hold the bitumen in place.

Nailable decks require a base sheet to prevent hot bitumen from flowing through the deck joints. A base sheet also allows for easy removal of the vapor retarder during reroofing without damaging the deck. The following is a general procedure for installing a vapor retarder on a nailable deck using hot-mopped bitumen:

Step 1 Make sure the deck is clean, dry, and in good condition.

Substrates, Decks, and Roof Insulation

Step 2   Mechanically fasten a base sheet using a manufacturer-approved fastener.

Step 3   Apply a layer of hot-mopped bitumen.

Step 4   Roll out two layers of felt, continuing to apply hot-mopped bitumen to the felts.

Installing Plastic Vapor Retarders

Plastic vapor retarders are best installed over continuous substrates to protect against punctures during construction. While plastic vapor retarders are uncommon, it is important to have a basic understanding of how they are installed. They may be installed in either of the following ways:

- *Loose-laid* — The vapor retarder is laid in place on a substrate installed on the roof deck.
- *Adhered* — Using an adhesive applied in a serpentine pattern to adhere the plastic to a substrate installed on the deck.

Installing a plastic vapor retarder directly on a steel deck increases the possibility of puncturing the vapor retarder. The gaps created by the flutes in a steel deck do not fully support the sheet, making it susceptible to puncturing (*Figure 10*). For this reason, cover board is usually installed over the steel deck prior to installing a plastic vapor retarder.

> **CAUTION**
>
> Do not use hot bitumen in installations with plastic vapor retarders. For example, using hot bitumen to adhere insulation would result in bitumen seeping through the joints of the insulation, melting the plastic and destroying the vapor retarder.

2.2.3 Steep-Slope Ventilation Systems

Steep-slope roof assemblies may be located over attics or rafter spaces where ceiling materials have been fastened directly to roof rafters. Insulation may be located on the attic floor or below the roof deck inside the rafter spaces, or spray foam insulation may be applied to the underside of the roof deck.

Depending on the ceiling configuration and insulation type and location, steep-slope roof assemblies may be ventilated or unvented. In ventilated assemblies, the ventilation space is typically adjacent to the underside of the roof deck. In conditioned, unvented attic assemblies, thermal insulation is applied directly to the underside of the roof deck, and the space directly below becomes conditioned interior space. Steep-slope roof assemblies do not typically include above-deck insulation.

Air intake and exhaust vents are used in ventilating steep-slope roof assemblies to provide a means of allowing outside air to enter and exit attic spaces and rafter spaces.

Intake vents are typically located along a roof's lowest eave, at or near soffits or eaves. They are usually installed by a siding or other contractor.

Air exhaust vents are used to allow air in attic spaces and rafter spaces to exit to the exterior. Exhaust vents are usually placed at or near a roof's ridge or high point. Roofers may need to cut holes where exhaust vents will go, install them, and flash around them.

The following are some common types of exhaust vents:

- Ridge vents (*Figure 11*)
- Static vents (*Figure 12*)
- Gable end vents
- Turbine vents
- Powered vents

2.3.0 Insulation and Cover Boards

Insulation and cover boards are often installed on top of the deck in low-slope systems to insulate the top of the building and provide a substrate for applying the roof membrane. Common types of insulation boards used for roofing are shown in *Figure 13*.

Insulation resists heat transfer. All insulation has an R-value, which is a measurement of its thermal resistance, or the amount of resistance it has against heat transfer. The larger the R-value, the better the insulation (*Figure 14*).

Figure 10 The flutes in a steel deck can lead to punctures in a vapor retarder.

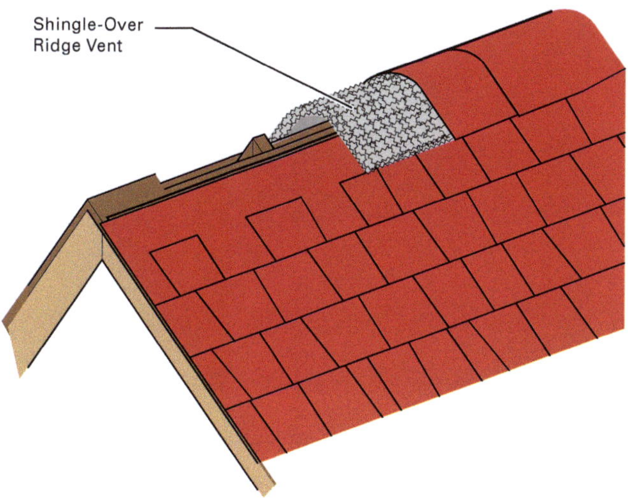

Figure 11 Example of a shingle-over ridge vent.

2.3.1 Insulation Layout and Installation

Insulation must be laid out in a specific pattern on a low-slope roof. The entire field must be covered without gaps. This allows for a stable temperature inside a building, as well as a stable substrate for installing the membrane.

Joints from the first layer of insulation and the second layer should not be lined up. This creates a greater change for heat transfer through the joints. Overlapping the joints from the first layer with the second layer reduces the change for heat transfer (*Figure 15*).

This module covers installation for new construction and assumes the substrate is the roof deck. When installing over an existing system, it is likely that you'll install a layer of cover board first to create a good substrate.

Insulation is attached to the roof in two possible ways:

- *Mechanically attached* — Mechanically attaching means using a fastener, such as a screw or nail, to connect the insulation to the substrate.
- *Adhered* — Adhering means using an adhesive to stick the insulation to the substrate. Common substances used to adhere insulation include low-rise foam, hot-applied asphalt, and cold-applied adhesive.

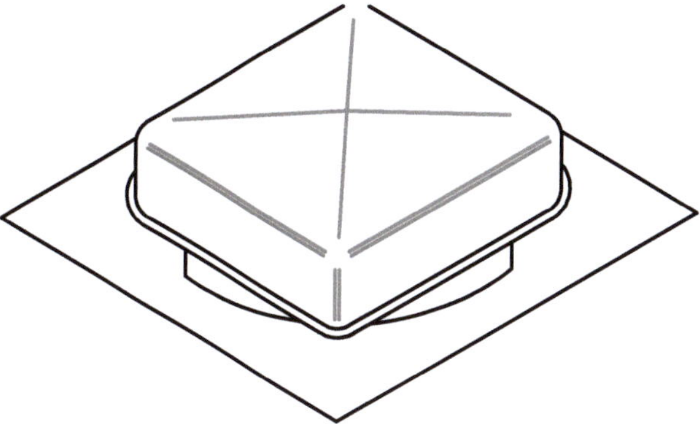

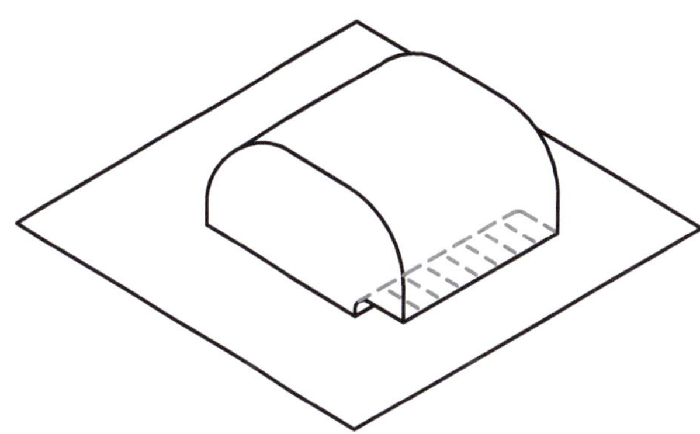

Figure 12 Static vents.

Some insulation is better suited to be mechanically attached or adhered because of the material. However, how insulation is installed often has more to do with the deck material than the insulation. For example, screws and nails cannot be used on non-nailable decks such as concrete; an adhesive must be used instead. Fasteners may also compromise the structural stability of a cementitious wood fiber deck.

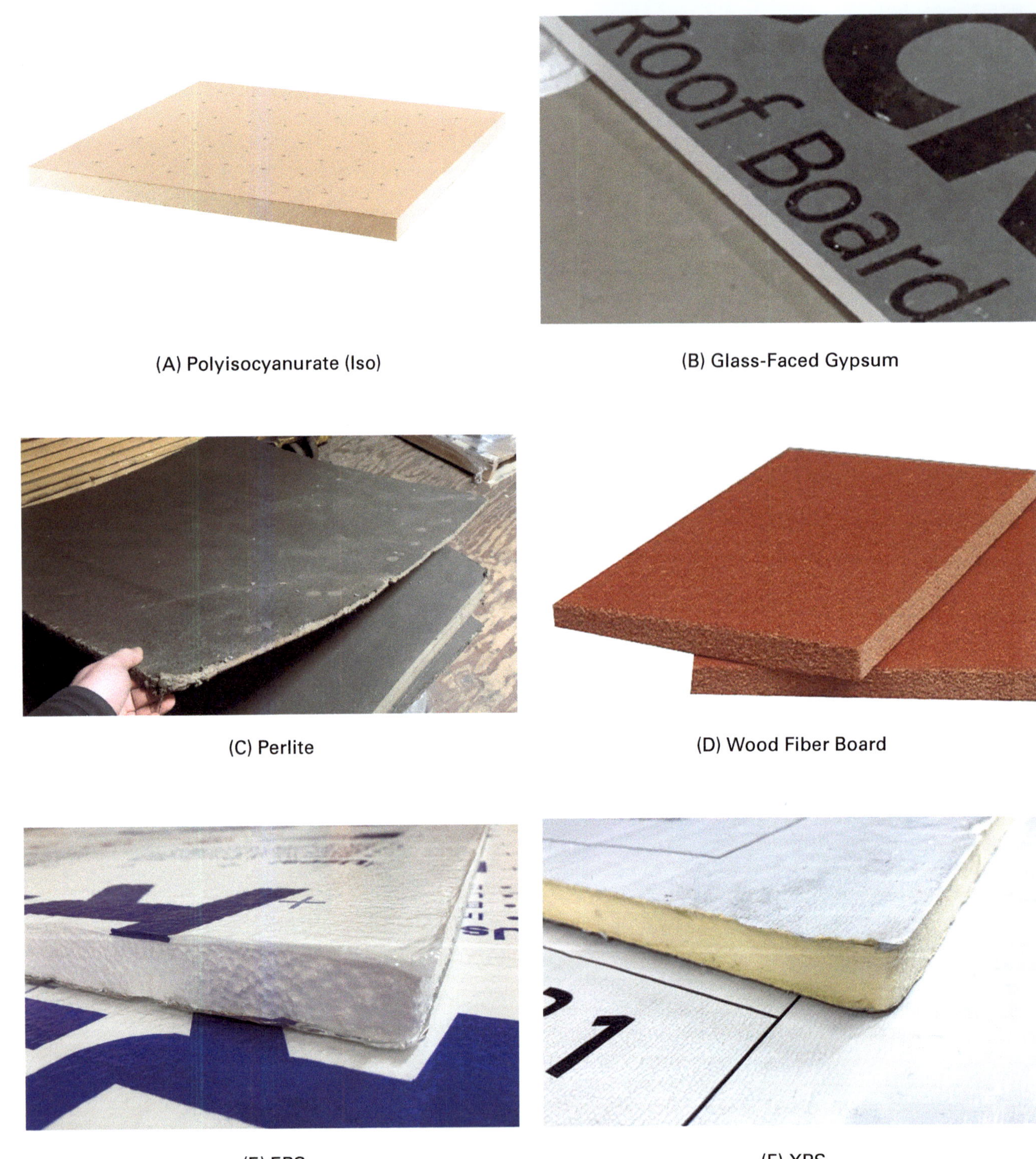

Figure 13 Common types of rigid board insulation used for roofing.

NCCER – *Roofing*

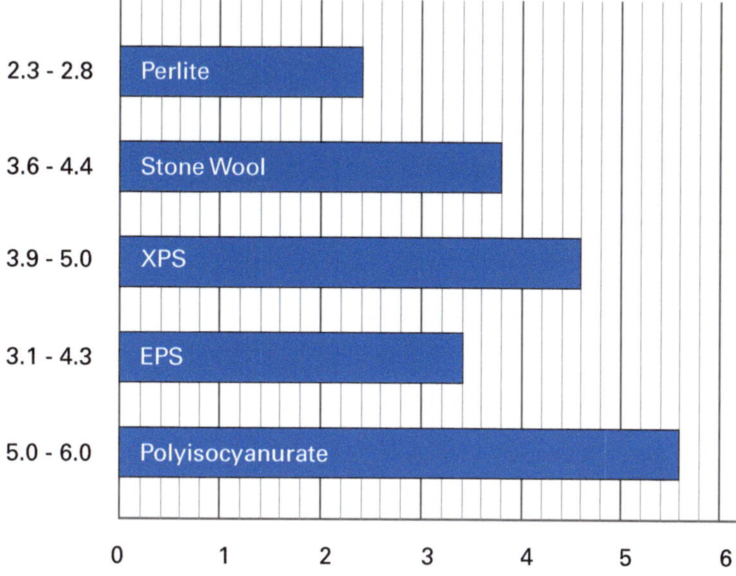

Figure 14 Typical R-values for common insulation materials (per inch of thickness).

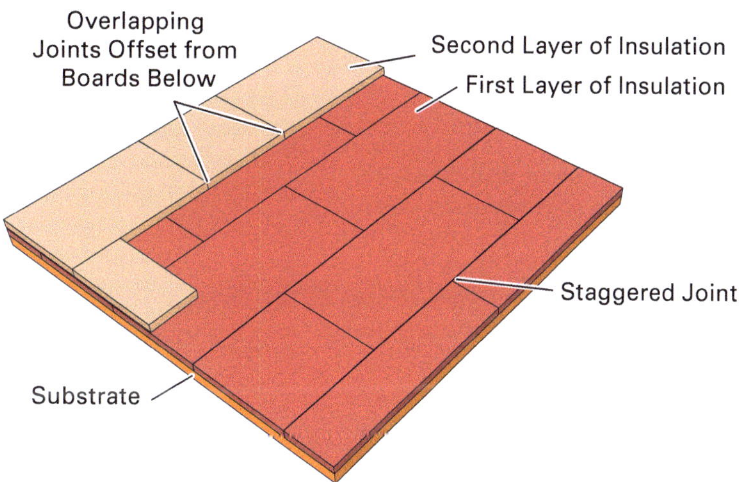

Figure 15 Insulation boards laid out with staggered and offset joints.

Often, on nailable decks such as wood and steel, the first layer of insulation is mechanically attached to the deck, and the second layer is adhered to the first layer.

Mechanically Attaching

Mechanically attaching means using the proper fasteners to attach the insulation to the substrate. Each fastener usually comes with a plate to help spread the load and keep the insulation from pulling up over the fastener (*Figure 16*). For example, the board in *Figure 17* (A) is secured better than the one in *Figure 17* (B). Pulling hard enough on the board without the plates would easily pull the fastener heads through. When a similar force is applied to the board with the plates, it will remain in place.

Adhering with Elastomeric Adhesives

When using adhesives to adhere insulation, always make sure the substrate is clean, dry, and free of contaminants before installing. Adhesives will not stick if either surface is dirty. Concrete surfaces must be cured or dry prior to adhering. Adhesives are not effective on moist surfaces.

There are two kinds of elastomeric adhesives used for roofing:

- *One-component* — One-component adhesives cure when exposed to air. This means roofers must wait between applying the adhesive and installing the insulation. Manufacturers will provide the number of minutes to wait before laying in the boards.

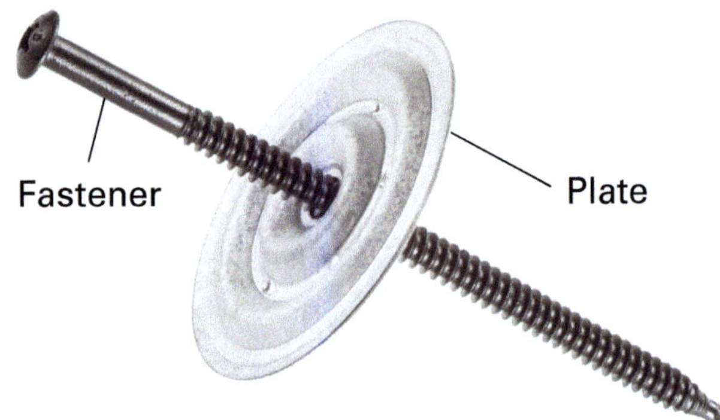

Figure 16 Fastener and plate for mechanical attachment of membrane.

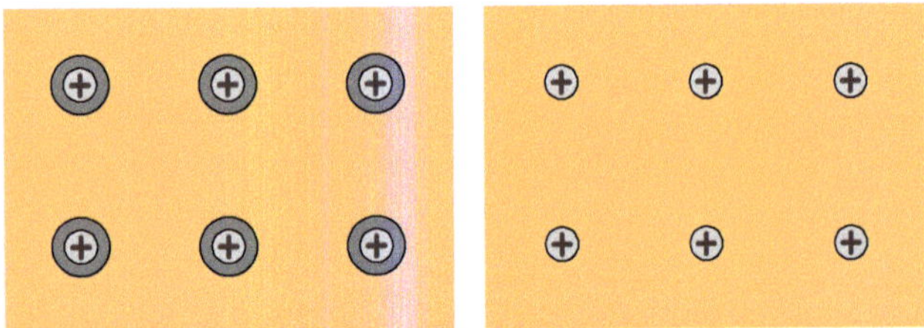

Figure 17 Insulation boards fastened with and without plates.

- *Two-component* — Two-component adhesives, as the name implies, consist of two separate components that are packaged in cartridges. The cartridges are often labeled *Part 1* and *Part 2* (*Figure 18*). The components come together as the adhesive is applied, and the adhesive begins to cure as soon as both components are mixed. This means boards can be laid in immediately. Application tools for two-component adhesives include handheld applicators that operate like caulk guns, wheeled applicators, and mixing carts.

Foam adhesive is applied in ribbons specified by manufacturer instructions. An example of manufacturer's instructions for ribbon spacing is shown in *Figure 19*. Field applications are typically spaced 12" on center (O.C.), meaning there are 12" between the center of one ribbon and the center of the next ribbon. Recommendations for ribbon spacing in perimeter and corner roof areas range from 4" to 6" on center, with 6" on center recommended most often.

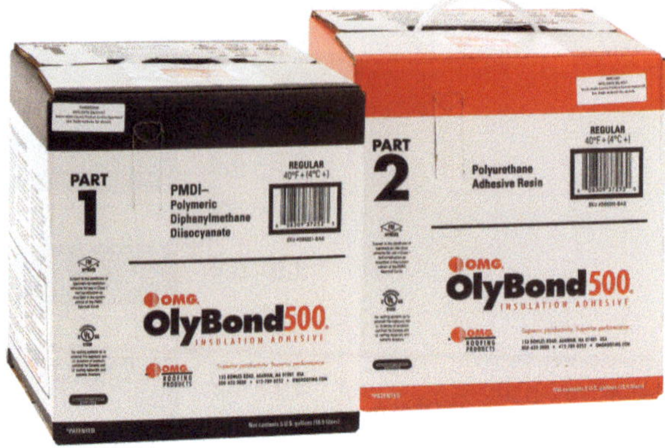

Figure 18 Two-component adhesive.

Adhering with Bitumen

A common method of installing insulation is by adhering it with bitumen. Bitumen can either be hot-mopped (*Figure 20*) or cold-applied (*Figure 21*). This section focuses on hot-mopped bitumen because of the many safety considerations involved.

Bitumen is heated into molten liquid in a kettle. The hot bitumen is pumped from the kettle, through an insulated supply pipe, and up to the roof where it is poured into luggers or mop buckets to be used.

WARNING!

Operating a kettle requires special training and special attention to safety.

Hot bitumen is applied to the roof surface using mops. Bitumen can be mopped into a solid coat to provide 100-percent coverage of the roof surface, or it can be spot mopped.

Hot-bitumen installations are usually done in a team, where one worker mops while others drop the insulation boards behind them. Once the insulation boards have been dropped in, someone will step on them. Depending on the substrate materials, hot asphalt application may require priming beforehand.

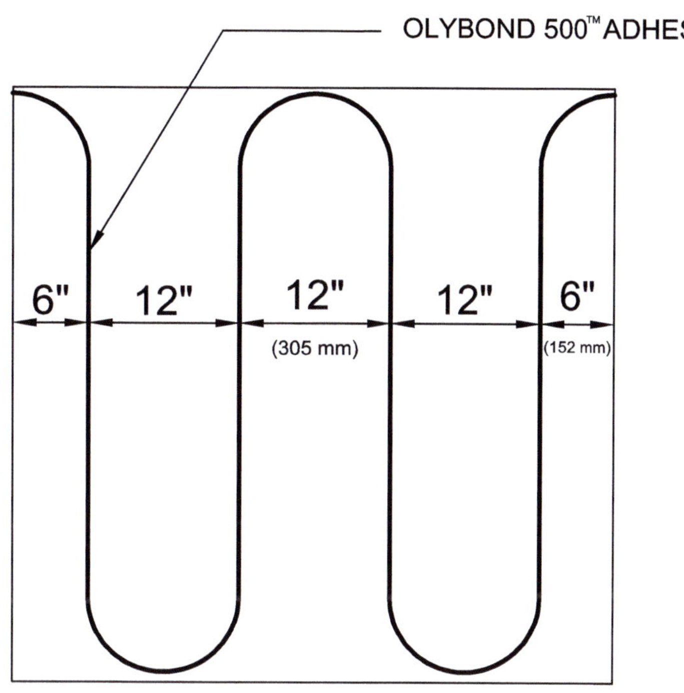

Figure 19 Example of a manufacturer's adhesive fastening pattern.

Figure 20 Adhering insulation with hot-mopped bitumen.

Figure 21 Cold-applied bitumen.

Observe the following safety considerations when working with hot bitumen:

- Always have two fully charged fire extinguishers nearby.
- Use PPE to protect your skin, including gloves and protective clothing. Heated bitumen is around 500°F (260°C) and can cause severe burns.
- If you get heated bitumen on your skin, seek help immediately.
- Be very careful when moving a lugger full of molten liquid. Full luggers are very heavy, making them difficult to move and easy to spill. Always push the lugger instead of pulling.
- Be careful to prevent spills when using wheeled or handheld buckets to transport hot bitumen.

2.3.2 Tapered Insulation

Tapered insulation is prefabricated to be thicker on one end to create a slope when placed on a flat surface. It can be used to move water toward drains.

It comes in pieces, each labeled according to where they should be placed in sequence. They are shaped to fit together and create a smooth slope.

Figure 22 shows an example of a tapered insulation set that includes letters AA, A, B, and C. Pieces labeled AA are the thinnest, and pieces labeled C are the thickest. The thinnest pieces are placed closest to drainage, and the thickest pieces are farthest from drainage.

When tapered insulation is needed, the roofing contractor typically will reach out to the manufacturer for a tapered insulation plan. They will send in the full set of construction drawings (including the roof plan) to indicate the intended slope, low points, and locations of gutters, scuppers, and drains. The manufacturer uses special software to convert this information into a tapered insulation plan and determine the quantity of tapered insulation panels needed in the system. An example of a tapered insulation plan is shown in *Figure 23*.

Manufacturers typically generate this plan and quote for free. Then, the roofing contractor purchases the tapered insulation materials needed from the manufacturer. Manufacturers may also provide training for installing the tapered insulation.

The tapered insulation plan shows where to install the insulation pieces. Roofers will have to snap chalk lines in locations where slopes will intersect and then cut the insulation to fit. Like insulation boards, tapered insulation should be installed with staggered joints for increased stability and thermal resistance.

When drainage or drains are not at the edge of a slope, tapered insulation may be used to create a counter-slope to prevent ponding below the drain, as shown in *Figure 24*. Tapered insulation is also used to create crickets and saddles that divert water out of potential ponding areas.

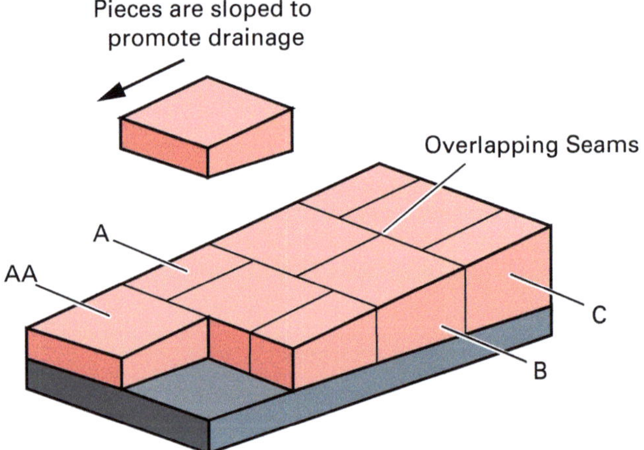

Figure 22 Example of tapered insulation.

2.3.3 Cover Board Installation

Cover board is a thin layer that serves as the membrane substrate. It is good practice to install cover board over insulation before installing a membrane.

Many cover boards are also insulation boards. However, this layer is considered to be separate from the insulation layer. There are several types of suitable cover boards, including the following:

- Glass-faced gypsum board
- Fiber-reinforced gypsum board
- Stone wool
- Perlite board
- High-density polyisocyanurate board
- Wood fiber board

The type of cover board used will be determined and specified by the roof system designer. Cover board serves the following purposes in a roof system:

- Provides a smooth substate and protects the roof membrane from uneven areas or irregularities in the insulation
- Protects insulation from impact damage, such as hail or dropped tools
- Increases the roof system's resistance to wind uplift
- Acts as a fire barrier in the event of external fires
- Allows for offsetting and staggering of board joints, especially if only a single layer of insulation was installed

Cover boards may be fastened to the deck or adhered to insulation with adhesive or hot asphalt. Just like insulation layers, always stagger the joints when installing cover board. Avoid aligning cover board joints with the joints of the top layer of insulation. This will help minimize heat flow through board joints and help evenly distribute mechanical stresses in the roof system.

> **NOTE**
> Always follow manufacturer instructions to get the correct adhesive or number and spacing of fasteners.

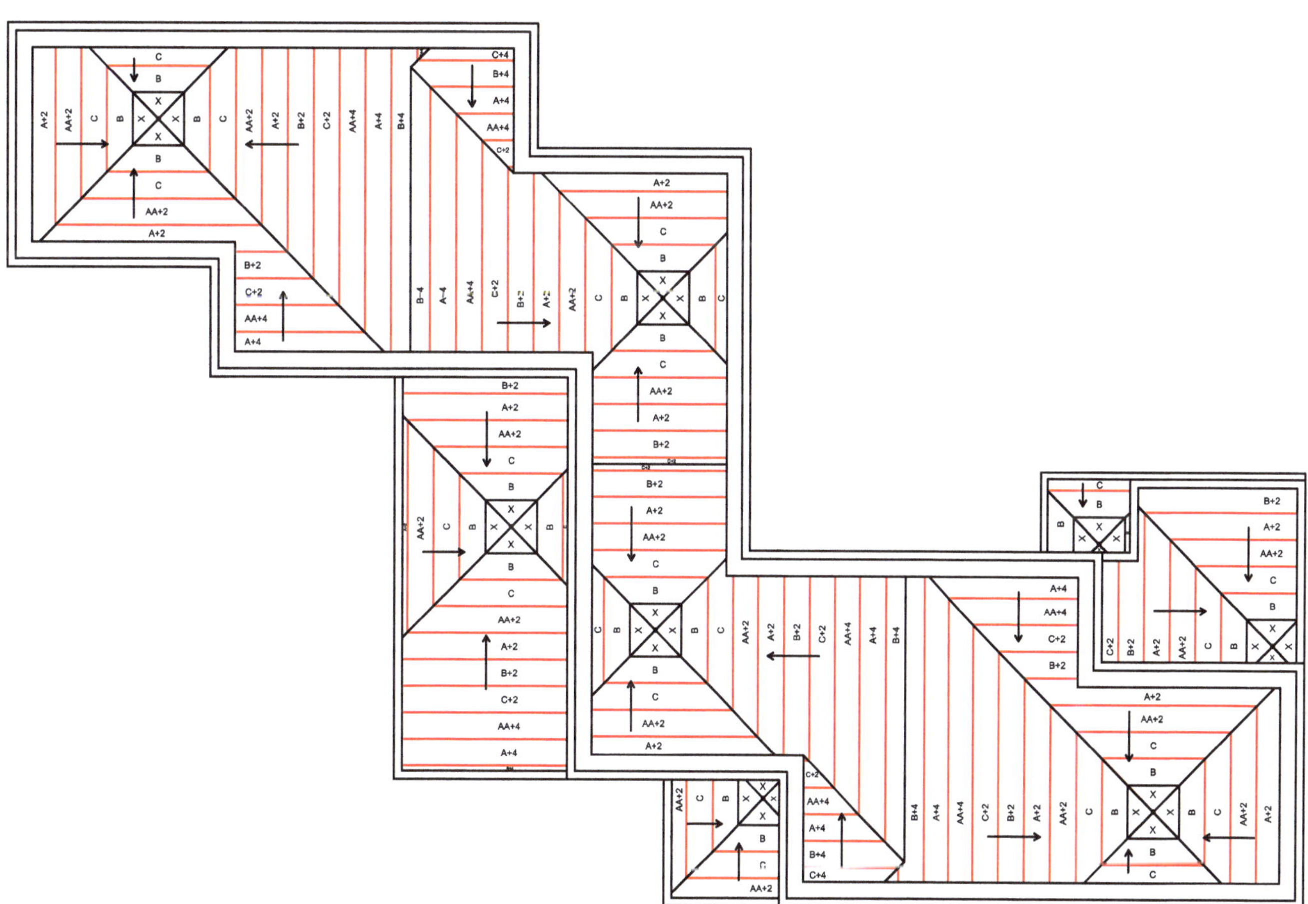

Figure 23 Tapered insulation plan.

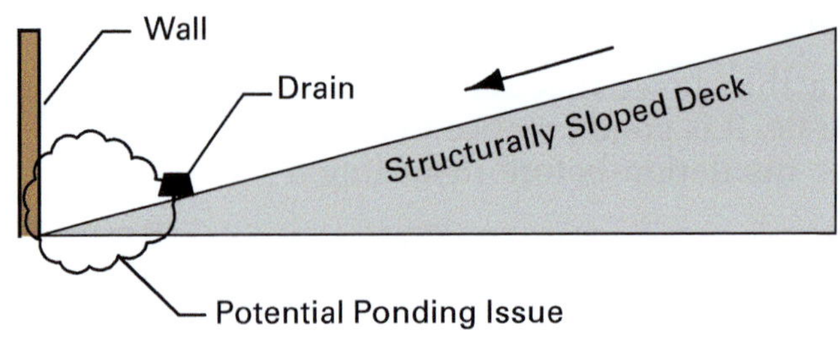

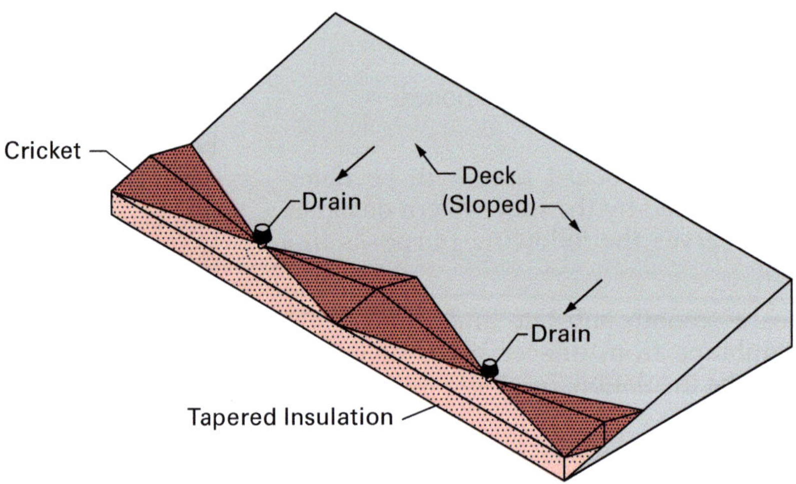

Figure 24 Tapered insulation used for a counter-slope.

2.3.4 Handling and Storage

Insulation boards must be handled and stored with care to prevent them from breaking, collecting moisture, becoming damaged, or blowing away in the wind.

Pallets of insulation are usually transported to the roof with a hoist. Only specially trained riggers can accept loads on a rooftop.

Because insulation is easy to damage, all roofers must be careful to protect the sides and edges during transportation.

Be prepared after unwrapping a load of insulation boards. They can easily catch the wind and fly away. If you are carrying a board, it can act as a sail and possibly force you off the roof. Wind can also blow a board and cause it to hit an unsuspecting person working on the roof. It can also blow off the roof and hit people on the ground or land in traffic.

Take the following precautions when storing insulation overnight on the roof or on the ground:

- Keep insulation on the shipping pallets and in the manufacturer's shipping wrap.
- Cover insulation with a moisture-resistant tarp and weigh the stack down.

1. The transfer of heat through movement of warm air to a cooler place is called _____.

    a. infrared
    b. conduction
    c. convection
    d. radiation

2. Which of the following is used to protect roof insulation from moisture inside the building?

    a. Cover board
    b. Radiation
    c. Expansion joints
    d. A vapor retarder

3. Which of the following roof system components is used to resist heat transfer?

    a. Insulation
    b. Membrane
    c. Asphalt
    d. Bitumen

1. A substrate is _______.
   a. any surface upon which a membrane is applied
   b. on top of the roofing membrane
   c. always underneath the roof deck
   d. rated for its permeability

2. Which of the following deck types is non-nailable?
   a. Wood panels
   b. Wood planks
   c. Steel
   d. Structural concrete

3. Which of the following is the *most* common material for steep-slope decks?
   a. Concrete
   b. Cementitious wood fiber panels
   c. Steel
   d. Wood

4. Which of the following is used to minimize the effects of deck movement due to weather and temperature changes?
   a. Cover boards
   b. Vapor retarders
   c. Expansion joints
   d. OSB

5. In buildings with steep-slope roofs, insulation is usually _______.
   a. below the roof deck
   b. under the shingles
   c. sprayed over the deck
   d. installed by roofers

6. Water is solid at what temperature?
   a. Below 32°F (0°C)
   b. 50°F (10°C)
   c. 122°F (50°C)
   d. Above 212°F (100°C)

7. When installing a vapor retarder on a nail-able deck using hot-mopped bitumen, you should _______.
   a. use only one layer of felt
   b. fasten a base sheet before hot mopping a layer of bitumen
   c. fasten a base sheet after hot mopping a layer of bitumen
   d. apply a primer to the deck surface

8. Two-component adhesives _______.
   a. should not be used on concrete decks
   b. are not recommended for roofing
   c. take a long time to cure
   d. cure when both components are mixed

9. Which of the following is thicker on one end to create a slope when placed on a flat surface?
   a. Tapered insulation
   b. Glass-faced gypsum
   c. Underlayment
   d. Low-rise foam

10. A tapered insulation plan is *usually* generated by a _______.
    a. roof installer
    b. roofing contractor
    c. manufacturer
    d. building owner

Fill in the blank with the correct term learned from your study of this module.

1. The natural change of liquid to vapor at a temperature below its boiling point is called ________________.

2. An individual capable of identifying hazards and has the authorization to correct them is called a(n) ________________.

3. Water in gas form that is diffused into the atmosphere is called ________________.

4. A type of visible light energy that is emitted by all objects is ________________.

5. A craftworker who constructs, assembles, installs, and repairs structures made of wood is called a(n) ________________.

6. The temperature at which air becomes saturated with water vapor and has a relative humidity of 100 percent is called the ________________.

7. The conversion of water vapor to liquid as the temperature drops or atmospheric pressure rises is called ________________.

Trade Terms

Carpenter
Competent person
Condensation
Dew point

Evaporation
Infrared
Water vapor

Cornerstone of Craftsmanship

Thomas R. Shanahan

VP of Enterprise Risk Management and Executive Education
National Roofing Contractors Association (NRCA)

How did you choose a career in the industry?

I came to work for NRCA because it is a not-for-profit business. My background placed me as its risk manager dealing with health and safety issues. I have worked at NRCA for 32 years now, mainly because the roofing industry is a wonderful family that takes care of its members and allows for career growth in ways you wouldn't think possible unless you were a part of it. Best decision ever!

What types of training have you been through?

NRCA has supported my development just like roofing contractors support their teams. That's why I've completed two graduate programs.

How important is education and training in construction?

It is extremely important. Roofing has developed into a sophisticated operation due to the variety and complexity of roof systems installed in today's market. It is both exciting and challenging.

What kinds of work have you done in your career?

In my role, I do a lot of safety and health and leadership training for the industry.

What do you enjoy most about your job?

Working with the wonderful professionals in the roofing industry.

What factors have contributed most to your success?

An environment where learning and growing are encouraged if that is what you want for your career.

What advice would you give to those new to the field?

Be curious and work hard—there are a lot of different career opportunities you will find in this dynamic industry.

Interesting career-related fact or accomplishment?

I've been a part of a number of teams that have developed unique, industry-specific educational programs where so much good has come out of them. This includes programs where accidents have been reduced, students become better managers and leaders, and—most importantly—people go home safe.

How do you define craftsmanship?

Through study and experience, the learned ability to create. In roofing, it is the ability to affect a professional roofing installation that enhances the beauty of the building and the safety of its occupants and their possessions.

Trade Terms Introduced in This Module

Carpenter: A craftworker who constructs, assembles, installs, and repairs structures made of wood or other materials.

Competent person: As defined by OSHA, an individual who is capable of identifying existing and predictable hazards in the surroundings or working conditions which are unsanitary, hazardous, or dangerous to employees, and who has the authorization to take prompt corrective measures to eliminate such hazards.

Condensation: The conversion of water vapor or other gas to liquid phase as the temperature drops or atmospheric pressure rises.

Dew point: The temperature at which air becomes saturated with water vapor and has a relative humidity of 100 percent.

Evaporation: The natural change of liquid to vapor at a temperature below its boiling point.

Infrared: A type of invisible light energy that is emitted by all objects. This type of energy is most effective at producing radiant heating when absorbed.

Water vapor: Water in a vapor (gas) form, especially when below the boiling point and diffused in the atmosphere.

Additional Resources

This module presents thorough resources for task training. The following reference material is suggested for further study.

GAF Roofing. "Installing a Tapered System | Roofing it Right with Dave & Wally by GAF." YouTube video. 4:35. April 11, 2020. **www.youtube.com**.

GAF Roofing. "Staggered Insulation Board Layout Detail | TPO Commercial Roofing | GAF Drawing 120." YouTube video. 1:01. March 19, 2019. **www.youtube.com**.

Hunter Panels Roof and Wall Polyiso. "Tapered 101." YouTube video. 10:07. June 20, 2017. **www.youtube.com**.

National Roofing Contractors Association (NRCA), **www.nrca.net**.

Figure Credits

Section Review Answer Key

SECTION 1.0.0

Answer	Section Reference	Objective
1. c	1.0.0	1
2. b	1.2.0	1b

SECTION 2.0.0

Answer	Section Reference	Objective
1. c	2.1.2	2a
2. d	2.2.2	2b
3. a	2.3.0	2c

This page is intentionally left blank.

NCCER CURRICULA — USER UPDATE

NCCER makes every effort to keep its textbooks up-to-date and free of technical errors. We appreciate your help in this process. If you find an error, a typographical mistake, or an inaccuracy in NCCER's curricula, please fill out this form (or a photocopy), or complete the online form at **www.nccer.org/olf**. Be sure to include the exact module ID number, page number, a detailed description, and your recommended correction. Your input will be brought to the attention of the Authoring Team. Thank you for your assistance.

Instructors – If you have an idea for improving this textbook, or have found that additional materials were necessary to teach this module effectively, please let us know so that we may present your suggestions to the Authoring Team.

NCCER Product Development and Revision

13614 Progress Blvd., Alachua, FL 32615

Email: curriculum@nccer.org
Online: www.nccer.org/olf

❏ Trainee Guide   ❏ Lesson Plans   ❏ Exam   ❏ PowerPoints   Other _______________________

Craft / Level: ____ Copyright Date: _______________

Module ID Number / Title: ____

Section Number(s): ____

Description: ____

____

____

____

____

Recommended Correction: ____

____

____

____

____

Your Name: ____

Address: ____

____

Email: ____ Phone: _______________________

This page is intentionally left blank.

Sheet Metal in Roofing

OVERVIEW

Most roof systems contain essential components that are made from sheet metal. These components may be provided by a system's manufacturer, custom-fabricated in a shop, or created on-site during a roofing project. Roofing professionals must be familiar with types and characteristics of sheet metal used in roof systems. They must also be able to identify metal components in roof systems and tools used to create and install them.

Module 16107

Trainees with successful module completions may be eligible for credentialing through the NCCER Registry. To learn more, go to **www.nccer.org** or contact us at 1.888.622.3720. Our website, **www.nccer.org**, has information on the latest product releases and training.

Your feedback is welcome. You may email your comments to **curriculum@nccer.org**, send general comments and inquiries to **info@nccer.org**, or fill in the User Update form at the back of this module.

This information is general in nature and intended for training purposes only. Actual performance of activities described in this manual requires compliance with all applicable operating, service, maintenance, and safety procedures under the direction of qualified personnel. References in this manual to patented or proprietary devices do not constitute a recommendation of their use.

16107 V1.0

From *Roofing, Trainee Guide*. NCCER.
Copyright © 2021 by NCCER. Published by Pearson. All rights reserved.

SHEET METAL IN ROOFING

Objectives

Successful completion of this module prepares you to do the following:

1. Explain the basic properties of metals and identify metal components found in roof systems.
   a. Understand basic physical properties of metals.
   b. Identify types of sheet-metal components commonly found in roof systems.
2. Describe tools and basic installation methods used in sheet-metal roofing.
   a. Identify tools and equipment used for sheet-metal roofing components.
   b. Explain some common procedures for handling sheet-metal roofing components.

Performance Tasks

Under supervision, you should be able to do the following:

1. Use aviation snips to cut a 6" (15 cm) circular hole into a piece of sheet metal.

   - Measure the hole.
   - Draw a circle with a compass.
   - Cut the hole with aviation snips.

2. Size, fit, and cut two pieces of L-style metal edge.

Trade Terms

Cleats	Mil
Corrosion	Oil canning
Diameter	Oxidize
Galvanic corrosion	Patina
Joinery	Thermal expansion

Industry Recognized Credentials

If you are training through an NCCER-accredited sponsor, you may be eligible for credentials from NCCER's Registry. The ID number for this module is 16107. Note that this module may have been used in other NCCER curricula and may apply to other level completions. Contact NCCER's Registry at 1.888.622.3720 or go to **www.nccer.org** for more information.

Contents

This page is intentionally left blank.

1.0.0 METAL IN ROOF SYSTEMS

Objective

Explain the basic properties of metals and identify metal components found in roof systems.

- a. Understand basic physical properties of metals.
- b. Identify types of sheet-metal components commonly found in roof systems.

Trade Terms

Cleats: Continuous metal strips, or angled pieces, used to secure metal components.

Corrosion: The breakdown or destruction of a material, especially metal, caused by chemical reactions. The most common form of corrosion is rust, which occurs when iron combines with oxygen to create iron oxide.

Galvanic corrosion: A form of corrosion (rusting) resulting from the creation of an electric current flowing between dissimilar metals in contact with each other.

Mil: A unit of measurement equal to one-thousandth of an inch, or 0.001" (0.0254 mm).

Oil canning: Physical distortions in the flatness of metal. This condition only effects the appearance of the metal and does not have negative effects on structural integrity.

Oxidize: To combine with oxygen, such as in burning (rapid oxidation) or rusting (slow oxidation).

Patina: A green or brown oxide film that develops on the surface of copper and copper alloys.

Thermal expansion: The increase in the dimension or volume of a body because of temperature variations.

M any essential components and accessories in roof systems are made with sheet metal (*Figure 1*). These components may be premade in a shop before roofing work begins. Roofing professionals must be able to recognize these components, understand how they fit together, and be able to adjust and install them in a roof system. When working with sheet metal components, you must be familiar with different types of metals, their basic characteristics, and how they interact with one another and the environment.

1.1.0 Characteristics of Metal

Metal roof-system components may be made out of a variety of metals, including steel, stainless steel, aluminum, copper, lead, and zinc. All metals oxidize when exposed to weather. Sometimes this oxidization protects the metal, and other times it results in unwanted corrosion and damage.

In addition to reacting to air, weather, and temperature, metals also react to other metals. When roofing professionals combine different types of metal in the same system, components must be carefully selected and matched to avoid corrosion.

1.1.1 Thickness and Gauge

Metal is usually considered sheet metal when it is less than $\frac{1}{4}$" (6 mm) thick. Thicker metal is considered plate metal rather than sheet metal.

The tool used to measure the thickness of sheet metal is called a *sheet metal gauge* (*Figure 2*). On one side, the gauge number is next to each slot on the gauge. Decimal and fractional values for each

Figure 1 Steel sheets.

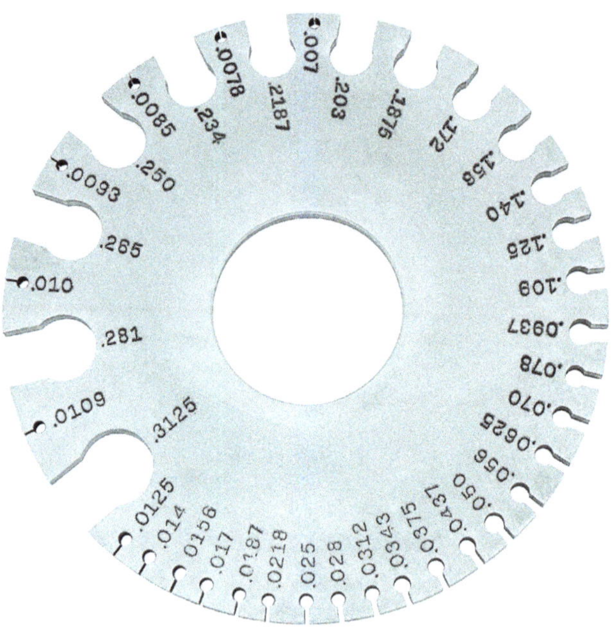

Figure 2 US standard sheet metal gauge.

thickness are indicated on the other side. To measure sheet metal, find the slot on the gauge that the sheet fits into best (snug, but not too tight or loose).

For steel, sheet metal thickness is typically indicated by gauge numbers. Each assigned gauge number represents a specific metal thickness. The lower the gauge number, the thicker the sheet. A table showing thickness and weight in pounds per square foot (lb/ft^2) for each type of sheet metal, by its gauge, is provided in *Appendix A*. In general, 24-gauge sheet metal is the most common type used in roofing components.

> **NOTE**
>
> When measuring coated metals, the actual gauge of the underlying sheet metal is generally one gauge thinner than that indicated on the measuring tool.

The thickness of some sheet metal materials, such as aluminum and zinc, is measured in mils. A **mil** is a unit of measurement equal to one-thousandth of an inch, or 0.001" (0.0254 mm). The thickness of these materials is measured with a micrometer (*Figure 3*).

1.1.2 Thermal Expansion

Metal expands and contracts when exposed to changing temperatures. This is referred to as **thermal expansion**. Some metals expand more than others. The expansion coefficient of a metal is a number indicating how much the metal expands per degree of change in temperature.

If not taken into consideration, thermal expansion can cause serious problems in roof systems. Thermal expansion must be considered in a system's fastening techniques and expansion joints. Expansion and contraction of sheet metal components cause movements that may lead to tears in the roof membrane, loosening of fasteners, or other roof-system damage. Some methods used to reduce the harmful effects of thermal expansion include the following:

- Securing flashings with clips that are designed to allow thermal movement (*Figure 4*)
- Using stiffening ribs (*Figure 5*), V-grooves, or cross breaks

1.1.3 Corrosion and Dissimilar Metals

Metals oxidizing when exposed to water is a form of corrosion. When two different metals come in contact and are exposed to water, one of

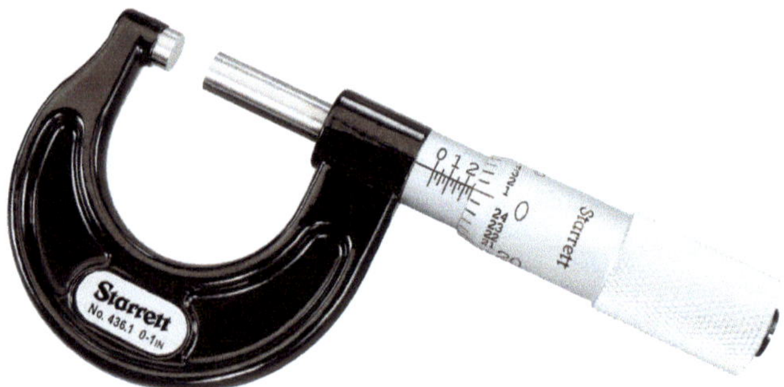

Figure 3 Micrometer.

NCCER – *Roofing*

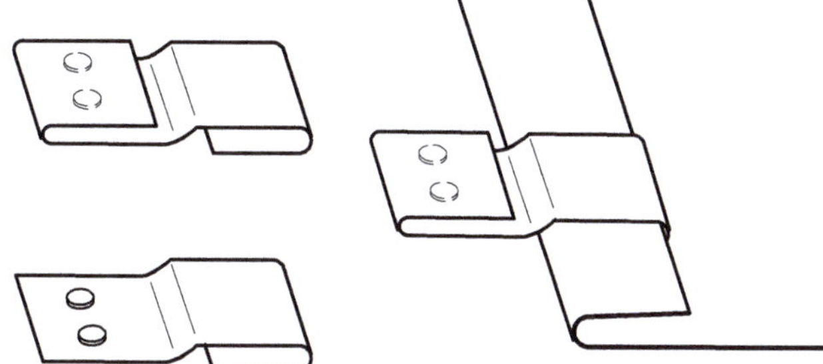

Figure 4 Metal clips.

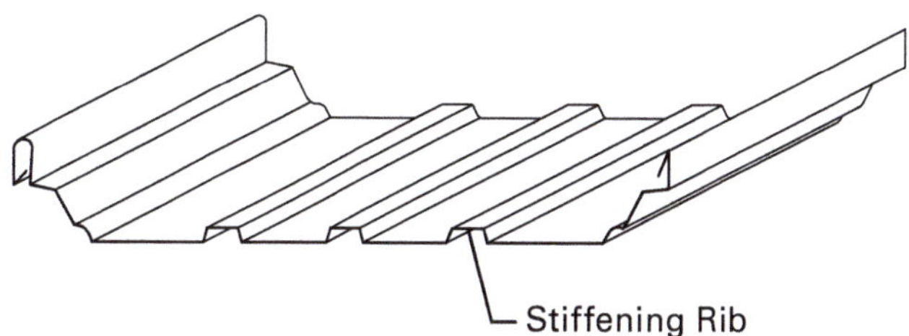

Figure 5 Stiffening ribs.

the metals can undergo increased corrosion while the other metal shows decreased corrosion. This reaction is called galvanic corrosion. *Figure 6* shows corrosion on a metal roof.

When different metals must be used together in a roof system, they must be matched carefully to avoid galvanic corrosion. For example, avoid pairing steel and aluminum together since steel causes aluminum to corrode quickly. If these metals are used together in a roof system, the structural integrity of that system could be compromised.

1.1.4 Protective Coatings

Sheet metal is often coated to protect it from rust and corrosion. Steel that is coated with a protective layer of zinc is called *galvanized steel*. When the coating is properly applied, it can generally protect the base metal from corrosion for 15 to 30 years or more, depending on the environment.

Paints or coatings are sometimes used to protect various types of base metals, as well as to enhance the metal's appearance. Paints and coatings may be applied in a factory or in the field. It is important to apply any necessary pretreatment material before painting or coating.

1.1.5 Types of Metal

Sheet metal comes in a variety of different base metals (*Figure 7*). Sheet metal is selected based on its cost as well as its compatibility with the roof system, local building codes, environment, and other materials used. Common types of sheet metal used in roofing include the following:

- *Galvanized steel* — Galvanized steel is steel coated with zinc. It is one of the oldest and most common metallic-coated metals. The zinc coating serves as a protective layer that adds weatherability. Galvanized steel oxidizes at different rates depending on coating weight and the environment. Environmental factors include moisture and humidity, the salt content of the moisture (proximity to the ocean), and industrial pollution. The more severe the conditions, the more rapid the loss of zinc, which means a more rapid loss of protection.
- *Stainless steel* — Stainless steel is a corrosion-resistant material, often used in harsh environments. It generally does not stain adjoining metals. Stainless steel is available in many finishes, from dull matte to highly polished. Stainless steel is designated by gauge.

Figure 6 Corrosion on a metal roof.

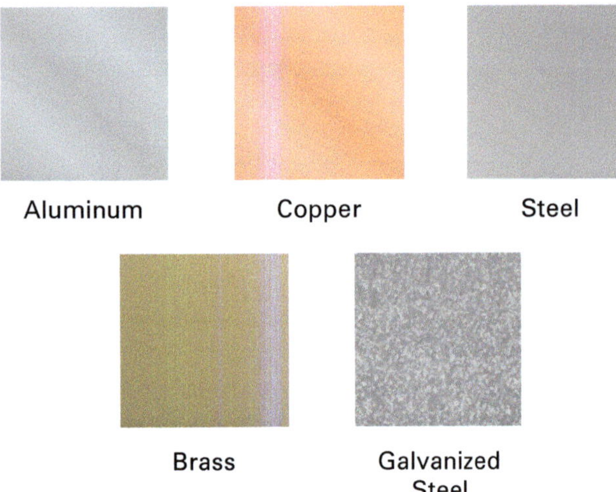

Figure 7 Examples of sheet-metal types.

- *Aluminum* — Aluminum is a lightweight, easily formed metal that does not require a protective coating in most cases. Aluminum develops a thin oxide film on the surface that is impermeable to many airborne contaminants. Aluminum has a high expansion coefficient compared to most roofing metals.

- *Copper* — Copper does not require a protective finish. Over time, copper develops a bronze, brown, or blue-green color, depending on geographic location. This color results from a protective layer of copper sulfate, referred to as **patina**, that forms on copper when it is exposed to the elements. Water runoff from copper can produce a patina-colored stain on light-colored building materials like concrete, so it is often recommended that water runoff from copper be directed to gutters and downspouts.

- *Lead* — Lead is used as a roofing accessory metal because of its durability and workability. Lead is extremely soft and can be formed by hand, making it useful for flashing irregular shapes and junctures. For example, lead is common in low-slope bituminous roof systems as a flashing material for internal roof drains. Sheet lead is designated by weight, not gauge. A weight range of 2 lb/ft^2 to 4 lb/ft^2 is typical for roof system applications, but heavier weights may be needed when soldering is required.

- *Zinc* — Zinc is a self-healing material that weathers to a blue-gray patina that protects it from corrosion. It is both durable and flexible. Zinc is used for drainage components, roof coverings, and wall cladding. To avoid adverse reactions, it should not be paired with copper or come into contact with water runoff from copper or certain wood species. Zinc is specified by gauge.

Galvanic Series

Galvanic corrosion occurs when an electrical current generates between two different metals (an anodic metal and a cathodic metal). This causes an oxide layer, such as rust, to form more quickly on the surface of the anodic metal.

A common method of predicting the effects of galvanic corrosion between metals is to list the metals in order of their corrosion potentials (from more anodic to more cathodic). This list is called a *galvanic series*. By observing the relative position of the two metals in the series, it is possible to predict which of the metals will experience more corrosion.

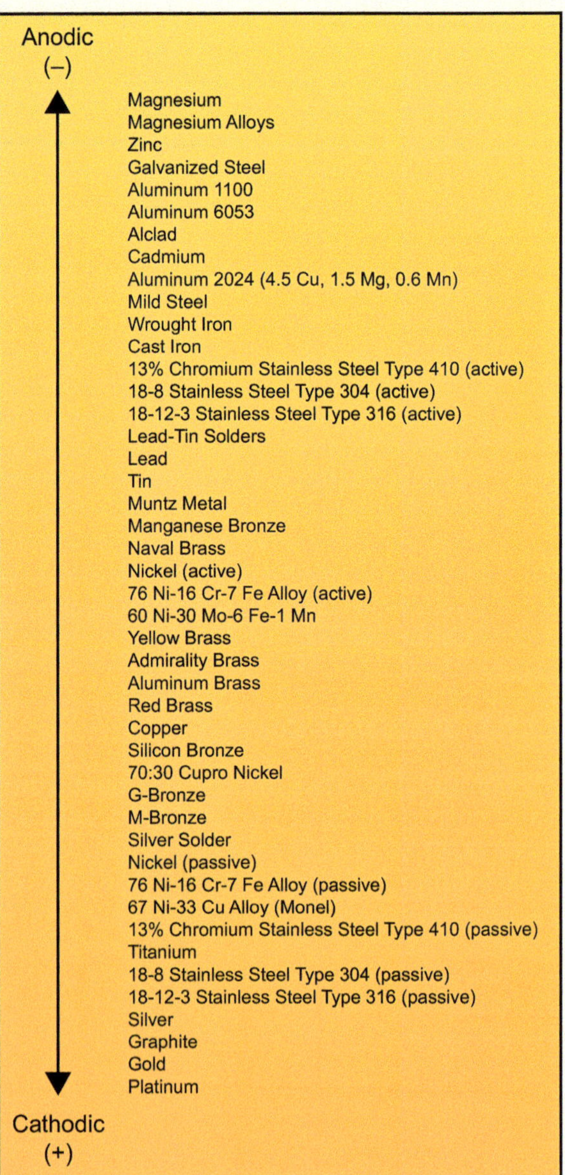

Did You Know?

Metal Cladding

Cladding bonds together two dissimilar metals. Cladding may be coated with PVC or TPO for weatherproofing purposes. Other trade professionals sometimes provide and install metal-cladding components on a roof system.

1.2.0 Metal Components in Roofing

Metal components in roof systems include copings, gutters, downspouts, scuppers, leader heads, and many types of flashings. Flashings are installed at roof edges, walls, and penetrations or interruptions in the roof field to divert water and prevent leaks. Flashings are often made of sheet metal, although they can also be made of flexible membrane material (such as rubber or PVC).

All roof systems are designed to prevent water from entering the building through the roof. If water pools or gets into seams on the roof, it can cause damage and corrosion to the roof system, leading to leaks and even structural damage.

Water always runs downhill. Anything that gets in its path is vulnerable to damage, corrosion, or leaks. Metal drainage components are designed and configured to ensure water is moved off the roof to protect the building from leaks. Drainage components installed by roofers include scuppers, drains, gutters, and downspouts.

1.2.1 Perimeter Edge Flashings

In steep-slope roof systems, drip edge metal is used to provide a continuous finished edge along the outer edges of a roof system. It typically provides a termination point for underlayment materials and metal panels.

The most common shapes of drip edge metal are T- and L-type metals (*Figure 8*). A T-type drip edge is shown in *Figure 9*.

In low-slope roof systems, perimeter edge metal is used to provide a continuous finished edge along the outer edges of a roof system. It provides a termination point and often secures the perimeter edge of the roof membrane. It can also help hold loose gravel or ballast at the edge of the roof.

Examples of perimeter edges are shown in *Figure 10*. The A-type profile, commonly called a *gravel stop*, is often used in low-slope applications. The height of the cant dam (the part that extends up vertically from the roof edge) varies depending on the type and thickness of ballast used. The L-type profile is commonly used in roof systems without an aggregate surfacing or when a smooth surface is desirable. The T-type profile is often used with metal panel roof systems. T-type edge metal is typically overlapped.

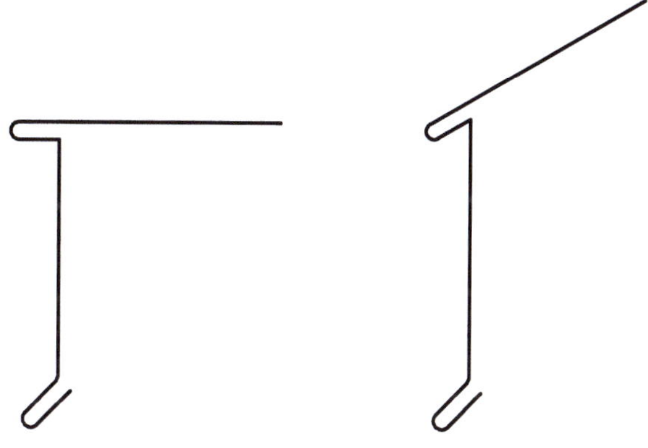

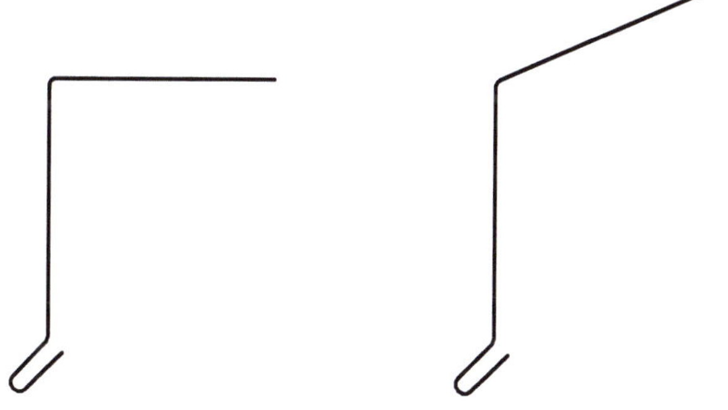

Figure 8 Common drip edge metal profiles.

Figure 9 T-type drip edge metal.

Perimeter edge or fascia metal with a large, flat face may be subject to **oil canning**. Stiffening ribs or heavier-gauge sheet metal can reduce this possibility.

Cleats are continuous metal strips, or angled pieces, used to secure metal components. Cleats may be used in the installation of the perimeter edge in combination with coping. *Figure 11* shows an example of a low-slope perimeter coping and cleat.

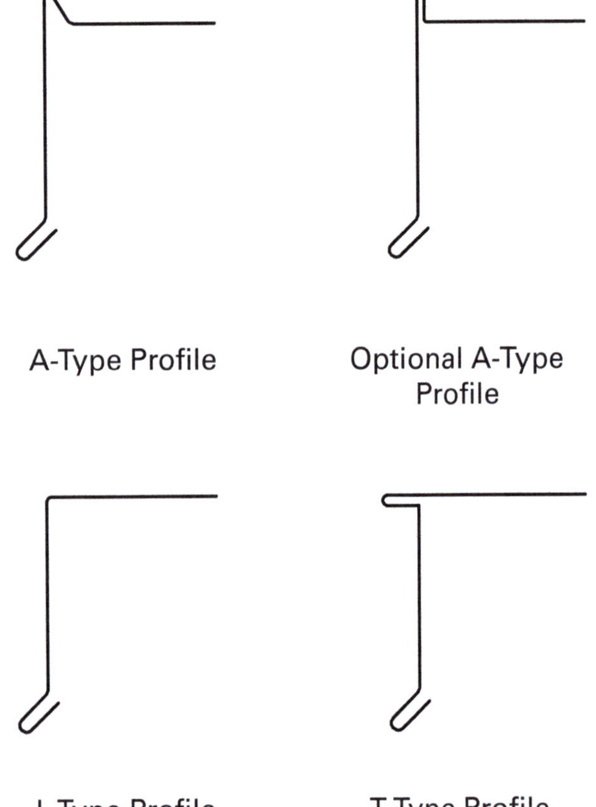

A-Type Profile

Optional A-Type Profile

L-Type Profile

T-Type Profile

Figure 10 Common perimeter edge metal profiles.

1.2.2 Base Flashings and Counterflashings

Metal base flashings rest on the deck at horizontal-to-vertical intersections to direct the flow of water onto the roof covering (*Figure 12*). They cover the edge of the field membrane and extend up the vertical surface.

Counterflashings are typically made from metal or membrane. They are secured on or into a wall or curb or to another component (such as a vent, mechanical unit, or chimney) to cover and protect the upper edge of the base flashing and its fasteners.

> Two-piece counterflashings provide better detail and make future roofing repairs more accessible. It is recommended to use two-piece counterflashings when possible to increase the overall quality of the roof system.

1.2.3 Valley Metal

A valley is created at the internal intersection of two sloping roof planes (*Figure 13*). Water runoff from the sloping planes collects in the valley, making this area especially vulnerable to leakage. Valleys must be weatherproofed to have a clear, unobstructed drainage path so water can run off quickly.

Valleys are installed only after the necessary layer(s) of underlayment and any specified valley-lining materials have been applied to the deck.

1.2.4 Penetrations

Penetrations are interruptions in the surface of the roof system. Penetrations on steep-slope roofs include chimneys, skylights, vent pipes, and exhaust fans. Low-slope roof penetrations include supply lines, electrical conduit, and roof-mounted equipment such as HVAC units. Examples of steep-slope and low-slope penetration flashings are shown in *Figure 14* and *Figure 15*.

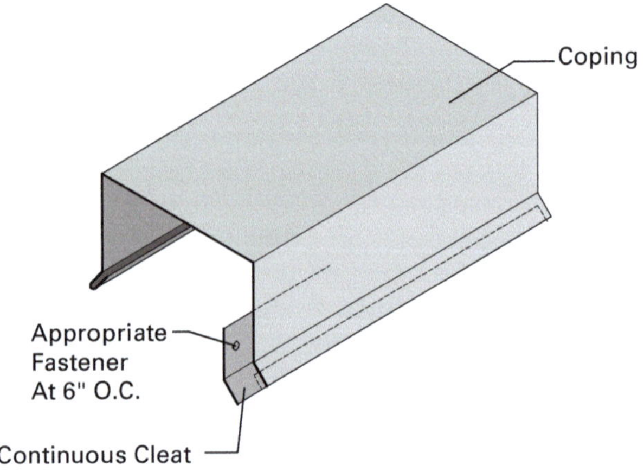

Figure 11 Low-slope perimeter coping and cleat.

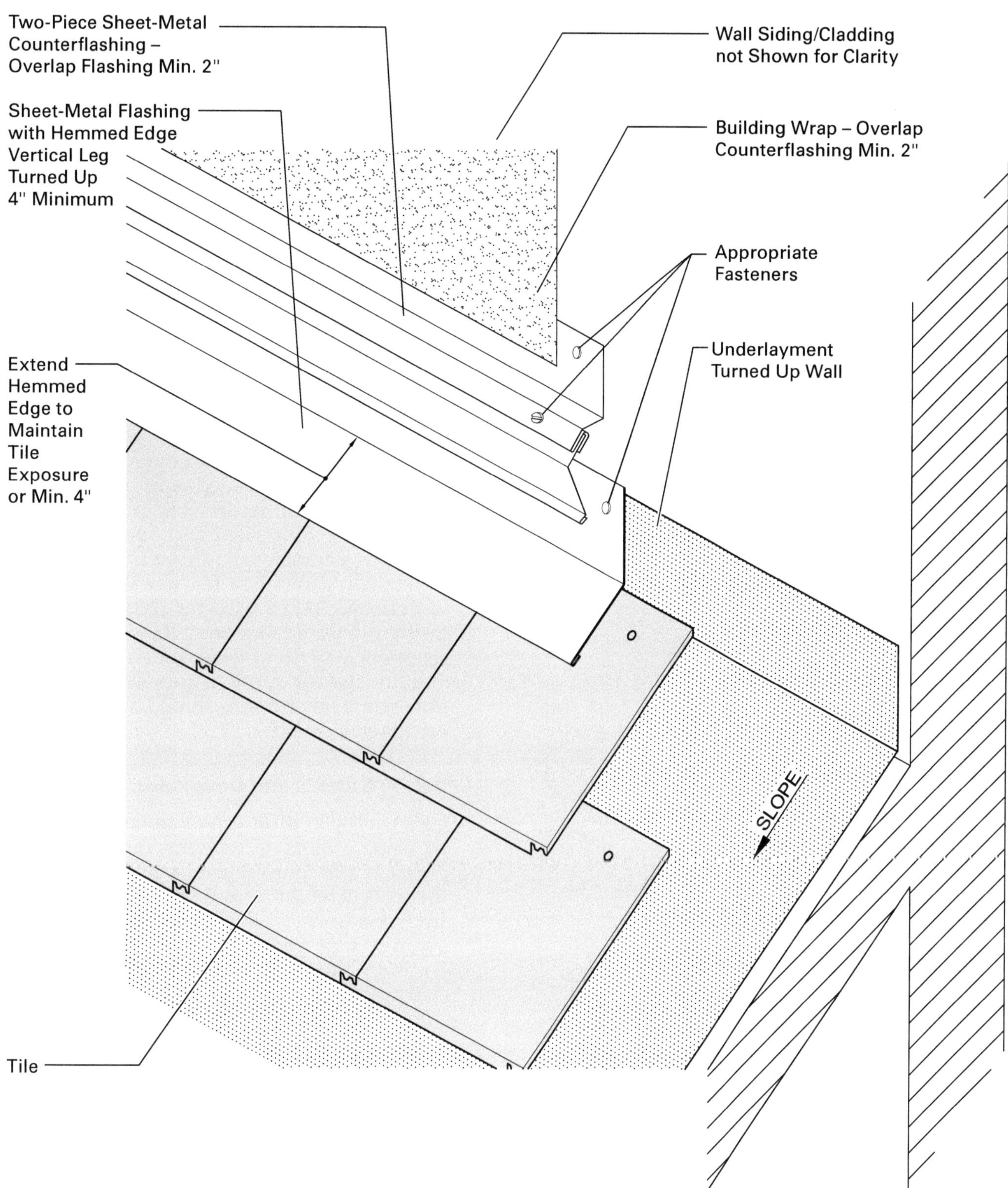

Figure 12 Base flashing and two-piece counterflashing.

Figure 13 Example of a valley flashing.

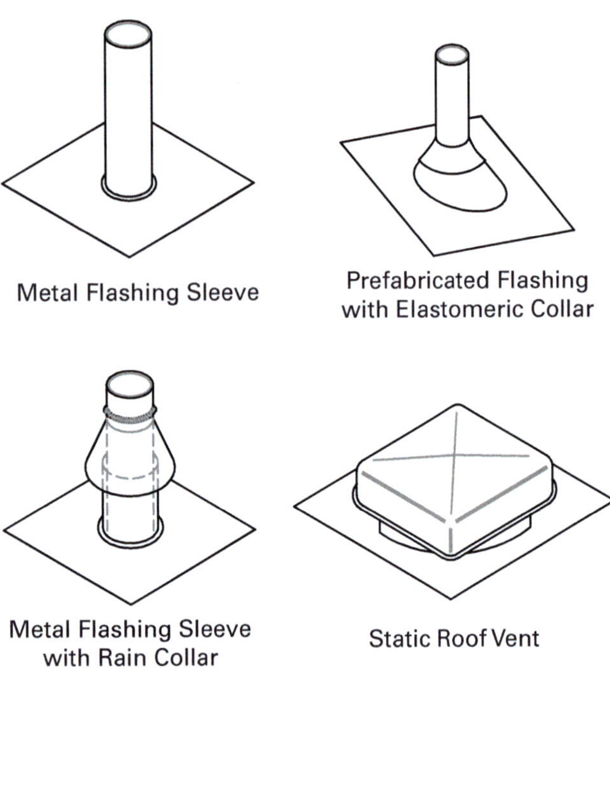

Figure 14 Examples of steep-slope penetration flashings.

1.2.5 Copings

Parapet walls must be adequately protected from moisture intrusion to prevent deterioration of the wall and damage to the roof system and building. Copings are used to cover the top of a parapet wall and provide a weatherproof cap, which seals and protects the wall from moisture intrusion. An example of a coping corner section is shown in *Figure 16*. Copings also use cleats (*Figure 17*) to aid installation and improve roof-system quality.

Copings sometimes require the use of a nailer. Wood nailers provide additional system protection and a substrate for anchoring flashing materials. Nailers must be properly fastened to a roof deck, wall, or structural supports to prevent wind uplift.

1.2.6 Scuppers

Scuppers are installed at edges of low-slope roofs. There are two general types of scuppers: through-wall scuppers and open scuppers. Open scuppers are three-sided and installed within a raised edge design (*Figure 18*). Through-wall scuppers are four-sided and installed through a perimeter parapet wall (*Figure 19*).

Scuppers can be a primary or secondary source of drainage. When used for secondary drainage, they are referred to as *overflow scuppers*. Primary scuppers may drain water into a conductor head and downspout or into a gutter. Secondary overflow scuppers or drains should be placed in obvious, open locations.

1.2.7 Gutters and Downspouts

One way to control drainage of roof systems is by using gutters and downspouts (*Figure 20*). A gutter is a channeled component installed along the perimeter of a roof to carry runoff water from the roof to the drain leaders or downspouts. Some gutters use scuppers to drain water through conductor heads into downspouts (*Figure 21*). Gutters must be appropriately sized to ensure they can properly drain the amount of water runoff on a given roof. Many types of sheet metal can be used to fabricate gutters and downspouts. Some common types include aluminum, copper, and steel. There are two types of gutters: externally attached (hanging) gutters and built-in gutters. Built-in gutters are uncommon, and they are used more often on steep-slope systems.

A downspout is joined to a gutter with an outlet tube inserted through a hole in the bottom of the gutter. This hole is often cut on the jobsite by the roofer. The size of the hole must be equal to the outside dimension of the outlet tube.

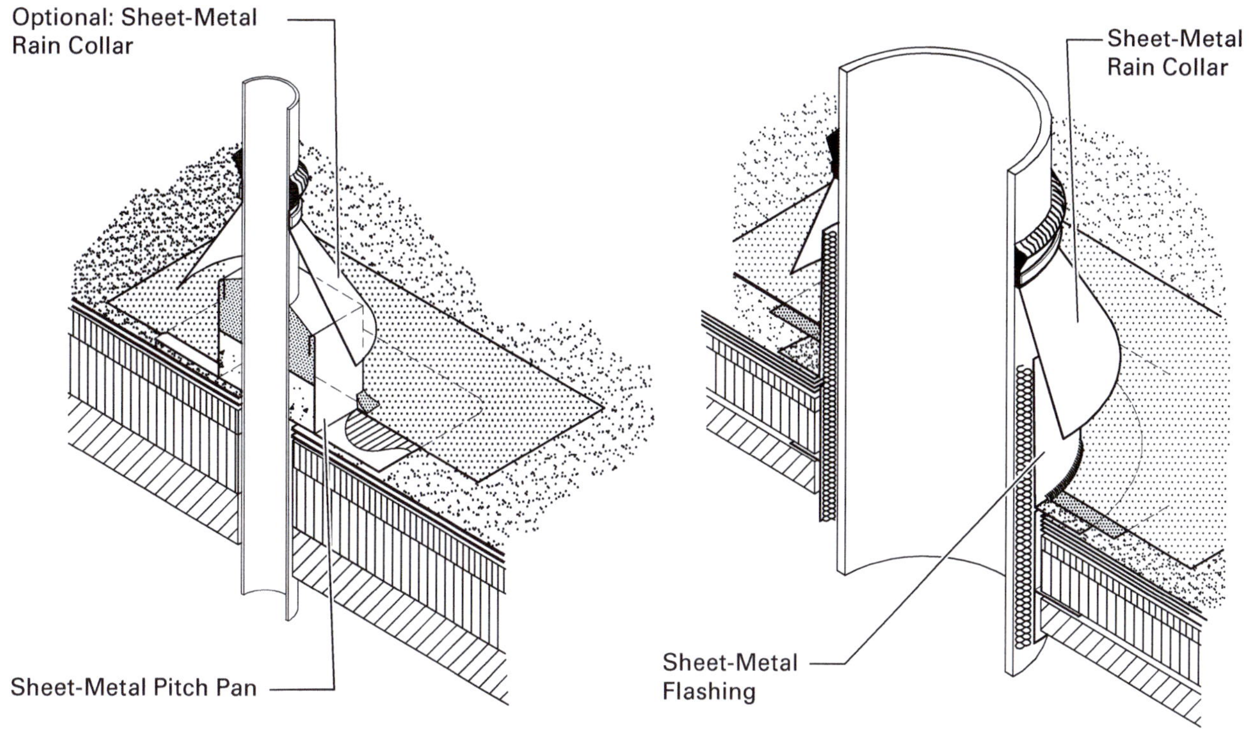

Figure 15 Examples of low-slope penetration flashings.

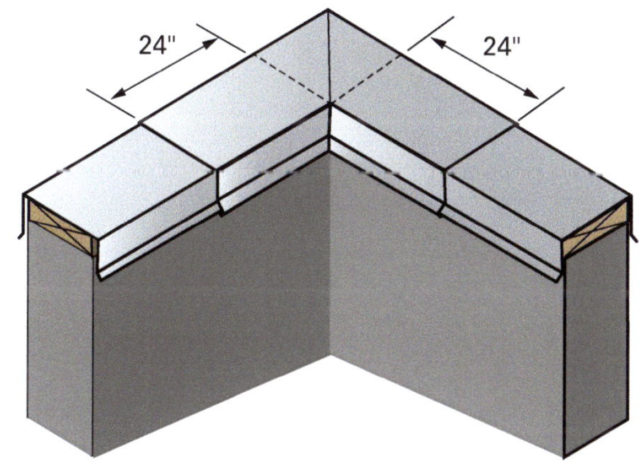

Figure 16 Coping corner.

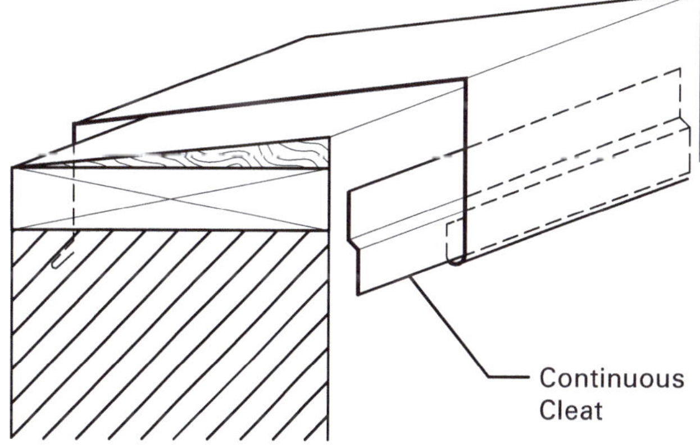

Figure 17 Coping cleat.

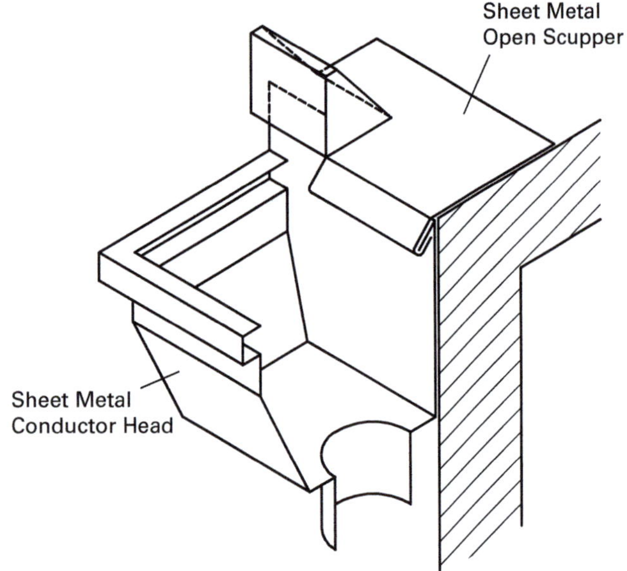

Figure 18 Open scupper.

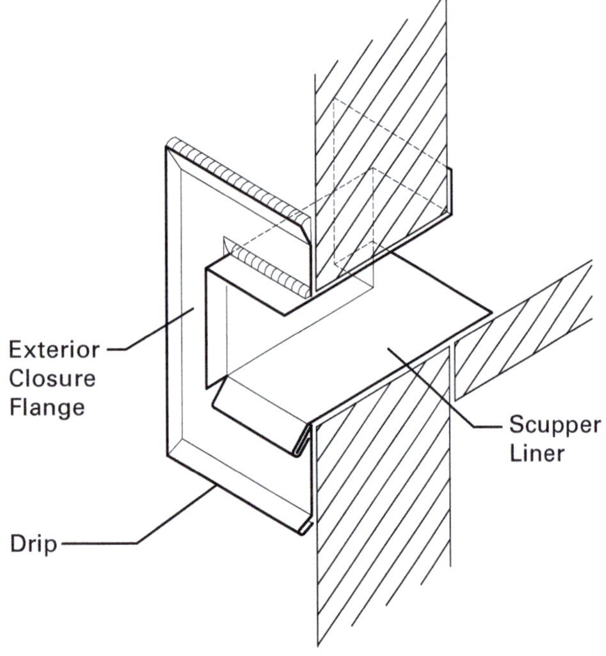

Figure 19 Through-wall scupper.

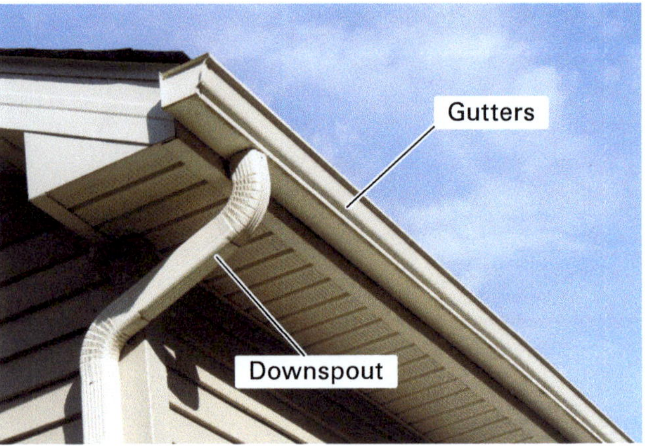

Figure 20 Gutter and downspout.

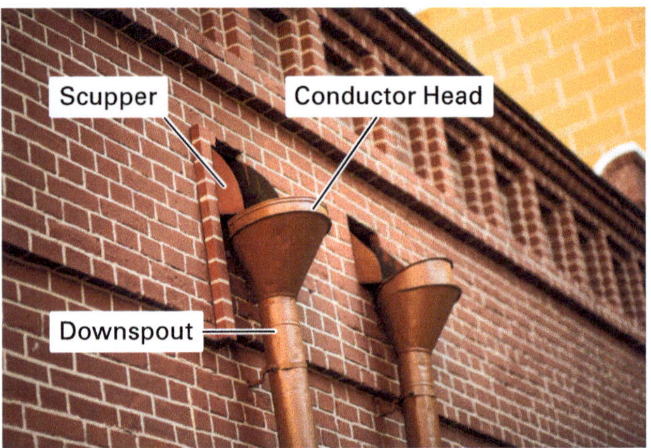

Figure 21 Scupper, conductor head, and downspout.

NCCER – *Roofing*

1. What happens to *all* metals once they are exposed to the weather?

    a. Corrosion
    b. Oxidation
    c. Flashing
    d. Gauging

2. What are continuous metal strips used to secure metal components?

    a. Mils
    b. Patinas
    c. Micrometers
    d. Cleats

2.0.0 SHEET METAL TOOLS AND EQUIPMENT

Objective

Describe tools and basic installation methods used in sheet-metal roofing.

a. Identify tools and equipment used for sheet-metal roofing components.
b. Explain some common procedures for handling sheet-metal roofing components.

Performance Tasks

1. Use aviation snips to cut a 6" (15 cm) circular hole into a piece of sheet metal.
   - Measure the hole.
   - Draw a circle with a compass.
   - Cut the hole with aviation snips.
2. Size, fit, and cut two pieces of L-style metal edge.

Trade Terms

Diameter: An invisible, straight line segment starting on one end of a circle, passing through the center point, and reaching the other end of the circle.

Joinery: The process of joining two or more pieces of sheet metal.

There are a variety of tools and procedures used for measuring, cutting, forming, and installing sheet metal. This section describes many of the tools that roofers will use when installing sheet-metal components. It also overviews some basic installation procedures used in sheet-metal roofing.

> **NOTE**
>
> Basic tools used in the construction industry are covered in detail in NCCER Module 00103, *Introduction to Hand Tools* and NCCER Module 00104, *Introduction to Power Tools*.

2.1.0 Tools

Layout is the measuring and marking of a piece of sheet metal to prepare it for cutting, bending, and forming into a roof component. Roofing

professionals must be familiar with the following layout and measuring tools:

- *Punches* — Used to put an indent in metal to mark it.
- *Scribes* — Used to mark sheet metal.
- *Combination squares and framing squares* — Used to measure and mark angles.
- *Rulers* — Used to measure sheet metal and has a straight edge to mark it.
- *Compasses* — Used to draw circles to cut into sheet metal, such as for downspouts. Also used in the shop to cut vent bases.

In order to measure and outline a correctly sized hole to cut out of sheet metal, use the following tools and steps.

Step 1 Locate the center of the hole you will cut out by using a ruler.

Step 2 Place one end of your compass on the center point of the circle you wish you draw.

Step 3 Calculate half of the width of the **diameter** of your circle cutout.

- For example, if the circle cutout will need to be 6" (15 cm), you must measure 3" (7.5 cm) away from the circle's center point to make a 6" (15 cm) circle with a compass.

Step 4 Use your measurement of half the diameter of the circle to place the other end of your compass that distance away from the center point.

- For example, if you would like to cut out a 6" (15 cm) circle, place the second end of your compass 3" (7.5 cm) from the center of the circle.

Step 5 Keeping the center of your compass still, rotate the outside piece of the compass all the way around the circle center point until you mark a complete circle in your sheet metal.

In addition to layout and measuring tools, there are a variety of additional tools used for cutting and forming sheet metal. These are important for roofers to be familiar with as they handle sheet metal components.

> **WARNING!**
>
> Take care when using sharp tools and handling sheet metal. Be sure to wear appropriate PPE, such as cut-resistant gloves, respirators, and eye protection.

Aviation Snips

A special type of snips, aviation snips, got their name in the early 20th century when they were used for aircraft manufacturing. Aviation snips are commonly just referred to as *snips* in the roofing industry. However, aviation snips specifically describe the type of color-coded snips specially designed to make left and right cuts.

Figure Credit: iStock@frankpeters

2.1.1 Snips

Snips are used like scissors to cut thin, soft metal. Snips can typically cut metal up to 18-gauge thick.

Snips have long handles, providing added leverage. It is important not to exert excessive force on the handles. If a cut is forced by extending the handles, applying body weight on the blades, or pounding the blades, the snips will likely be permanently damaged. This typically forces the blades apart, leaving too much clearance between them. Once the tips no longer meet, even making a simple notch cut will not be possible. Keep snips in good operating condition and do not force them beyond their stated capacity. If the metal is too thick or too hard to cut with snips, choose another, more appropriate tool.

Snips are available in straight, left-cut, and right-cut models (*Figure 22*). These directionally oriented snips are sometimes referred to as *aviation snips*. Right-cut snips cut tight curves to the right, while left-cut snips cut tight curves to the left. The blades are serrated so that they grip the metal tightly.

The three types are easy to tell apart based on their jaw design. If the upper blade is on the right, it is a right-cut snip. If the upper blade is on the left, it is a left-cut snip. However, manufacturers also color-code the handles to make it even easier. Straight-cut snips have yellow handles, left-cut snips have red handles, and right-cut snips have green handles.

2.1.2 Tongs

Tongs (*Figure 23*) are also referred to as *hand seamers* or *hand brakes*. They are used to form and flatten metal when brakes are not available, or when it is impossible to use a brake to form the seam. The top jaw of the tongs is often ruled off in $\frac{1}{4}$" (6 mm) increments to help the user make accurate bends and flanges. Tongs are not effective on heavy-gauge metal since they are a hand-operated tool.

2.1.3 Folding Tools and Brakes

A folding tool (*Figure 24*) is a very simple device. To form bends in sheet metal, insert the metal into the top slot then use the tool to make a fold. This hand tool is effective on lighter-gauge metals.

Brakes are metalworking machines that form bends in sheet metal. Brakes use a clamping bar to hold the material firmly as the metal is bent automatically or through the use of a hand lever or foot pedal.

Figure 22 Snips.

Figure 23 Tongs.

2.1.4 Nibblers and Shears

A nibbler (*Figure 25*) may be electrically, manually, or compressed-air powered and can be either bench-mounted or portable. It can make straight or circular cuts. A hardened steel cutter punches out metal notches against a die as either the sheet metal is moved over the nibbler or the portable nibbler is applied to the sheet metal. Nibblers can cut irregular shapes and around tight corners; however, they don't usually leave a straight, finished edge. Chips produced by cutting are ejected downward, away from the worker. The width of the cut is usually greater than $\frac{1}{8}$".

Ground-fault circuit interrupter (GFCI) protected cords or receptacles must be used for power tools on the job at all times. GFCI devices are designed to interrupt power to protect the user in the event of an electrical ground fault.

Single-cut and double-cut shears are precise tools used to cut sheet metal. While single-cut shears use one blade to cut cleanly through metal, double-cut shears use two separate cutters to remove a thin metal strip as the tool slices the sheet metal. Double-cut shears are not recommended for making curved cuts. There are hand-powered and power-tool version of both single-cut and double-cut shears.

Figure 24 Folding tool.

2.1.5 Reciprocating Saws

Reciprocating saws (*Figure 26*) can make cuts in a wide variety of materials. They are used to cut irregular shapes and holes and are handy for cutting in tight spaces. They are well suited for demolition work.

The straight, metal saw blades are interchangeable and generally measure up to 12" (30 cm) long × 1" (2.5 cm) wide. The blades cut by moving back and forth, with the cut being made on the backstroke. Note that this is opposite of the hacksaw's cutting stroke. On the forward stroke, a cam lifts the blade slightly to prevent it from rubbing against the metal. Many reciprocating saws offer variable speed operation. A low-speed setting is usually best for metal work.

2.2.0 Sheet-Metal Procedures

Metal roof system components are cut, seamed, joined, and fastened in a variety of ways. Many metal components provide a drainage path for water. Water-shedding and weatherproofing must always be considered when installing these components.

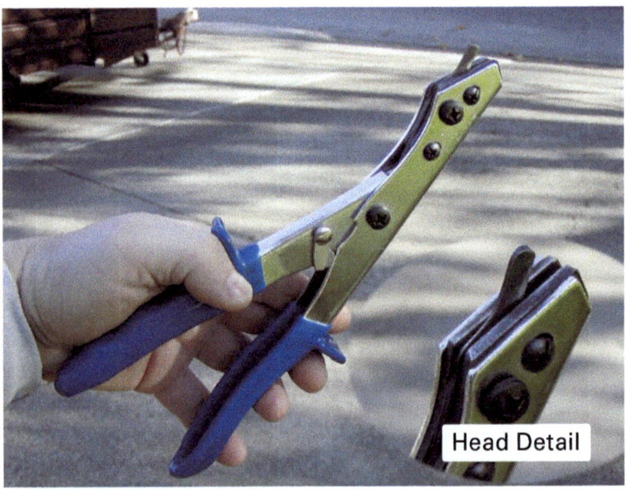

Figure 25 Manual nibbler.

Figure 26 Reciprocating saw.

2.2.1 Cutting

Before cutting sheet metal, be sure to have all the appropriate tools and PPE for the job. Cutting sheet metal exposes you to many potential safety hazards, such as sharp edges and loose metal fragments. Be sure to consult your safety data sheet and supervisor for safety-related guidance.

It is helpful to measure, layout, and mark your metal before you make cuts to ensure a smoother process. If it is challenging to use snips or other hand tools for cutting metal, a different tool may be required.

Sometimes you may need to cut a hole through the center of a larger piece of sheet metal—for example, to make room for a pipe penetration in a metal-roof panel. The following are general procedure steps involved in cutting out a centerpiece of sheet metal with snips.

Step 1 Don proper PPE and ensure snips are the appropriate tool to use for the job.

Step 2 Plan, measure, and mark the cut you will make in the sheet metal.

Step 3 Create a pilot hole with the appropriate hand or power tool.

Step 4 Insert snips through the open space and resume carefully cutting your desired shape from the sheet metal.

Be sure to only create pilot holes in areas of the sheet metal you will be removing from the roof system.

If you are cutting out larger or thicker metal pieces, use power shears for straight cuts and nibblers for curved cuts. As is the case with hand tools, maintain a slow and steady pace when using power tools to cut sheet metal.

2.2.2 Joinery

Joinery is the process of joining two or more pieces of sheet metal. Joints can be made with mechanical fasteners, interlocking seaming methods, sheet-metal cover plates, sealants, soldering and welding, or a combination of these methods.

The type of joinery used usually depends on the function and location of the sheet-metal joint, its exposure to water, and the type of metal. Some joints are constructed to be water-shedding while others require a more weatherproof seal. Methods of joining sheet metal include the following:

- *Mechanical fasteners* — A simple and common way to join sheet metal is using mechanical fasteners, such as screws, bolts, and rivets. However, this method of joining can be difficult to weatherproof. Sealants, gaskets (rubber seals), or solder may be used with mechanical fasteners to provide a weatherproof joint.
- *Crimping and locking* — Pre-painted metals are commonly joined without the use of soldering because the coating does not allow for it. Crimping and locking is a typical method used to join laps in sheet-metal components. For example, a standing seam is a locking type of seam. Another example is the outer face of a through-wall scupper—where the scupper overlaps the flange, the two pieces of metal are crimped and locked together. Copings and cleats are joined in the same fashion.
- *Soldering and welding* — Welded or soldered joints are a bond between two metal shapes. A welded joint is a structural joint formed when two metals are fused together using high heat and a filler material. A soldered joint is not a structural joint. While it has some structural characteristics, it is only a method of bonding two metals. Whether a metal can be welded or soldered primarily depends on the metal type, thickness of the material, equipment, and skill.

- *Mitering* — A miter joint is created by cutting two metal parts to form a 45-degree angle at a 90-degree joint.

NOTE

Welding is a separate skilled profession and specialty area. NCCER has a four-year Welding Program that prepares trainees to be knowledgeable, safe, and effective welders. For more information, visit **www.nccer.org**.

The following are examples of seam types commonly used in roof systems:

- *Overlap standing seam* — A minimum $\frac{7}{8}$" (22 mm) high female leg should be installed over the adjoining metal shape's male leg and crimped to complete the seam. Sometimes, sealant or butyl tape is installed in the female leg before installation to provide additional weatherproofing capabilities. The seam is secured by the button-punch method or with rivets. See *Figure 27* (A).
- *Single-lock standing seam* — A minimum $\frac{7}{8}$" (22 mm) high female leg should be installed over the adjoining metal shape's male leg. The female leg is folded over the male leg, and both legs are bent 90 degrees. Sometimes, sealant or butyl tape is installed in the female leg before to provide additional weatherproofing capabilities. See *Figure 27* (B).
- *Double-lock standing seam* — A minimum $\frac{7}{8}$" (22 mm) high female leg should be installed over the adjoining metal shape's male leg. The legs are folded over themselves twice to complete the double-lock seam. Double-lock standing seams are weatherproof without the addition of sealants or sealant tapes. See *Figure 27* (C).
- *Capped standing seam* — A capped standing seam is composed of two single-lock standing seams constructed with two adjacent vertical legs and a cap. It is generally used where free movement is essential, such as expansion joint and ornamental purposes. See *Figure 27* (D).
- *Flat lock seam* — Flat lock seams should be at least $\frac{1}{2}$" (12 mm) wide on copper up to 20 ounces and at least $\frac{3}{4}$" (19 mm) wide with heavier weights. See *Figure 27* (E).
- *Double flat lock seam* — A double flat lock seam is a double-lock standing seam bent flat and can be soldered or unsoldered (for metals that cannot be soldered). See *Figure 27* (F).
- *Drive cleat seam* — A minimum 2" (5 cm) wide folded drive cleat should be installed over the folded ends of two metal shapes. The space between the ends of the metal shapes will vary depending on the type of material used, gauge thickness, and building locale. See *Figure 27* (G).
- *"S" cleat seam* — A minimum 2" (5 cm) wide "S" cleat should be installed between the unhemmed ends of two metal shapes. Sealant is recommended to be installed in the female legs before installing the metal shape to provide additional weatherproofing capabilities. The space between the ends of the metal shapes will vary depending on the type of material used, gauge thickness and building locale. See *Figure 27* (H).
- *Double "S" cleat seam* — A double "S" cleat seam is similar to a drive cleat seam but with the cleat having two female legs on each side to accept the unhemmed end of a metal shape. See *Figure 27* (I).
- *Cover plate seam* — A minimum 6" (15 cm) wide exposed joint cover plate should be installed, evenly divided and placed over the two edges of adjacent metal pieces. The space between the ends of the adjacent metal pieces will vary depending on the type of material used, gauge thickness and building locale. Each side of the space requires sealant; a minimum of two beads of sealant should be installed under the cover plate. See *Figure 27* (J).
- *Concealed (backer plate seam)* — A minimum 6" (15 cm) wide concealed joint plate should be installed, evenly divided and placed under the two edges of adjacent metal pieces. The space between the ends of the adjacent metal pieces will vary depending on the type of material used, gauge thickness and building locale. Each side of the space requires sealant; a minimum of two beads of sealant should be installed over the backer plate. See *Figure 27* (K).
- *Cover with backer plate seam* — A minimum 6" (15 cm) wide concealed joint plate and exposed joint plate should be installed, evenly divided and placed between the two edges of adjacent metal pieces. The space between the ends of the adjacent metal pieces will vary depending on the type of material used, gauge thickness, and building locale. Each side of the space requires sealant; a minimum of two beads of sealant or butyl tape should be installed under the cover plate. See *Figure 27* (L).

- *Lap seam* — Simple joinery consists of overlapping the adjacent ends of two pieces of sheet metal. Sometimes mechanical fasteners or rivets are used to secure the metal in place. Succeeding pieces of metal, lapped about 3" (7.5 cm) over the preceding pieces, should be installed. A notch can be cut into the return hem along the bottom edge of the metal that receives and locks in place the preceding piece. For weatherproof installations, sealant or solder is required in the seam. See *Figure 27* (M).

- *Pittsburgh lock seam* — A Pittsburgh lock seam is used at corners and involves joining two pieces of metal. The horizontal metal piece has a turned-down flanged edge. The vertical piece is formed to have a vertical pocket, and the flanged edge is inserted into the pocket. The extended edge of the vertical piece is folded on top of the horizontal surface to complete the lock. See *Figure 27* (N).

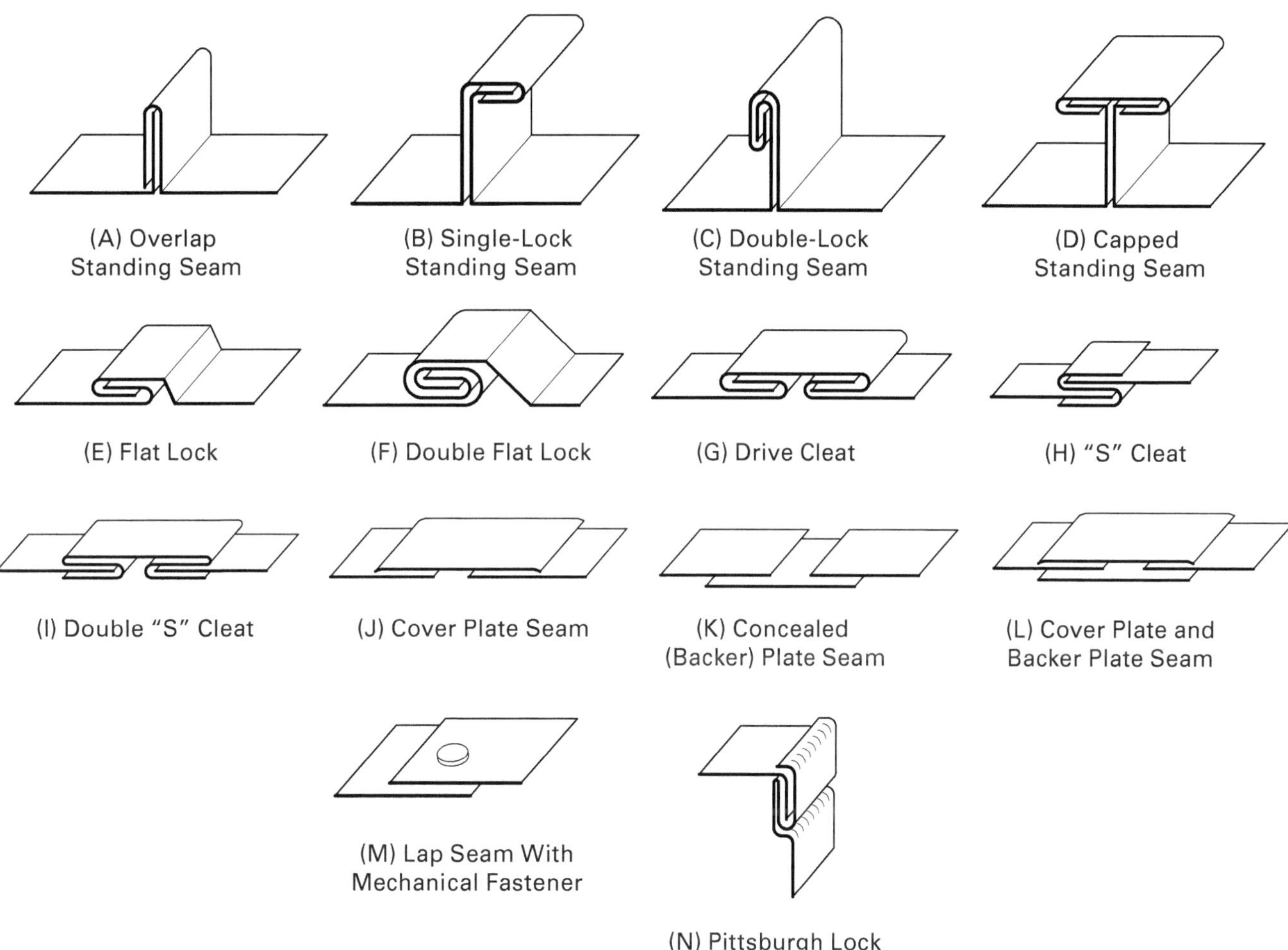

Figure 27 Types of seams.

1. Which of the following layout and measuring tools is used *primarily* to mark sheet metal?

    a. Punches
    b. Scribes
    c. Rulers
    d. Compasses

2. Which of the following seams is a double-lock standing seam bent flat that can be soldered or unsoldered?

    a. Flat lock seam
    b. Double flat lock seam
    c. "S" cleat seam
    d. Cover with backer plate seam

Review Questions

1. Metal that is less than $\frac{1}{4}$" (6 mm) thick is considered _____.
   a. sheet metal
   b. plate metal
   c. aluminum
   d. steel

2. What tool is used to measure the thickness of sheet metal?
   a. Corrosion
   b. Sheet metal ruler
   c. Sheet metal gauge
   d. Patina

3. The expansion and contraction of metal when it is exposed to changing temperatures is called _____.
   a. expansion coefficient
   b. thermal expansion
   c. patina
   d. corrosion

4. Which of the following is used to provide a continuous finished edge along the outer edges of a roof system?
   a. Valley metal
   b. Scuppers
   c. Downspouts
   d. Drip edge

5. A vent pipe is an example of a roof _____.
   a. valley metal
   b. coping
   c. penetration
   d. base flashing

6. To provide a substrate for anchoring flashing materials, copings sometimes require the use of a _____.
   a. nailer
   b. scupper
   c. gutter
   d. downspout

7. Which of the following layout and measuring tools is used to put an indent in metal to mark it?
   a. Punches
   b. Scribes
   c. Rulers
   d. Compasses

8. What is half the diameter of a circle with a width of 8" (20 cm)?
   a. 2" (5 cm)
   b. 3" (7.5 cm)
   c. 4" (10 cm)
   d. 6" (15 cm)

9. Typically, right-cut snips have _____.
   a. green handles
   b. yellow handles
   c. red handles
   d. blue handles

10. Tongs are also referred to as _____.
    a. aviation snips
    b. nibblers
    c. hand brakes
    d. shears

Fill in the blank with the correct term learned from your study of this module.

1. _______________ is a form of rusting resulting from the creation of an electric current flowing between dissimilar metals in contact with each other.

2. _______________ are continuous metal strips, or angled pieces, used to secure metal components.

3. A green or brown oxide film that develops on the surface of copper and copper alloys is _______________.

4. _______________ is the increase in the dimension or volume of a body because of temperature variations.

5. _______________ is an invisible, straight line segment starting on one end of a circle, passing through the center point, and reaching the other end of the circle.

6. The process of joining two or more pieces of sheet metal is _______________.

7. A unit of measurement equal to one-thousandth of an inch, or 0.001" (0.0254 mm) is a(n) _______________.

8. _______________ is physical distortions in the flatness of metal.

9. _______________ is the breakdown or destruction of a material, especially metal, caused by chemical reactions. The most common form is rust, which occurs when iron combines with oxygen to create iron oxide.

10. _______________ means to combine with oxygen, such as in burning or rusting.

Trade Terms

Cleats	Galvanic corrosion	Oil canning	Thermal expansion
Corrosion	Joinery	Oxidize	
Diameter	Mil	Patina	

Cornerstone of Craftmanship

Mike Silvers
Subject Matter Expert
NRCA

How did you choose a career in the industry?

My father was working for a roofing contractor in Indiana. He and my mother moved the family to St. Petersburg, Florida in 1961 to start their own company. I was 8 years old and along for the ride. I walked to our business every day after school. A broom in my hand came very quickly. The broom gave way to snips, hammers, and wrenches. So, at a very early age I was up to my rear end in the roofing business. Since then, I have run my own company and have been a Florida State Certified Roofing Contractor for nearly 40 years.

Who inspired you to enter the industry?

I did not need to look any further than both of my parents. They both worked in our businesses and demonstrated daily how to properly conduct oneself in that environment and beyond. I was also very fortunate as a young man to be encouraged by another contractor to attend our local roofing association meetings which led me to the Florida Roofing and Sheet Metal Contractors Association (FRSA). My father had passed and the mentorship I received from competitors was both selfless and priceless.

What types of training have you been through?

The daily exposure to roofing company operations was the source of most of my early training. I took a liking to and had a knack for fabrication, so I took all the shop classes that I could. The most influential of these was sheet metal class in high school (Thank you, Mr. Nelson). I have both taken and taught many trade-related courses and seminars as part of my professional development.

How important is education and training in construction?

As my career developed, I came to realize just how much I drew on the information that I had been exposed to through education, training, and working. From formal education and classroom training to the many other ways we learn through exposure, all forms of education allow us to expand our critical skills and hopefully allow us to become a conduit to help others learn.

What kinds of work have you done in your career?

I have been fortunate to have some incredible employees whose skills at a particular task are much better than my own, but I am proud of the fact that I can do anything that we do. Most of my experience has been in roofing, but it only begins with roofing. We have to understand sales, geometry, estimation, procurement, human resources, logistics, and possibly mechanics or a little welding before we even head to the job. OSHA and CDL requirements are next. Then, you add the complexities of roofing, the multitude of different types, sizes, shapes, and locations of buildings that we encounter, all of the different types of roofing systems that are used and the different techniques used to remove and install them.

Tell us about your present job.

I have stepped back from running my roofing company to work with FRSA as the Technical Director. My primary focus is on the Florida Building Code and other technical issues. I field technical questions from contractors, manufacturers, building officials, architects, engineers, and consumers. I also get to provide instruction through seminars for almost all of the aforementioned on these subjects.

What do you enjoy most about your job?

When someone comes to you with a complex problem, and you lead them to the answer and see the light go on. It has also helped to quench a lifelong thirst for knowledge and information. Stay thirsty, my friends.

What factors have contributed most to your success?

Curiosity, perseverance, empathy, a willingness to take on a challenge, and an ongoing thirst for knowledge.

Would you suggest construction as a career to others? Why?

Yes, as long as they are up for a challenge. There is great satisfaction to physically building or improving the structures that we live and work in and that we depend on for our way of life, as well as the inner pride one gets from a job well done. It has also allowed me and so many others to gain financially for our efforts.

What advice would you give to those new to the field?

Do not underestimate the willingness of roofing professionals to share their knowledge. Work hard to expose yourself to this information.

Interesting career-related fact or accomplishment:

After watching and responding to a vehicular accident, removing the driver from the overturned and burning vehicle, and then extinguishing the fire, a bystander asked me if I was a firefighter or a paramedic. I responded, "No, sir, I'm a roofer." A somewhat puzzled look followed. The experience gained through a lifetime of training and exposure can serve one in unexpected, but often very important ways.

How do you define craftsmanship?

A craftsman brings the mental and physical skills to complete a given task in a functional and aesthetically pleasing way. They also understand the importance of a job well done and conduct themselves in a manner that reflects well on their chosen trade.

Trade Terms Introduced in This Module

Cleats: Continuous metal strips, or angled pieces, used to secure metal components.

Corrosion: The breakdown or destruction of a material, especially metal, caused by chemical reactions. The most common form of corrosion is rust, which occurs when iron combines with oxygen to create iron oxide.

Diameter: An invisible, straight line segment starting on one end of a circle, passing through the center point, and reaching the other end of the circle.

Galvanic corrosion: A form of corrosion (rusting) resulting from the creation of an electric current flowing between dissimilar metals in contact with each other.

Joinery: The process of joining two or more pieces of sheet metal.

Mil: A unit of measurement equal to one-thousandth of an inch, or 0.001" (0.0254 mm).

Oil canning: Physical distortions in the flatness of metal. This condition only effects the appearance of the metal and does not have negative effects on structural integrity.

Oxidize: To combine with oxygen, such as in burning (rapid oxidation) or rusting (slow oxidation).

Patina: A green or brown oxide film that develops on the surface of copper and copper alloys.

Thermal expansion: The increase in the dimension or volume of a body because of temperature variations.

Appendix A

Thicknesses and Weights of Metals by Gauge

For convenience, a chart showing decimal and fractional dimension equivalents in inches is provided in *Appendix B*.

Galvanized			Stainless				Aluminum		Copper	
Gauge	Mfr's Thickness for Steel + 0.0037"	Weight per Square Foot, Pounds	Gauge	US Standard Approximate Decimal Parts of an Inch Thickness	Pounds per Square Foot — Chrome	Nickel	Gauge	Ounces per Square Foot	Ounces per Square Foot	Approximate Thickness in Inches
									96	—
							—	0.0403	88	$\frac{1}{8}$ 0.1250
							9	0.0359	72	$\frac{7}{64}$ 0.1080
							—	0.0320	64	$\frac{3}{32}$ 0.0972
10	0.1382	5.781	10	0.140	5.793	5.906	10	$\frac{1}{32}$ 0.0313	56	0.0863
							—	0.0253	48	$\frac{5}{64}$ 0.0647
12	0.1084	4.531	12	0.109	4.506	4.593	12	0.0226	44	$\frac{1}{16}$ 0.0647
							—	0.0201	40	0.0593
							13	0.0179	36	0.0539
14	0.0785	3.281	14	0.078	3.218	3.281	14	0.0159	32	$\frac{3}{64}$ 0.0485
							—	$\frac{1}{64}$ 0.0156	28	0.0431
							15	0.0142	24	0.0377
16	0.0635	2.656	16	0.062	2.575	2.625	16	0.0126	20	$\frac{1}{32}$ 0.0323
							—	0.0113	18	0.0270
							—	0.0100	16	0.0243
18	0.0516	2.156	18	0.050	2.060	2.100			15	0.0216
20	0.0396	1.656	20	0.037	1.545	1.575			14	0.0202
22	0.0336	1.406	22	0.310	1.287	1.312			13	0.0189
24	0.0276	1.156	24	0.250	1.030	1.050			12	0.0175
26	0.0217	0.906	26	0.018	0.772	0.787			11	$\frac{1}{64}$ 0.0162
28	0.0187	0.781	28	0.015	0.643	0.656			10	0.0148
30	0.0157	0.656	30	0.012	0.515	0.525			9	0.0135
									8	0.0121

Appendix B

Fraction	Decimal	Fraction	Decimal	Fraction	Decimal	Fraction	Decimal
$\frac{1}{64}$	0.015625	$\frac{17}{64}$	0.265625	$\frac{33}{64}$	0.515625	$\frac{49}{64}$	0.76525
$\frac{1}{32}$	0.03125	$\frac{9}{32}$	0.28125	$\frac{17}{32}$	0.53125	$\frac{25}{32}$	0.78125
$\frac{3}{64}$	0.046875	$\frac{19}{64}$	0.296875	$\frac{35}{64}$	0.546875	$\frac{51}{64}$	0.796875
$\frac{1}{16}$	0.0625	$\frac{5}{16}$	0.3125	$\frac{9}{16}$	0.5625	$\frac{13}{16}$	0.8125
$\frac{5}{64}$	0.078125	$\frac{21}{64}$	0.328125	$\frac{37}{64}$	0.578125	$\frac{53}{64}$	0.828125
$\frac{3}{32}$	0.09375	$\frac{11}{32}$	0.34375	$\frac{19}{32}$	0.59375	$\frac{27}{32}$	0.84375
$\frac{7}{64}$	0.109375	$\frac{23}{64}$	0.359375	$\frac{39}{64}$	0.609375	$\frac{55}{64}$	0.859375
$\frac{1}{8}$	0.125	$\frac{3}{8}$	0.375	$\frac{5}{8}$	0.625	$\frac{7}{8}$	0.875
$\frac{9}{64}$	0.140625	$\frac{25}{64}$	0.390625	$\frac{41}{64}$	0.640625	$\frac{57}{64}$	0.890625
$\frac{5}{32}$	0.15625	$\frac{13}{32}$	0.40625	$\frac{21}{32}$	0.65625	$\frac{29}{32}$	0.90625
$\frac{3}{16}$	0.1875	$\frac{7}{16}$	0.4375	$\frac{11}{16}$	0.6875	$\frac{15}{16}$	0.9375
$\frac{13}{64}$	0.203125	$\frac{29}{64}$	0.453125	$\frac{45}{64}$	0.703125	$\frac{61}{64}$	0.953125
$\frac{7}{32}$	0.21875	$\frac{15}{32}$	0.46875	$\frac{23}{32}$	0.71875	$\frac{31}{32}$	0.96875
$\frac{15}{64}$	0.234375	$\frac{31}{64}$	0.484375	$\frac{47}{64}$	0.734375	$\frac{63}{64}$	0.984375
$\frac{1}{4}$	0.250	$\frac{1}{2}$	0.500	$\frac{3}{4}$	0.750	1	1.000

Additional Resources

This module presents thorough resources for task training. The following reference material is suggested for further study.

National Roofing Contractors Association (NRCA), **www.nrca.net**.
NCCER Module 00103, *Introduction to Hand Tools*.
NCCER Module 00104, *Introduction to Power Tools*.
NCCER Module 04102, *Sheet Metal Tools and Equipment*.
NCCER Module 04307, *Architectural Sheet Metal*.
Sheet Metal and Air Conditioning Contractors' National Association (SMACNA), **www.smacna.org**.

Figure Credits

Shutterstock.com/Blaze986, Module Opener
iStock@SimoneN, Figure 1
Courtesy of the L.S. Starrett Company, Figures 2–3
Courtesy of the National Roofing Contractors Assoc., Figures 4–5, 7–15, 17–19, 27
iStock@Wendy Townrow, Figure 6
iStock@db_beyer, Figure 20
iStock@DENIS Starostin, Figure 21
Image property of Stanley Black & Decker. Used with permission., Figures 22, 23–24, 26
"Nibbler2.jpg" by Hydrargyrum is licensed under the Public Domain., Figure 25

Section Review Answer Key

Section 1.0.0

Answer	Section Reference	Objective
1. b	1.1.0	1a
2. d	1.2.1	1b

Section 2.0.0

Answer	Section Reference	Objective
1. b	2.1.0	2a
2. b	2.2.2	2b

This page is intentionally left blank.

NCCER CURRICULA — USER UPDATE

NCCER makes every effort to keep its textbooks up-to-date and free of technical errors. We appreciate your help in this process. If you find an error, a typographical mistake, or an inaccuracy in NCCER's curricula, please fill out this form (or a photocopy), or complete the online form at **www.nccer.org/olf**. Be sure to include the exact module ID number, page number, a detailed description, and your recommended correction. Your input will be brought to the attention of the Authoring Team. Thank you for your assistance.

Instructors – If you have an idea for improving this textbook, or have found that additional materials were necessary to teach this module effectively, please let us know so that we may present your suggestions to the Authoring Team.

NCCER Product Development and Revision

13614 Progress Blvd., Alachua, FL 32615

Email: curriculum@nccer.org
Online: www.nccer.org/olf

❏ Trainee Guide ❏ Lesson Plans ❏ Exam ❏ PowerPoints Other __________________________

Craft / Level: ____ Copyright Date: ________________

Module ID Number / Title: ____

Section Number(s): ____

Description: ____

____

____

____

Recommended Correction: ____

____

____

____

Your Name: ____

Address: ____

____

Email: ____ Phone: __________________________

This page is intentionally left blank.

Rigging Practices

OVERVIEW

Rigging is the preparation of a load for movement, as well as preparation of the hardware and other components used to connect the load to a crane. Rigging is associated with all types of cranes, and rigging skills are also required to move and position equipment inside buildings and other areas where cranes are not involved. This module will provide insight into rigging hardware, lifting slings and their proper use, and various types of rigging equipment.

Module 38102

RIGGING PRACTICES

Objectives

When you have completed this module, you will be able to do the following:

1. Identify and describe various types of rigging hardware.
   a. Identify and describe various hooks, shackles, eyebolts, and clamps.
   b. Identify and describe various lugs, turnbuckles, plates, and spreader beams.
2. Identify and describe various types of slings and sling hitches.
   a. Identify and describe wire-rope slings and their proper care.
   b. Identify and describe synthetic slings and their proper care.
   c. Identify and describe chain slings and their proper care.
   d. Explain the significance of sling angles and describe common hitches.
   e. Describe how to properly rig and handle piping materials and rebar.
   f. Identify and describe how to use taglines and knots for load control.
   g. Identify common rigging-related safety precautions.
3. Identify and describe how to use various types of hoisting and jacking equipment.
   a. Identify and describe how to use manual and powered hoisting equipment.
   b. Identify and describe how to use jacks.

Performance Tasks

Under the supervision of your instructor, you should be able to do the following:

1. Inspect various types of rigging components and report on the condition and suitability for a task.
2. Configure a sling to produce a single-wrap basket hitch.
3. Configure a sling to produce a double-wrap basket hitch.
4. Configure a sling to produce a single-wrap choker hitch.
5. Configure a sling to produce a double-wrap choker hitch.
6. Select the correct tagline for a specified application.
7. Tie specific instructor-selected knots.
8. Select, inspect, and demonstrate the safe use of the following rigging equipment:
   - Block and tackle
   - Chain hoist
   - Ratchet-lever hoist
   - One or more types of jack

Trade Terms

Basket hitch
Bird caging
Blind hole
Bridle hitch
Center of gravity (CG)
Choker hitch
Equalizer beams
Equalizer plates
Gantry
Hauling line
Independent wire rope core (IWRC)

Minimum breaking strength (MBS)
Parts of line
Rated load
Rigging links
Saddle
Sling angle
Spreader beams
Spur track
Tagline
Vertical hitch

Industry Recognized Credentials

If you are training through an NCCER-accredited sponsor, you may be eligible for credentials from NCCER's Registry. The ID number for this module is 38102. Note that this module may have been used in other NCCER curricula and may apply to other level completions. Contact NCCER's Registry at 888.622.3720 or go to **www.nccer.org** for more information.

Note

This module provides instruction and information about common rigging equipment and hitch configurations, but it does not provide any level of rigging certification. Any questions about rigging procedures and/or certification should be directed to an instructor or supervisor.

Contents

This page is intentionally left blank.

1.0.0 RIGGING HARDWARE

Objective

Identify and describe various types of rigging hardware.

a. Identify and describe various hooks, shackles, eyebolts, and clamps.
b. Identify and describe various lugs, turnbuckles, plates, and spreader beams.

Performance Task

1. Inspect various types of rigging components and report on the condition and suitability for a task.

Trade Terms

Blind hole: A hole that does not penetrate the material completely, leaving a hole with a bottom.

Bridle hitch: A type of hitch that consists of two or more slings that support the load attached to a common lifting point.

Center of gravity (CG): The point at which the entire weight of an object is considered to be concentrated, such that supporting the object at this specific point would result in its remaining balanced in position.

Equalizer beams: Beams used to distribute the load weight on multi-crane lifts. The beam attaches to the load below, with two or more cranes attached to lifting eyes on the top.

Equalizer plates: A type of rigging plate that has three or more holes, used to level loads when sling lengths are unequal.

Minimum breaking strength (MBS): The amount of stress required to bring a rigging component to its breaking point. The MBS is a factor in determining a component's rated load capacity.

Rated load: The maximum working load permitted by a component manufacturer under a specific set of conditions. Alternate names for rated load include *working load limit (WLL), rated capacity*, and *safe working load (SWL)*.

Rigging links: Links or plates with two holes used as termination hardware to appropriate lifting points.

Saddle: The portion of a hook directly below the center of the lifting eye.

Sling angle: The angle formed by the legs of a sling with respect to the horizontal plane when tension is placed on the rigging.

Spreader beams: Beams or bars used to distribute the load of a lift across more than one point to increase stability. Spreader beams are often used when the object being lifted is too long or large to be lifted from a single point, or when the use of slings around the load may crush the sides.

Hardware used in rigging includes items such as slings, hooks, shackles, eyebolts, spreader beams, equalizer beams, and blocks. There are also many unique pieces of hardware designed for specific applications. These hardware items must be carefully matched to the load to be lifted to ensure the safety of the load and all workers in the area. Careful inspection and maintenance of all lifting hardware is essential to safe and effective material movement. Rigging hardware should always be inspected before each use.

To understand rigging hardware and applications, it is important to be familiar with terms related to the capacity of an object to withstand weight or force. Many of these terms are used interchangeably. The term rated load for rigging components can be defined simply as the weight that a component can safely lift without fear of breaking, based on manufacturer testing. Other terms for rated load that are used by manufacturers in their product specifications include *rated capacity* and *working load limit (WLL)*.

> **NOTE**
>
> *ASME Standard B30.10*, which is devoted exclusively to hooks, uses the term *rated load* to describe the maximum amount of weight or force that can be safely applied to a component. For this reason, rated load will be used primarily throughout this module.

To determine the weight a component can lift without fear of breaking and ensure that rigging hardware provides safe service, testing is done to determine the breaking point. This weight is referred to as the minimum breaking strength (MBS). The rated load for a component is determined by applying a factor to the MBS. It is not uncommon for the rated load to be only 20 percent of the MBS, resulting in a safety factor of 5.

For example, if the MBS of a hook is 5,400 lbs (2,449 kg) and the factor applied is 20 percent, the rated load would be 1,080 lbs (490 kg). As a result, the proper application of rigging hardware means that the load imposed should always be well below the breaking point. Safety factors may vary, so it is important not to assume that the rated load should be $\frac{1}{5}$ of the MBS. Always work within the specified rated load of the component.

> **CAUTION**
>
> Always refer to the manufacturer's instructions for all types of rigging equipment and its proper application.

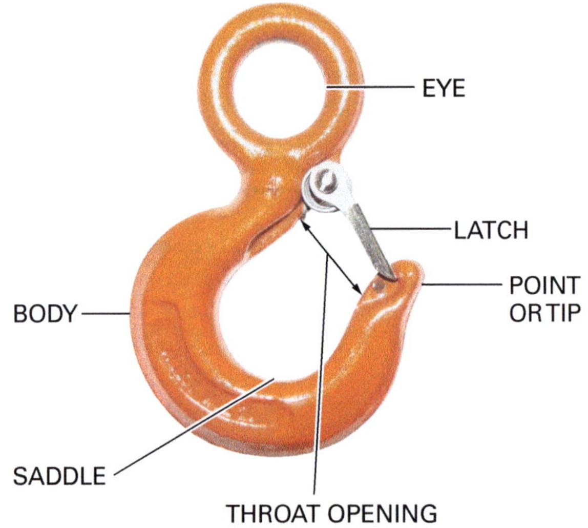

Figure 1 Typical eye hook.

1.1.0 Hooks, Shackles, Eyebolts, and Clamps

Hooks and shackles are essential items for rigging almost every load. Eyebolts and clamps are also commonly used components for lifts. These devices come in different types, and there are a variety of guidelines that must be observed for their safe operation.

1.1.1 Hooks

A rigging hook is typically used to attach a sling to a load. The eye hook (*Figure 1*) is the most common type. Hooks used for rigging must be equipped with safety latches to prevent a connection from slipping off of the hook if any slack develops in the sling. The capacity of a rigging hook is determined by its material of construction, size, and physical dimensions. Information about a specific hook's capacity is always available from the manufacturer or authorized distributor.

Always inspect hooks before each use. Look for wear in the saddle of the hook. Wear in a hook should not exceed 10 percent of the original hook

dimensions. *ASME Standard B30.10* also requires that the hook be removed from service if the throat opening has enlarged 5 percent from its original size, to a maximum of $\frac{1}{4}$" (6.4 mm). *Figure 2* shows some of the inspection points for a hook.

Some other problems that result in the hook being removed from service include the following:

- Missing or illegible manufacturer's identification or rated load information
- Any visually apparent amount of bending or twisting of the hook
- Cracks, nicks, or gouges that could compromise its strength
- Damaged or inoperative latch that does not properly close the throat of the hook
- Any sign of modifications such as grinding, drilling, or machining

Never use a sling on a hook if the sling eye is marginal in size and must be forced over the hook. The body diameter of the hook should fit easily into the sling eye. (Note that information regarding the proper fit of the sling eye to other

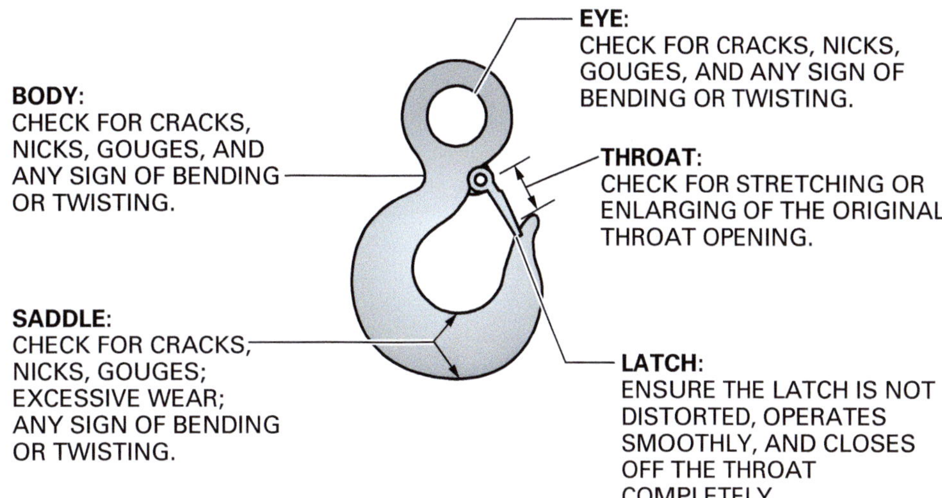

Figure 2 Rigging hook inspection points.

lifting hardware varies depending on the type of sling and hitch used. Slings and hitches will be discussed in more detail in later sections.)

The rated load of the hook is accurate only when the load is suspended from the saddle of the hook. The saddle is the portion of the hook that is directly beneath the center of the lifting eye. If the load is applied anywhere between the saddle and the hook tip, the rated load is reduced considerably, as shown in *Figure 3*. Point loading, also shown in the figure, is not acceptable.

1.1.2 Shackles

A shackle is used to attach an item to a load or to attach slings together. It can be used to attach the end of a wire rope to an eye fitting, hook, or other type of connector. Shackles used for lifting are made of forged steel and are sized by the diameter of the steel in the body, or bow section, of the shackle. However, the rated load capacity, pin size, and distance between the two lugs are often provided in the catalog data. Shackles are made with either screw pins or round pins as a means of safe closure, as shown in *Figure 4*. Screw-pin shackles, the most popular type, are threaded and have threaded pins that screw directly into the body of the shackle. No nut or cotter pin for security is required. A screw-pin shackle designed for synthetic web slings that allows the material to lie flat in the shackle throat is also shown in *Figure 4*.

A round-pin shackle body is not threaded. The round pin itself is threaded, but it is designed to pass completely through the shackle body. The threaded pin then receives a nut and a cotter pin to keep the nut from loosening. This type may also be referred to as a *safety shackle*, since the pin is more secure.

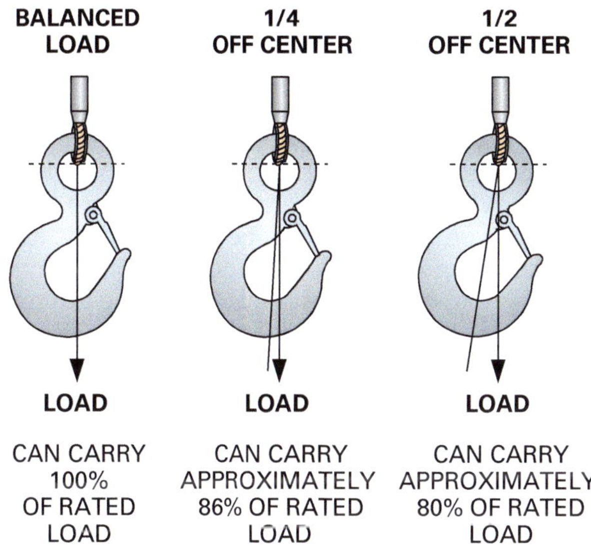

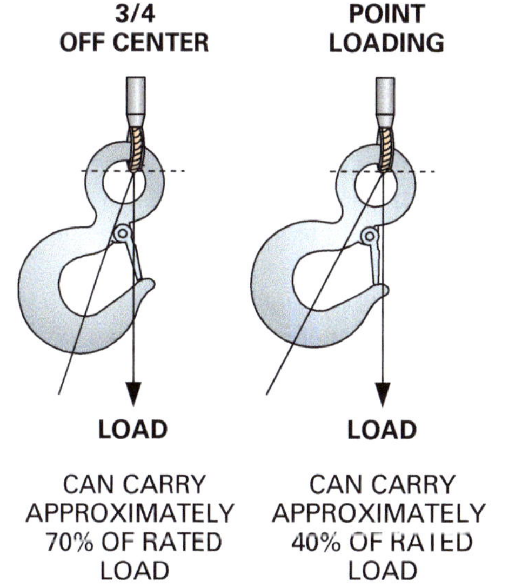

Figure 3 Balanced and off-center loads.

Figure 4 Screw-pin and round-pin shackles.

The threaded pin of a screw-pin shackle must be fully seated in order for the shackle to function at its full rated capacity. In the field, the threaded pin is often left loose to make it easier to remove later, since lifting the load tends to tighten the pin. When it is tight, a tool is needed to loosen it. However, capacity can be significantly affected by leaving the pin loose, and vibration may cause the pin to back out and fall away. The pin must be fully seated in all cases to be safe.

When using shackles, be certain that all pins are straight, all screw pins are completely seated, and cotter pins are used with all round-pin shackles. It is a good practice to replace cotter pins used with shackles after each use to ensure their integrity.

Shackle pins should never be replaced with a common bolt. Common bolts are not hard enough and cannot take the stress normally applied to a shackle pin. Shackles that are stretched, or that have crowns or pins worn more than 10 percent of their original size should be removed from service.

Only shackles with suitable load ratings can be used for lifting. Like hooks, shackles must have the rated load information on the body. When using a shackle on a hook, the pin of the shackle should be hung on the hook, while the load is placed on the bow of the shackle (*Figure 5*). Spacers, such as large washers, can be used on the pin, on each side of the hook, to keep the shackle centered on it. Never use a screw-pin shackle in a situation where the pin can roll as the load sways, as shown in *Figure 6*.

Hook Integrity

The crane hook is a crucial part of many lifts that is used over and over again. Hook failure can happen at any time and may be caused by a number of factors, including cumulative fatigue, overloading, and mechanical abuse such as a free fall to a hard surface.

Hooks can be visually inspected in the field, but they should also be periodically removed, disassembled, and tested to ensure their integrity. In addition to the thorough examination of areas that are not always visible to the operator, various nondestructive testing techniques such as dye-penetrant, magnetic-particle and magnetic-rubber tests can be conducted. These techniques can help detect fatigue and damage that lead to failure. This photo

Figure Credit Konecranes Americas, Inc.

shows a magnetic-particle test being conducted on a hook. Magnetic-particle, or magnaflux, testing can reveal surface flaws, as well as flaws slightly below the surface of ferrous metals.

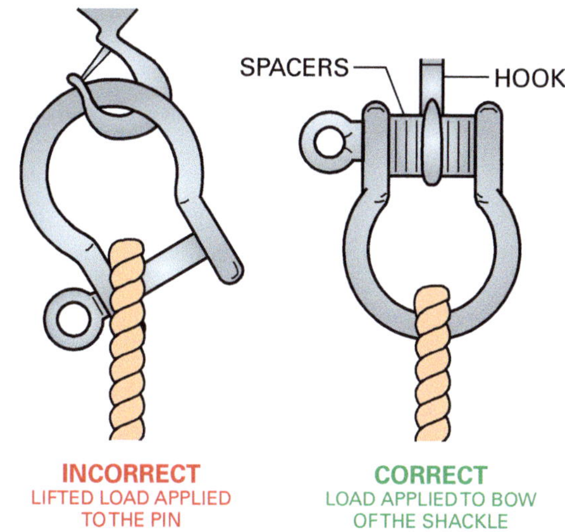

Figure 5 Shackle positioning.

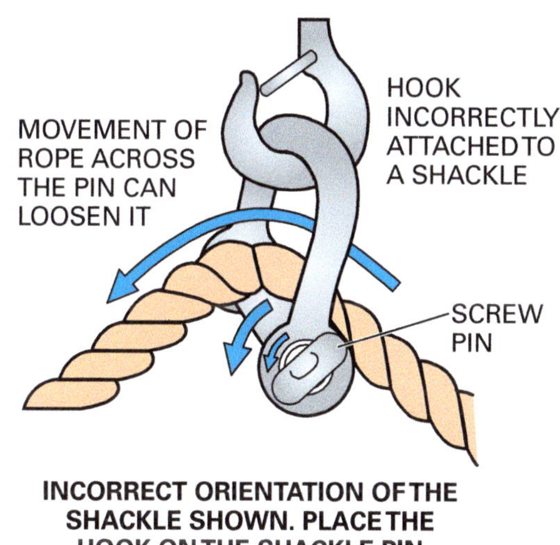

Figure 6 Incorrect attachment of shackle causing pin to roll.

1.1.3 Eyebolts

Eyebolts (*Figure 7*) are often attached to heavy loads to aid in their handling and hoisting. Eyebolts can either have shoulders or be shoulderless. The shouldered type is recommended for use in hoisting applications because it is stronger and resists bending when being pulled from an angle. The shoulder provides a broad surface area of contact between the bolt eye and the load, stabilizing the bolt. The shoulderless type is designed only for lifting a vertical load.

Regardless of the type used, the rated load of all eyebolts however, is reduced during angular loading. Loads should always be applied in the same plane as the eye to reduce the chance of bending. Aligning the eye of the bolt with the shackle or other connector is particularly important when a bridle hitch is used. Proper and improper orientation of the eyebolt is shown in the bottom left corner of *Figure 7.*

When installed, the shoulder surface of the eyebolt must make full contact with the load surface. Washers or other suitable spacers may be used to ensure that the shoulders are in firm contact with the working surface. A threaded blind hole used for an eyebolt should have a minimum depth of $1\frac{1}{2}$ times the bolt diameter to ensure sufficient thread engagement. However, this does not ensure that the threaded portion of the hole is deep enough to allow the shoulder to make firm contact with the surface. For shouldered eyebolts, a blind hole should be deep enough so that the threaded portion is deeper than the length of the threaded eyebolt shank. This prevents the eyebolt from bottoming out or reaching the end of the threads before it is properly seated.

Swivel hoist rings, also shown in *Figure 7*, may be used instead of eyebolts. These devices swivel to the desired lift position and therefore do not require any load-rating reduction due to an angular pull.

1.1.4 Beam and Plate Clamps

Beam clamps (*Figure 8*) are used to connect hoisting devices to beams so the beams can be lifted and positioned. Observe the following guidelines when using beam clamps:

- Do not use clamps unless they are designed, load tested, and stamped by an engineer. Homemade clamps should never be used.
- Ensure that the clamp fits the beam and has the necessary rated load capacity.
- Ensure that the clamp is securely fastened to the beam.
- Be careful when using beam clamps where angular lifts are required. Most are designed for straight vertical lifts only.
- Be certain the rated load appears on the beam clamp and is legible. Never load a beam clamp beyond its rated load capacity.
- Attach rigging to the beam clamp using a shackle; do not place a hoist hook directly in the beam clamp lifting eye.
- Examine the clamp for the following defects, and remove it from service if any are present:
  - Jaws of the beam clamp have been opened more than 15 percent of their normal opening
  - Lifting eye worn, bent, or elongated
  - Excessive rust or corrosion
  - Capacity and beam size information unreadable

Plate clamps (*Figure 9*) attach quickly to structural steel plates to allow for easier rigging attachment and handling of the plate. There are two basic types of plate clamps: serrated-jaw clamps and screw clamps.

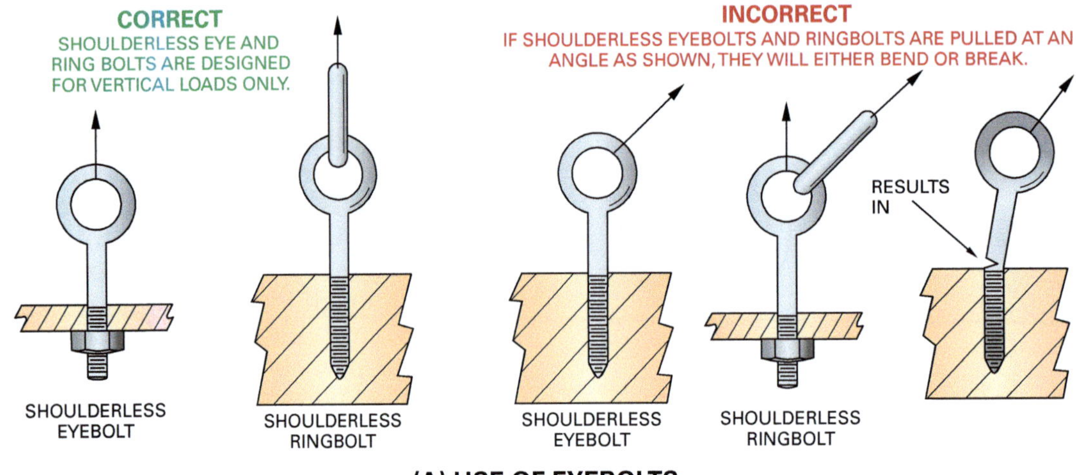

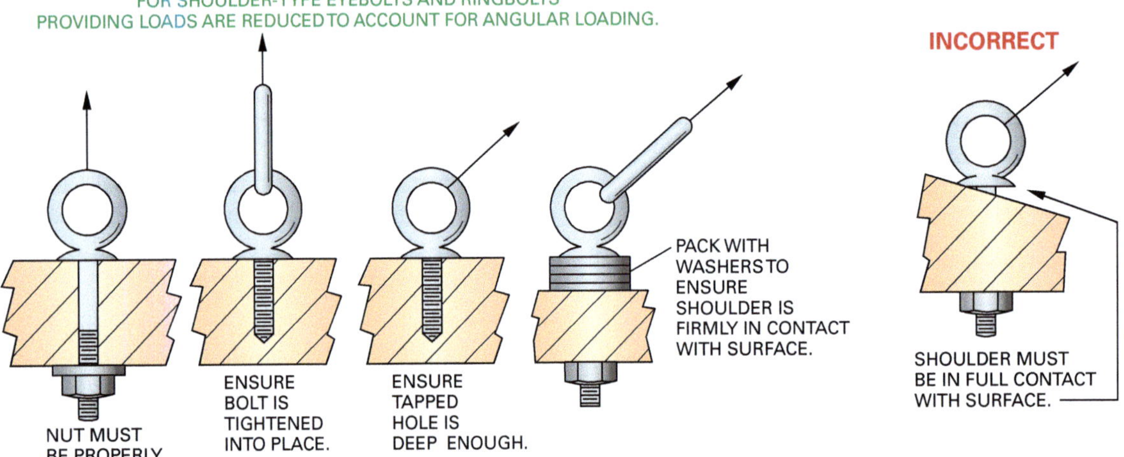

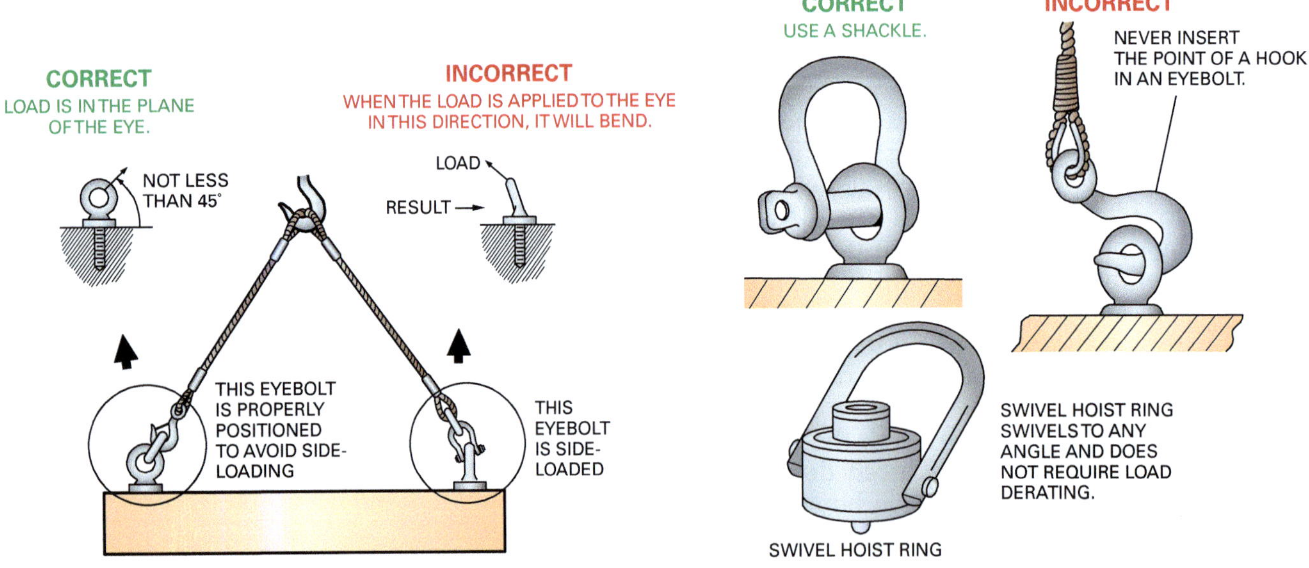

Figure 7 Eyebolt installation and lifting criteria.

Figure 8 Typical beam clamp application.

Serrated-jaw clamps are designed to grip a single plate for hoisting and are available with a locking device. The jaws of a serrated-jaw clamp are cam-operated; as the plate tries to escape the jaws, they rotate down slightly and press more tightly against the plate. Note that all plate clamps are designed to lift only one plate at a time. Always follow the manufacturer's recommendations for their use and remain within the rated load of the clamp.

Plate clamps must be examined before use and removed from service if any of the following defects are present:

- Identifying information and/or the rated load is absent or unreadable
- Distortion of the opening or wear of the jaw teeth
- Cracks in body
- Loose or damaged rivets

- Lifting eye worn, bent, or elongated
- Excessive rust or corrosion

1.2.0 Lugs, Turnbuckles, Plates, and Beams

There are many different types of hardware used for both common and unique rigging applications. Some of these include lifting lugs, turnbuckles, plates, and spreader beams. Riggers and crane operators must be familiar with different varieties of these components and know how to use them safely and effectively.

1.2.1 Lifting Lugs

Lifting lugs are often welded or bolted to an object by the manufacturer so that it can be lifted or moved more easily. A welded lifting lug is shown in *Figure 10*. They are typically designed and located to suspend a load near its center of gravity (CG) and support it safely. Lifting lugs should be used for straight, vertical lifting only. They are more likely to bend or fail if they are side-loaded. When lifting lugs are field-installed, consideration must be given to how they will be used in lifting and from what direction the stress will be applied.

1.2.2 Turnbuckles

Turnbuckles are available in a variety of sizes. They are used to adjust the length of rigging connections without twisting the cables or ropes. Two common types of turnbuckles use eyes and jaws as terminations (*Figure 11*). They can be used in any combination. Turnbuckles with hooks are also available but should not be used for rigging purposes as they can easily be disconnected and

Figure 9 Plate clamps.

Figure 10 Welded lifting lug on a spreader beam.

have lower rated load capacities than the other types. The rated load for turnbuckles is based on the diameter of the threaded rods.

Observe the following guidelines and considerations when selecting turnbuckles:

- Turnbuckles should be made of forged steel and should not be welded.
- Do not use turnbuckles with open hooks for rigging loads.
- When using turnbuckles with multi-leg slings, do not use more than one turnbuckle per leg.

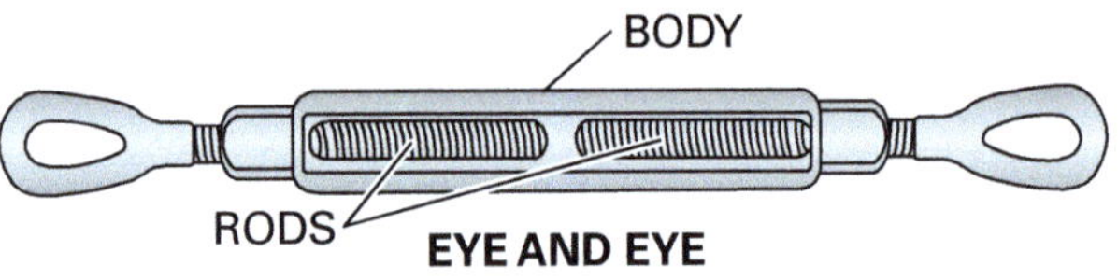

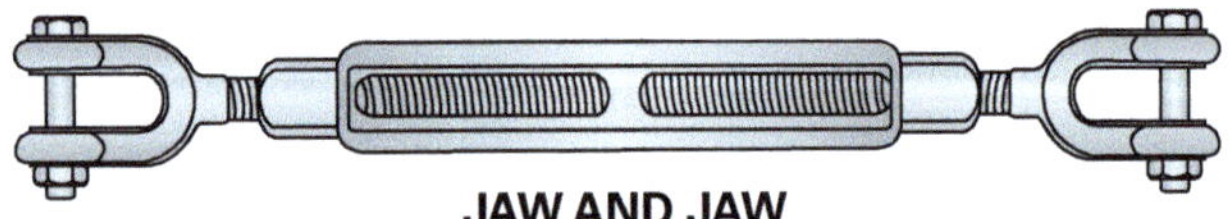

Figure 11 Turnbuckles.

- Do not use jamb nuts on turnbuckles that do not come equipped with them.
- Turnbuckles should not be overtightened. Perform tightening with a wrench of the proper size, using only as much force as a person can achieve by hand, without applying added leverage.

When inspecting turnbuckles, check for bent threaded rods and thread damage, and look for cracks in the body, threaded rods, and terminations.

1.2.3 Rigging Plates and Links

Rigging plates and links (*Figure 12*) are made for specific uses. The holes in the plates or links may be different sizes and may be placed in different locations in the plates. This creates a piece of connecting hardware that can be used for rigging the same type of objects repetitively. Plates with two holes are called **rigging links**. Plates with three or more holes are called **equalizer plates**. Equalizer plates can be used to level loads when the legs of a sling are unequal. Plates are attached to the rigging with high-strength pins or bolts.

Inspect rigging plates and links before using them, and remove them from service if any of the following are present:

- Cracks in body
- Worn or elongated lifting eye
- Excessive rust or corrosion

1.2.4 Spreader and Equalizer Beams

Spreader beams (*Figure 13*) are used to support and protect long loads. These devices prevent the load from tipping, sliding, or bending. They decrease the possibility of a low **sling angle** and help prevent the sling from crushing the load. Equalizer beams are used to balance the load on sling

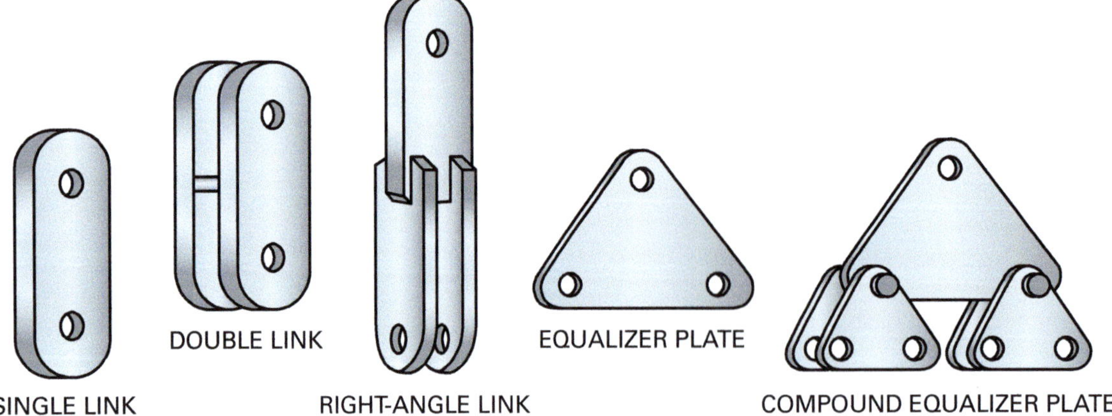

Figure 12 Rigging plates and links.

NCCER – *Basic Rigger*

legs and to maintain equal loads on dual hoist lines when making multi-crane, or tandem, lifts.

> Spreader and equalizer beams are discussed in more detail in NCCER's *Mobile Crane Operations Level Two* and *Mobile Crane Operations Level Three*.

Both types of beams are often fabricated to suit a specific application. They are commonly made of heavy pipe, I-beams, or other suitable material. Custom-fabricated spreader or equalizer beams must be designed by a qualified person (often an engineer) and have their capacity clearly stamped on the side. All such beams should be tested at 125 percent of their rated load. Information on any beams in use should be kept on file.

The rated load capacity of beams designed for use with multiple attachment points depends upon the distance between attachment points. For example, if the distance between the attachment points is doubled, the beam capacity is typically cut in half.

Before use, a spreader or equalizer beam should be inspected for the following:

Figure 13 Typical use of a spreader beam.

- Beam is clearly marked with the following information:
  - Manufacturer's name and address
  - Serial number
  - Device weight, if over 100 pounds (45 kg)
  - Rated load
  - *ASME Standard BTH-1* Design Category
  - *ASME Standard BTH-1* Service Class

Moving Materials Using Vacuum Principles

There are many devices used by cranes and other heavy equipment to grab, lift, and position construction materials. However, one method is a little more unusual than the others.

Vacuum lifters are equipped with vacuum pumps, usually powered by a small diesel engine, that draw out the air between the lift mechanism and the load. Irregular surfaces that cannot provide a tight seal are not candidates for vacuum lifting because integrity of the seal is essential for them to function. Vacuum lifters have been specifically designed for pipe, slabs, beams, and similar items. An advantage of this method is the elimination of a great deal of rigging hardware and the labor required to assemble it prior to a lift, saving both time and money. These devices and their design criteria are covered by *ASME Standard B30.20, Below-The-Hook Lifting Devices*, and *ASME Standard BTH-1, Design of Below-The-Hook Lifting Devices*. The power of a vacuum is surprising—rated load capacities for the strongest vacuum lifters exceed 20 tons (18 metric tons).

(A) PLATE LIFTER

(B) SLAB LIFTER

Figure Credit: Vaculift ™ , Inc. d.b.a. Vacuworx®

- Welds are free of cracks or other significant flaws.
- No cracks, nicks, gouges, or corrosion are present.
- Attachment points are not damaged or distorted.

Additional Resources

ASME Standard B30.5, Mobile and Locomotive Cranes. Current edition. New York, NY: American Society of Mechanical Engineers.

ASME Standard B30.10, Hooks. Current edition. New York, NY: American Society of Mechanical Engineers.

ASME Standard B30.20, Below-The-Hook Lifting Devices. Current edition. New York, NY: American Society of Mechanical Engineers.

ASME Standard BTH-1, Design of Below-The-Hook Lifting Devices. Current edition, New York, NY: American Society of Mechanical Engineers.

29 *CFR* 1926, Subpart CC, **www.ecfr.gov**

29 *CFR* 1926.251, **www.ecfr.gov**

29 *CFR* 1926.753, **www.ecfr.gov**

Mobile Crane Safety Manual (AEM MC-1407). 2014. Milwaukee, WI: Association of Equipment Manufacturers.

Willy's Signal Person and Master Rigger Handbook, Ted L. Blanton, Sr. Current edition. Altamonte Springs, FL: NorAm Productions, Inc.

North American Crane Bureau, Inc. website offers resources for products and training, **www.cranesafe.com**

1.0.0 Section Review

1. Which of the following statements about shackles is true?
    a. Shackles with pins worn more than 10 percent of their original size should be removed from service.
    b. The size of a shackle is based on its pin size.
    c. When connecting a shackle to a hook, the body of the shackle is hung on the hook.
    d. If a shackle is pulled at an angle rather than a straight pull, its rated load increases.

2. Which rigging devices require the *ASME Standard BTH-1* Design Category and Service Class on the labeling?
    a. Spreader beams
    b. Shackles
    c. Hooks
    d. Eyebolts

2.0.0 SLINGS AND HITCHES

Objective

Identify and describe various types of slings and sling hitches.

   a. Identify and describe wire-rope slings and their proper care.

   b. Identify and describe synthetic slings and their proper care.

   c. Identify and describe chain slings and their proper care.

   d. Explain the significance of sling angles and describe common hitches.

   e. Describe how to properly rig and handle piping materials and rebar.

   f. Identify and describe how to use taglines and knots for load control.

   g. Identify common rigging-related safety precautions.

Performance Tasks

1. Inspect various types of rigging components and report on the condition and suitability for a task.

2. Configure a sling to produce a single-wrap basket hitch.

3. Configure a sling to produce a double-wrap basket hitch.

4. Configure a sling to produce a single-wrap choker hitch.

5. Configure a sling to produce a double-wrap choker hitch.

6. Select the correct tagline for a specified application.

7. Tie specific instructor-selected knots.

Trade Terms

Basket hitch: A common hitch made by passing a sling around a load or through a connection and attaching both sling eyes to the hoist line.

Bird caging: A deformation of wire rope that causes the strands or lays to separate and balloon outward like the vertical bars of a bird cage.

Choker hitch: A hitch made by passing a sling around the load, and then passing one eye of the sling through the other. The one eye is then connected to the hoist line, creating a choke-hold on the load.

Independent wire rope core (IWRC): Wire rope with a core consisting of wire rope, as opposed to a fiber or single-stranded core; considered to be the most durable for rigging applications.

Spur track: A relatively short branch leading from a primary railroad track to a destination for loading or unloading. A spur is typically connected to the main at its origin only (a dead end).

Tagline: A rope attached to a lifted load for the purpose of controlling load spinning and swinging, or used to stabilize and control suspended attachments.

Vertical hitch: A simple hitch that uses one end of a sling to connect to a point on the load and the opposite end to connect to the hoist line. Also known as a *straight-line hitch*.

Slings are available in a variety of configurations. Some of these include wire rope, synthetic web, and chain slings, which are all presented in this section. Some slings are made from natural fibers or synthetic rope, but these are not typically used for rigging applications involving mobile cranes. All types of slings are likely to need protection to ensure they are not damaged by sharp edges, protrusions, and other potential sources of damage. In many cases, the load itself also benefits from such protection.

Like hooks used in rigging, an ASME standard is devoted exclusively to slings. *ASME Standard B30.9, Slings*, provides a wealth of information about slings and their proper application. In addition, slings are also the topic of 29 *CFR* 1910.184.

2.1.0 Wire-Rope Slings

Wire-rope slings (*Figure 14*) are made of high-strength steel wires formed into strands and wrapped around a supporting core. They are lighter and easier to handle than chain slings, and can withstand substantial abuse and relatively high temperatures. However, because wire-rope slings can slip, the use of synthetic slings is often preferred. Wire-rope slings are still being used, so it is still important to learn the design, characteristics, applications, inspection, and maintenance of wire-rope slings.

2.1.1 Characteristics and Applications

Wire-rope slings usually consist of six strands, with each strand containing an average of 19 wires (written as 6 x 19), laid in a specific pattern

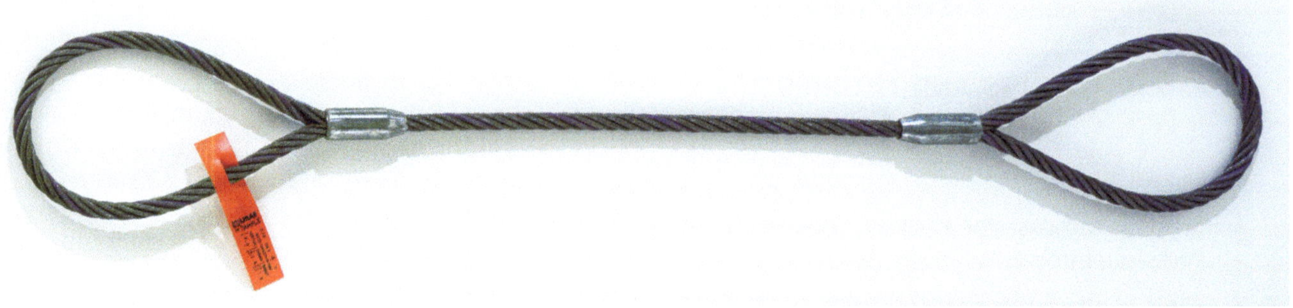

Figure 14 Typical wire-rope sling.

around a wire rope core. Cores other than wire are available, such as fiber cores, but independent wire rope core (IWRC) is the most common and considered the most durable. Wire rope will suffer the loss of only about 1 percent of its strength if a wire breaks. *Figure 15* shows examples of the most common wire-rope sling applications used in construction rigging.

Wire-rope slings should be protected where the slings are wrapped around the sharp edges of the object to be lifted (*Figure 16*). Of course, the load can often be damaged by the sling as well. Even if the edge of the load is a soft material that would not cut the sling, one or more individual wires may break if they are sharply bent. If wire-rope slings are kinked, the severe bending stress and displaced strands allow for unequal distribution of the live load, with some strands taking more than their normal load. Damage done to the rope by kinking is usually permanent, resulting in the disposal of many slings as well as some failures.

Protective material can include simple pieces of wood to separate the sling from the load. However, protective materials are also available in a variety of manufactured forms. *Figure 17* shows examples of some materials designed for the task. The plastic protectors shown here have magnets embedded in them, allowing them to remain in place on ferrous materials before the sling is applied. Others are made of metal alloys and are designed to slip around a wire-rope sling at any point along its length. Some type of protective material is often needed for any type of sling, depending upon the hitch used and the nature of the load being lifted. These protective materials are not used with wire rope only.

A special type of wire-rope sling, called a *braided-belt sling*, is made by braiding six or more small-diameter wire ropes together. This provides a sling with a wide, flat bearing surface of great strength and flexibility in all directions. They are especially well suited for a basket hitch or a choker hitch where sharp bends are encountered.

Rigging hardware placed into the eye of a sling must be appropriate to prevent failures and/or sling damage. *ASME Standard 30.9*, Section 2.10.4(p) states that "an object in the eye of a (wire-rope) sling should not be wider than one half the length of the eye nor less than the nominal sling diameter." For example, if the eye of wire-rope sling is 6" (15 cm) long, then nothing wider than 3" (7.5 cm) should be placed through the eye. (Half of 6" is 3", or half of 15 cm is 7.5 cm.) The second portion refers to the minimal width of hardware placed in the sling eye; it should never be smaller in diameter than the wire-rope sling itself. Sling manufacturers provide information about the application of their products that should be followed as well.

2.1.2 Storage and Inspection

Store slings in a rack to keep them off the ground. The rack should be in an area free of moisture and away from acid, acid fumes, or extreme heat. Both moisture and fumes can lead to corrosion. Never let slings lie on the ground in areas where heavy machinery may run over them, or where they can become filled with sand and other abrasives internally.

> **WARNING!**
> Broken wires in a wire-rope sling are extremely sharp and can easily cut or puncture the skin. Always wear gloves when handling wire-rope slings and when inspecting them by running your hand along their length.

Slings should be regularly inspected for broken wires, kinks, rust, or damaged fittings. A visual inspection should be made before each use. Inspections at the beginning of each shift or work day are required by *ASME Standard B30.9*. Any slings found to be defective should be removed from service for repair or disposal. Repairs should not be attempted by anyone other than the manufacturer or other qualified party.

NCCER – *Basic Rigger*

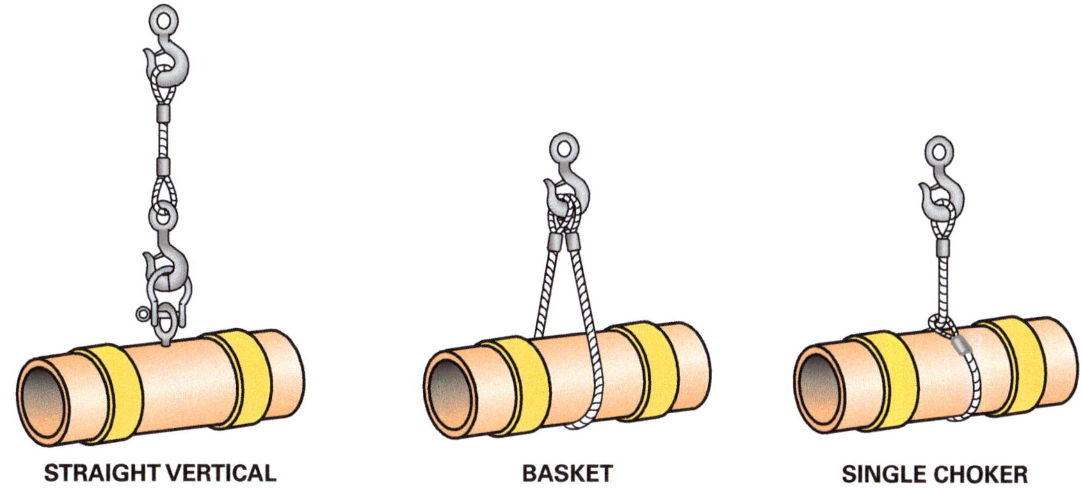

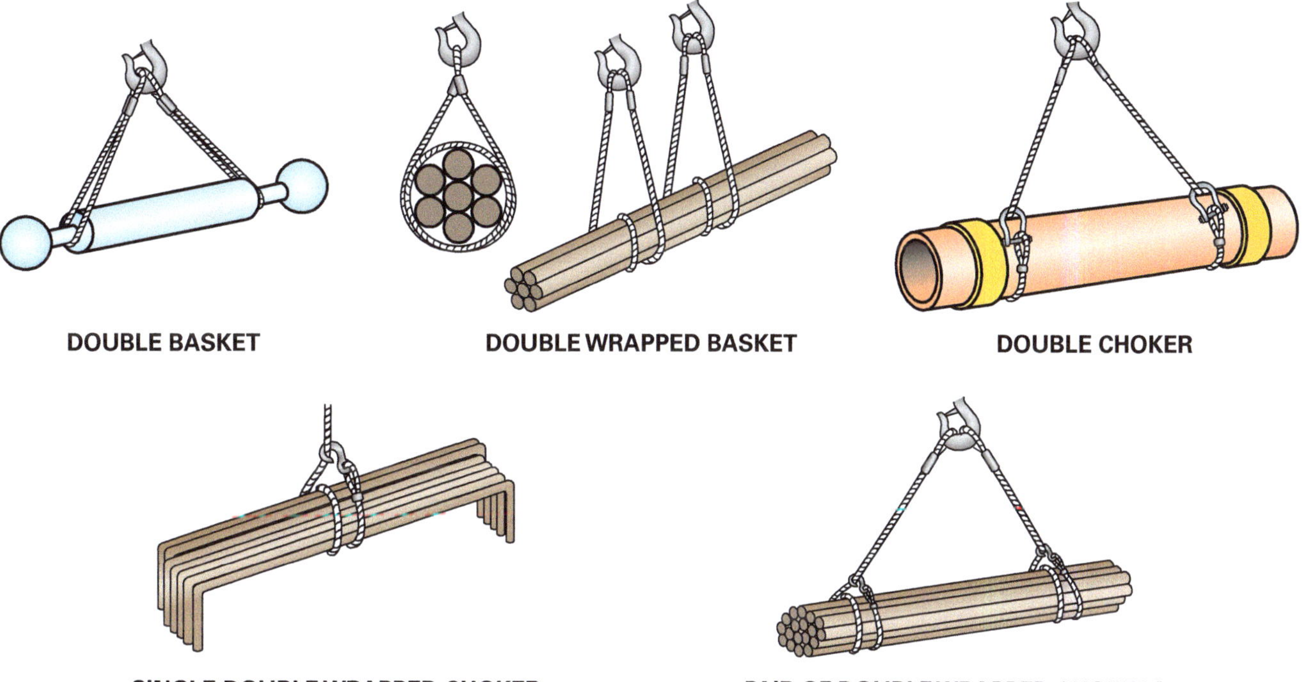

Figure 15 Examples of wire-rope sling applications.

Wire-rope slings must be properly tagged with specific information. This is required by both ASME and OSHA standards. *Figure 18* shows a typical wire-rope tag. A missing or illegible identification tag or data plate is a cause for sling disposal. Information that must be on the tag includes the following:

- Name or trademark of manufacturer or repair organization
- Rated load for at least one hitch and the angle upon which it is based
- Size/diameter
- Number of legs, if more than one

Wire-rope slings should also be removed from service if any of the following conditions are discovered; refer to *Figure 19* for examples:

- Localized abrasion or scraping that reduces the diameter of the sling by more than 5 percent of its original size
- Rope distortion, which includes kinking, crushing, and bird caging

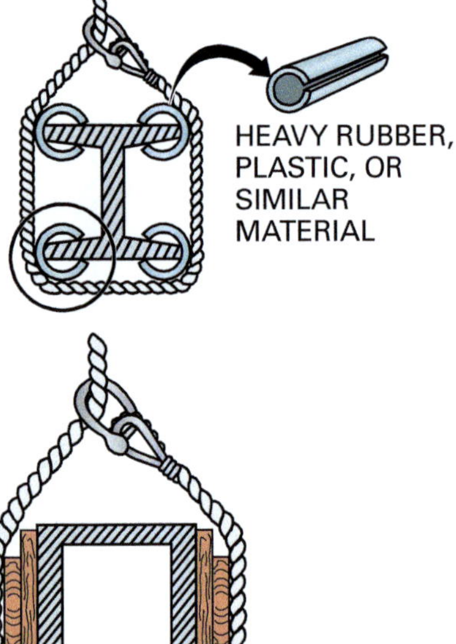

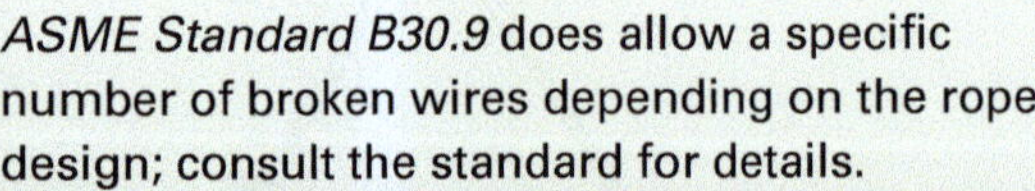

Figure 16 Use of sling protection.

- Evidence of heat damage, usually indicated by discoloration
- Damaged end fittings, such as cracks, deformation, and excessive wear
- Severe corrosion
- Any other type of damage that results in doubt about the integrity of the sling
- Broken wires

ASME Standard B30.9 does allow a specific number of broken wires depending on the rope design; consult the standard for details.

(A) UPPER AND LOWER WIRE-ROPE SADDLES

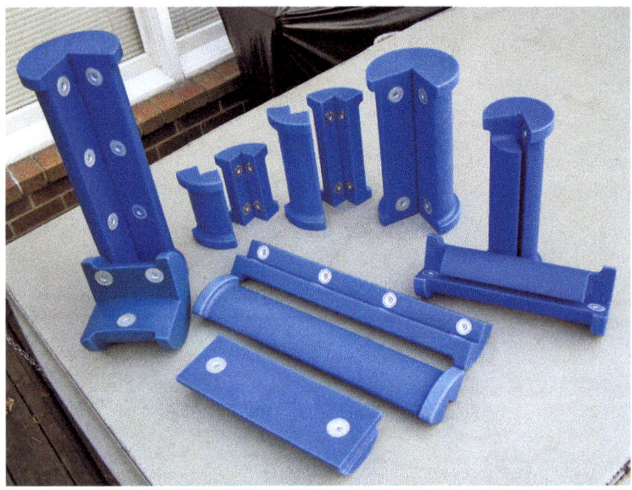

(B) PLASTIC SLING PROTECTORS FOR ALL SLING TYPES

Figure 17 Sling protection products.

Sling Maintenance

Slings are required to have periodic inspections. *ASME Standard B30.9* requires that documentation of the most recent periodic inspection be maintained. Someone is responsible for maintaining that documentation as well as the slings themselves.

Most sling manufacturers now offer radio-frequency identification (RFID) chips permanently attached to new slings. Passive RFID chips called transponders are scanned by devices that receive the signal and synchronize important data with a database based on the embedded serial number. RFID tags are extremely durable and use Bluetooth technology; the tag can be located and information downloaded from a reasonable distance and without a direct line of sight to the receiver. The technology is not only used for slings, but for many other rigging components such as hooks and shackles.

Figure Credit: Lift-All Company, Inc.

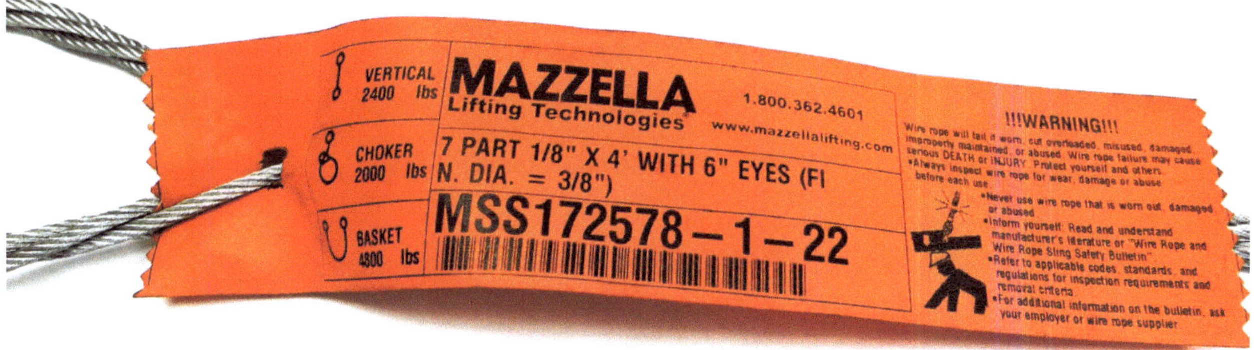

Figure 18 Wire-rope sling tag.

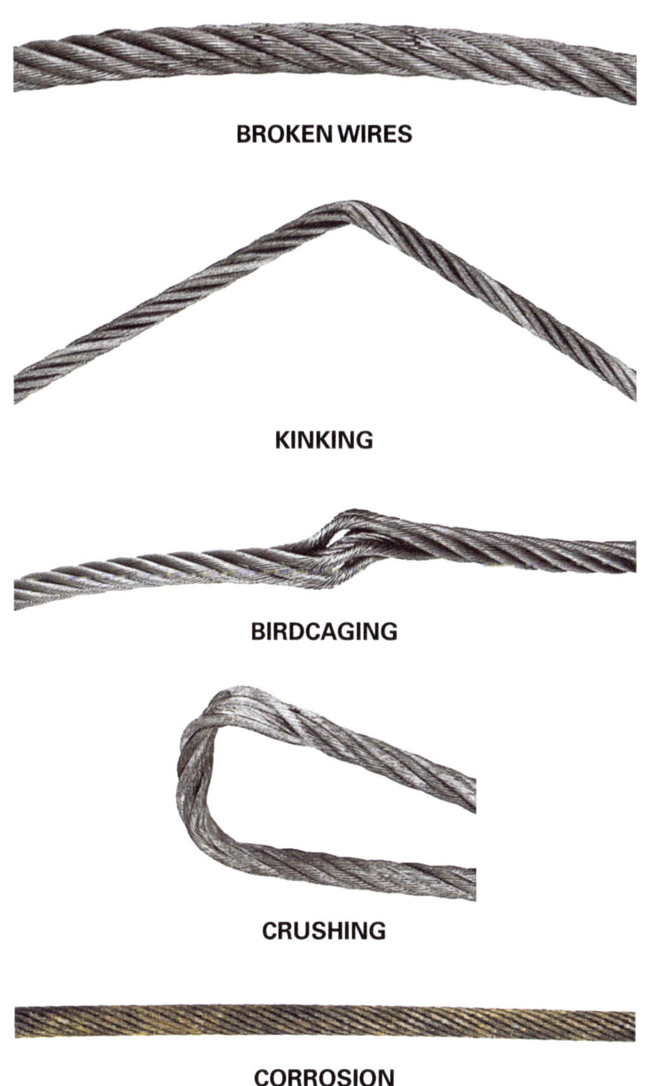

Figure 19 Common types of wire-rope damage.

2.2.0 Synthetic Web Slings

Synthetic slings are widely used to lift loads, and they are especially suitable for easily damaged ones. Common types of synthetic slings, as well as guidelines for their storage and inspection, are outlined in the following sections.

2.2.1 Characteristics and Applications

Synthetic web slings commonly used for construction rigging are made of polyester, nylon, or other high-performance synthetic materials. While synthetic slings are useful in many situations, there are some applications and environments that are likely to damage them.

Synthetic web slings are available in a number of configurations, with common types shown in *Figure 20*. These include the following:

- *Endless slings* – Endless slings are also referred to as *continuous loop slings* or *grommet slings*. The ends of a piece of webbing are overlapped and sewn together to form an endless loop. They can be used for a vertical hitch, bridle hitch, choker hitch, or basket hitch.
- *Standard eye-and-eye slings* – Webbing in these slings is sewn to form a flat body with an eye on each end. The eye is in the same plane as the sling body. They are used for the same purposes as wire-rope slings of similar design.
- *Round slings* – Round slings are available in endless and eye-and-eye styles. They are generally made from a continuous length of polyester filament yarn covered by a woven sleeve. The eye-and-eye style of round sling simply has a sleeve wrapped around the two sides, forming eyes on the end.
- *Twisted eye-and-eye slings* – The eyes in twisted eye-and-eye slings are sewn at right angles to the plane of the sling body.

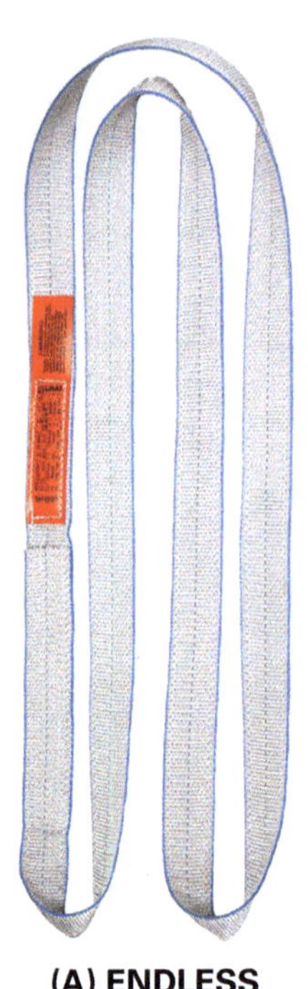

(A) ENDLESS

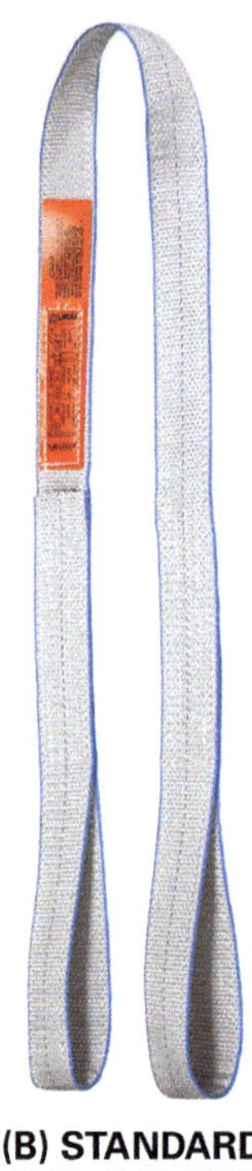

**(B) STANDARD
EYE-AND-EYE**

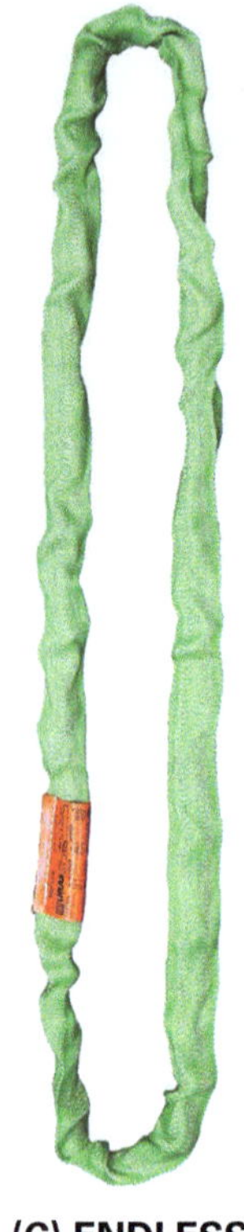

**(C) ENDLESS
ROUND**

**(D) TWISTED
EYE-AND-EYE**

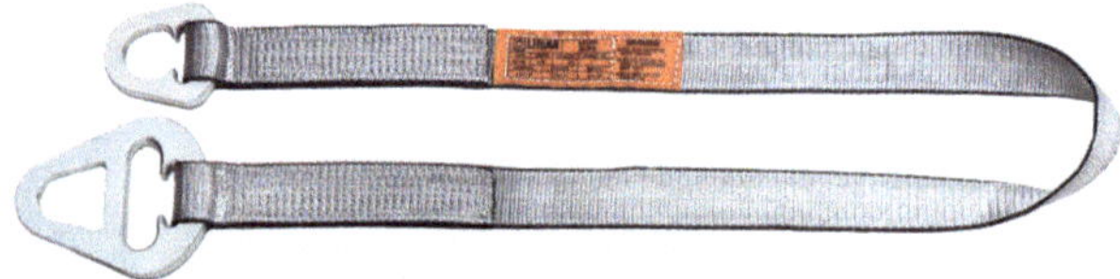

**(E) SLING WITH TRIANGLE
AND CHOKER HARDWARE**

Figure 20 Examples of synthetic web slings.

- *Slings with attached hardware* – Web slings are also available with triangular and choker hardware end fittings.

Synthetic web slings can be cut by sharp edges and corners or damaged by excessive sling angles. A razor sharp edge is not required for damage to be done; a sling stressed with weight against a seemingly dull edge can result in failure. The same type of protection used with wire-rope slings must also be applied to web slings. Advantages of synthetic web slings include the following:

- Their texture and width reduce marring or scratching of polished, finished, or soft surfaces.
- They are less likely to crush fragile surfaces.
- They mold themselves to the shape of the load.
- Moisture and a variety of chemicals do not damage them.

- As long as they remain clean and dry, they are nonsparking and nonconductive. (Excessive soil can also make a sling conductive.)
- They minimize twisting of the load during lifting.
- They do not stain ornamental materials.
- They are lightweight and soft, making them easier and safer to handle.
- They stretch somewhat under load, allowing them to withstand shock.

ASME Standard B30.9, Section 5.10.4(p) provides the following guidance for fitting hardware into the eye of the sling: "An object in the eye of a sling should not be wider than one-third the length of the eye." For example, if the eye of a synthetic sling is 9" (23 cm) long, the maximum width of a hook or other object placed through the sling eye is 3" (7.5 cm). This is smaller than the maximum width allowed for wire-rope sling eyes, which is half the length of the eye.

NCCER – *Basic Rigger*

2.2.2 Storage and Inspection

Nylon and polyester web slings should not be stored or allowed to contact objects at temperatures over 194°F (90°C) or below −40°F (−40°C). This temperature range is a requirement of *ASME Standard B30.9*; note that manufacturers may specify a more limited range for their products that would supersede the standard. It is unlikely that storage facilities would reach these temperatures, but the surface of a rigged load could in some cases. It is also important to limit long-term exposure to sunlight or ultraviolet (UV) light. Both can affect the strength of synthetic web slings. Sling manufacturers can assist with information related to sunlight and UV light exposure.

The strength of synthetic webbing slings can also be degraded by certain chemicals. This includes exposure in the form of solids, liquids, vapors or fumes. Consult the sling manufacturer for a specific list of the chemicals and their variations that can damage synthetic slings.

Like other types of slings, synthetic web slings should not be laid on the ground where they can be damaged or run over by heavy equipment. They also must be properly labeled with specific information. An unreadable or missing label is cause for removing a sling from service. However, a damaged label can be repaired by the manufacturer or other qualified party. When repairs are made, the repair organization must record their information on the label. Labels must include the following information:

- Name or trademark of manufacturer, or if repaired, the entity performing repairs
- Manufacturer's code or stock number
- Rated load for at least one hitch type and the angle upon which it is based
- Type of synthetic web material
- Number of legs, if more than one

Per *ASME Standard B30.9*, synthetic web slings should be removed from service if any of the following conditions are found during an inspection; refer to *Figure 21* for examples:

- Missing or illegible sling identification and rated load information
- Acid or caustic burns
- Melting or charring of any part of the sling
- Holes, tears, cuts, or snags
- Broken or worn stitching
- Excessive abrasion
- Knots in any part of the sling
- Discoloration and brittle or stiff areas on any part of the sling, which may indicate chemical or ultraviolet/sunlight damage has occurred
- For attached hooks, removal criteria as stated in *ASME Standard B30.10*
- For other attached rigging hardware, removal criteria as stated in *ASME Standard B30.26*
- Other conditions, including visible damage, that create doubt as to the continued use of the sling

2.3.0 Chain and Metal-Mesh Slings

Chain slings and metal-mesh slings are often used for lifts in high heat or rugged conditions. They are versatile because they can be adjusted over the center of gravity, and they are also very durable. However, because chain and metal-mesh are very heavy, they can be harder to inspect than other types.

2.3.1 Chain Slings

For some lifts, chains slings are more appropriate than wire-rope or web slings. For example, the use of chain slings is recommended when lifting rough castings that would quickly destroy wire or synthetic web slings. They are also used in high-heat applications or where wire-rope chokers are not suitable, and for dredging and other marine work because they withstand abrasion and corrosion better than wire rope. *Figure 22* shows some common configurations of chain slings and hooks.

Synthetic Slings and Rigging Incidents

Industrial Training International (ITI) compiled a webinar entitled "Rigging & Sling Failures: Case Studies & Solutions" in 2013. During ITI's research for the webinar, the results of which were supported by a poll taken of attendees during the webinar, it was found that over 80 percent of rigging accidents were related to synthetic slings. The incidents were generally due to cutting or severe abrasion resulting from a lack of sling protection used during lifting. As many rigging professionals have stated over the years, if you have a synthetic sling in your right hand, you should have sling protection in your left. All slings, regardless of their materials of construction, should be protected in use.

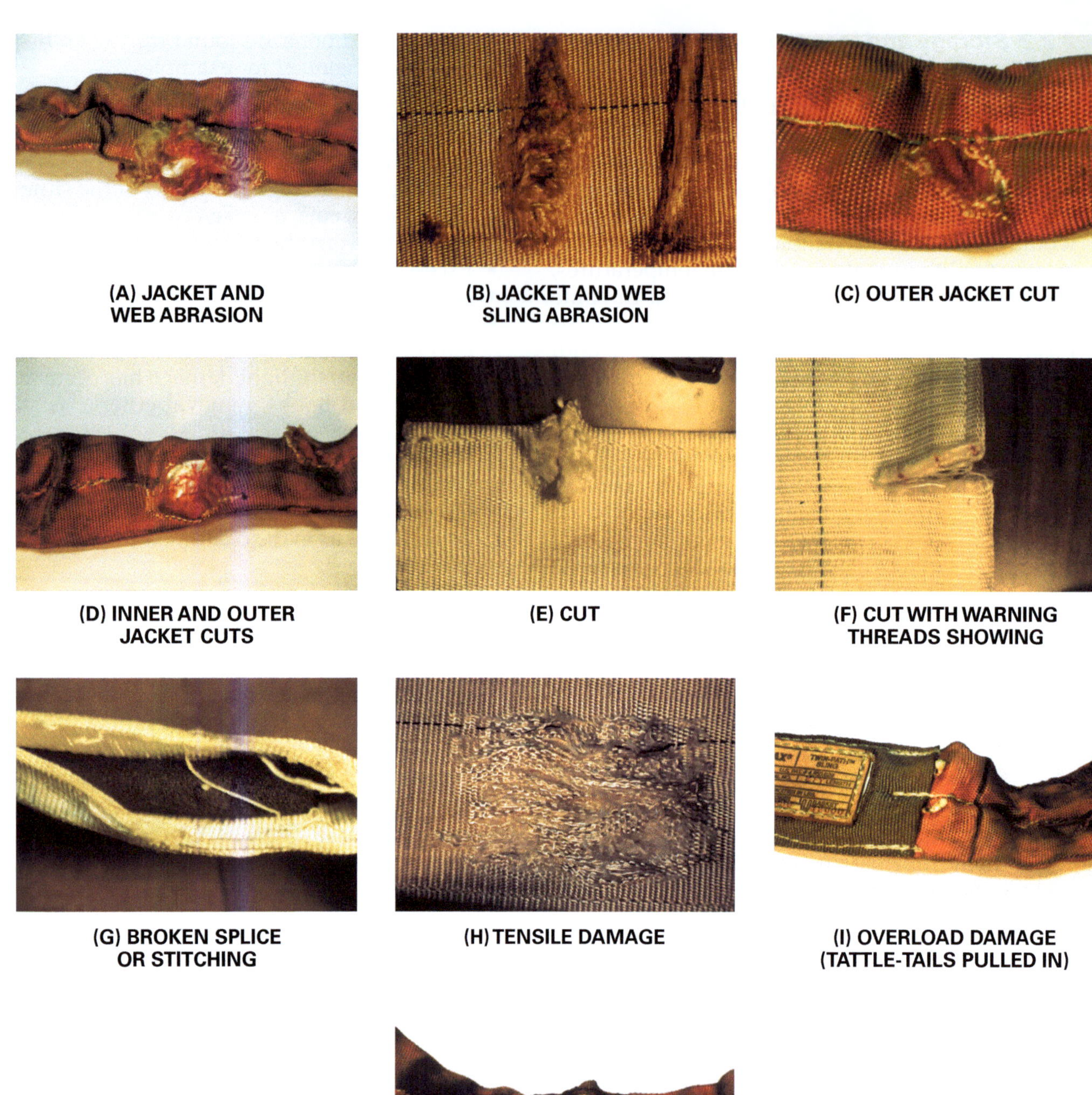

Figure 21 Examples of synthetic-web sling damage.

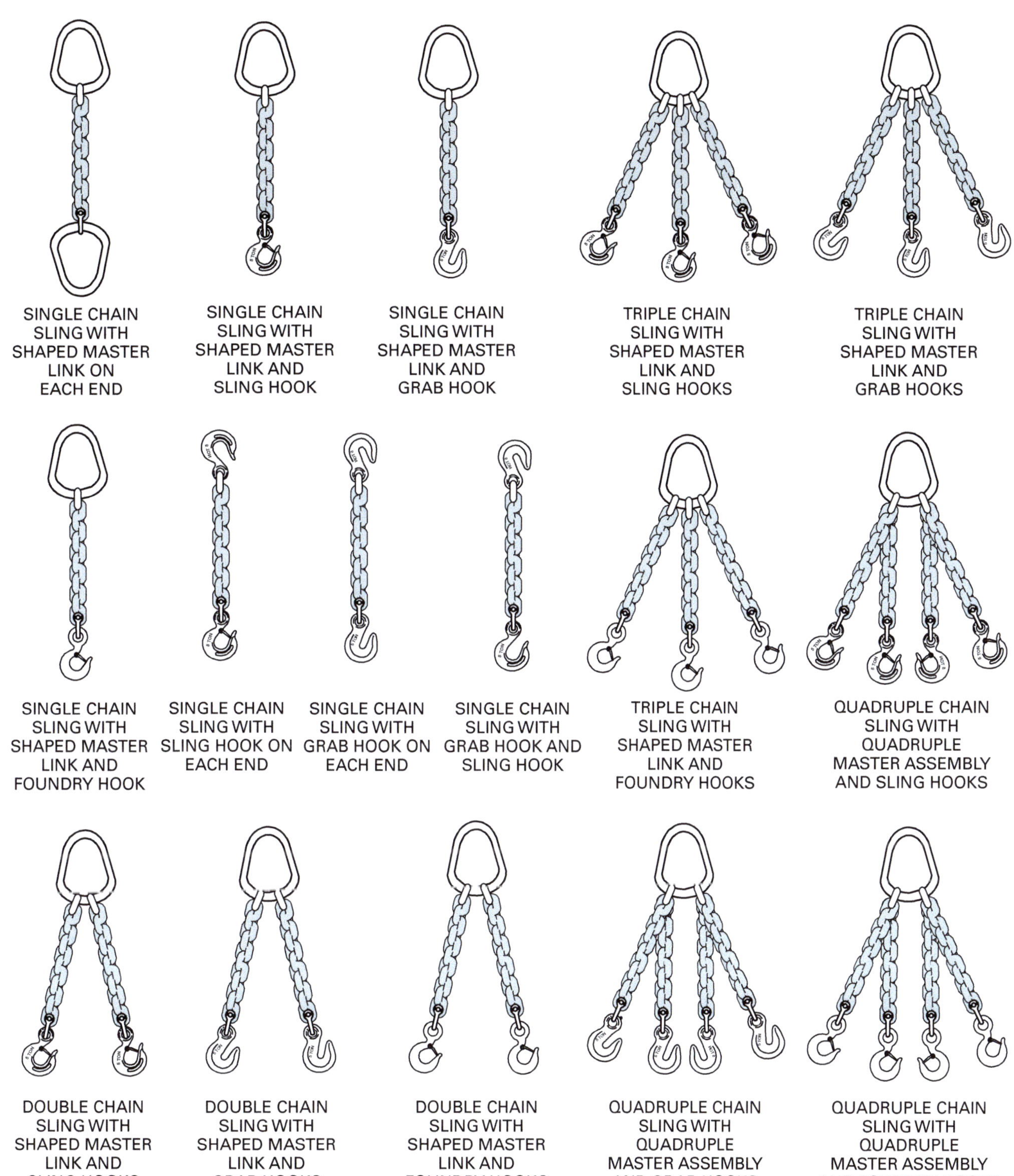

Figure 22 Common chain slings and hooks.

Chain links have two sides. Failure of either side causes the link to open and drop the load. Wire rope is frequently composed of as many as 114 individual wires, all of which must fail before the rope finally breaks. In other words, wire rope is more likely to experience a progressive failure than chain. Chains have less reserve strength and are more likely to fail quickly once the process begins.

Chains will stretch under excessive loading. This causes elongating and narrowing of the links until they bind on each other, giving visible warning. If overloading is severe, the chain will fail with less warning than a wire rope. When a chain link breaks, there is little or no warning.

2.3.2 Metal-Mesh Slings

Metal-mesh slings (*Figure 23*) are typically made of wire or chain mesh. They are similar in appearance and flexibility to web slings and are suited for some situations where other slings do not perform well. Metal-mesh slings have the following advantages:

- Resist abrasion and cutting
- Grip the load firmly without stretching
- Conform to irregular shapes
- Do not kink or tangle
- Can withstand high temperatures

ASME Standard B30.9 requires that the use of metal-mesh slings at temperatures above 550°F (288°C) requires manufacturer approval. These slings are available in several mesh sizes and can be coated with a variety of substances, such as rubber or plastic, to help protect the load. When used in high-temperature applications though, slings with coatings are not typically approved.

2.3.3 Storage and Inspection

Chain and metal-mesh slings must be stored inside a building or vehicle and hung on racks to reduce deterioration due to weather-related rust or corrosion. Never let chain slings lie on the ground in areas where heavy machinery can run over them.

Some manufacturers suggest lubrication of alloy chains while in use. However, slippery chains increase handling hazards. Chains coated with oil

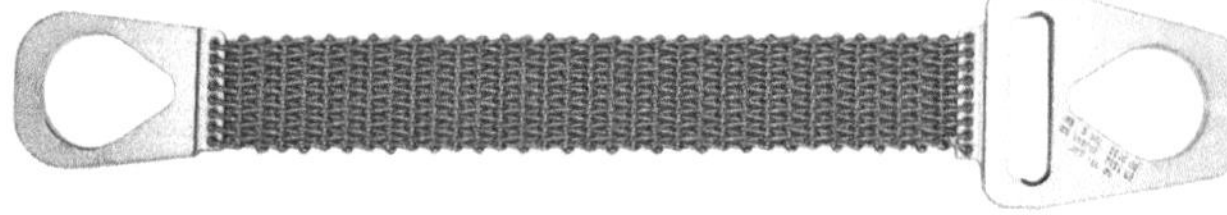

Figure 23 Metal-mesh sling.

or grease also attract dirt and grit that may cause abrasive wear. This is especially true of metal-mesh slings that are far more difficult to clean. Chain slings to be stored in exposed areas should be coated with a film of oil or grease for rust and corrosion protection.

Like all other slings, chain slings should be visually inspected before every lift. They should be removed from service if any of the following conditions are found during inspection:

- Missing or unreadable identification and/or rated load information
- Cracks or breaks
- Nicks, gouges, and excess wear
- Stretched, bent, twisted, or deformed links or end fittings;*ASME Standard B30.9*, Table 9-1.9.5.1 provides the minimum allowable thickness of any point on a link, based on the nominal size of the link.
- Evidence of overheating
- Excessive pitting or corrosion
- Lack of ability of chain or fittings to hinge freely
- Weld splatter
- Any other condition, including visible damage, that causes doubt about the continued use of the sling

Although metal-mesh slings are similar to chain slings, they are constructed in a very different way. Metal-mesh slings should be removed from service if any of the following conditions are found during inspection:

- Missing or unreadable identification tag
- A broken weld or brazed joint along the sling edge
- A broken wire in any part of the mesh
- A reduction in wire diameter of 25 percent due to abrasion, or 15 percent as the result of corrosion
- Lack of flexibility due to distortion of the mesh
- Distortion of the choker fitting so the depth of the slot is increased by 10 percent
- Distortion of either end fitting so the width of the eye opening is decreased by more than 10 percent
- Fittings that are pitted, corroded, cracked, bent, twisted, gouged, or broken
- A 15 percent reduction in the original cross-sectional area of metal at any point around
- Slings with individual spirals that are locked in place
- Any other condition, including visible damage, that causes doubt about the continued use of the sling

2.4.0 Sling Angles and Basic Hitches

Every sling, regardless of type and manufacturer, has a specified rated load capacity that should never be exceeded. Less obvious is that the load on a sling changes dramatically as the sling angle changes. What appears to be a light load for a sling can quickly become a very stressful load due to the additional stress placed on the sling as the sling angle changes. Sling angles and other significant factors in sling selection and use are explored in this section.

Since there are an infinite number of loads and load configurations, slings must be used in different ways to connect the load to the hoist line. An understanding of basic hitches and their performance characteristics help the crane operator understand how a reliable and properly made hitch should look.

2.4.1 Sling Capacity

Sling capacity depends on the sling material, sling construction, hitch configuration, number of slings, and the angle of the sling in the hitch used. This type of information, along with other relevant rigging information, is available from rigging equipment manufacturers and trade organizations in the form of easy-to-use pocket guides like the one shown in *Figure 24*. Sling capacity information is also provided in *ASME Standard B30.9* for various types of slings. *Table 1* is an example of a manufacturer's capacity table for wire-rope slings. Note that the capacity of a sling used in a simple basket hitch with a vertical pull is double the capacity for a straight vertical hitch. This is common, since two legs are being used to suspend the load instead of one. However, as the angle between the legs of the hitch widens, the rated load capacity of the sling is reduced. Remember that most anything you do with a sling that varies from a straight vertical pull is likely to negatively affect its rated load capacity. Choking a sling, for example, even for a vertical pull, significantly reduces the capacity of the sling.

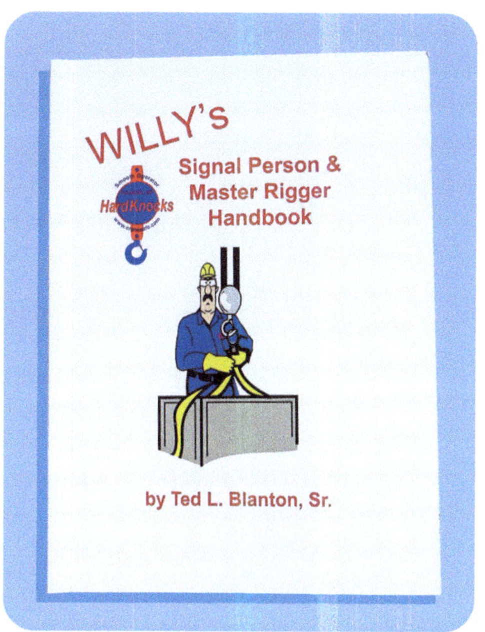

Figure 24 A rigging pocket guide.

2.4.2 Sling Angles

The angle formed by the legs of a sling, with respect to the horizontal plane, when tension is placed on the load is called the *sling angle*. The sling angle directly affects the tension applied to each sling. For this reason, maintaining acceptable sling angles is crucial to safe rigging. In addition, since the sling angle affects its rated load capacity, the rigger must ensure that even a sling at an acceptable angle remains within its rated load capacity.

To determine the effective weight placed on a sling at an angle, a sling angle factor is needed. The factor is applied to the actual load weight per sling to determine the tension on the sling. The proper way to calculate the sling angle factor is through some simple field measurements and math.

Figure 25 shows a load suspended from two slings in a vertical hitch. "L" represents the distance between the points of sling contact with the hook and the load. "H" represents the distance from where the sling contacts the hook down to

---

Convenient Rigging Information

Over the years, there have been quite a few rigging handbooks developed and marketed. Many are pocket-sized for the convenience of the rigger, allowing quick access to the needed information. As you might expect, rigging information has also gone electronic. Apps are now available from various manufacturers and trade organizations that are compatible with all popular smartphones, and more are sure to reach the market in the future. As OSHA and/or ASME standards change, it is much easier to update an app than it is to throw away and replace books.

Table 1 Wire-Rope Sling Capacity Table

CLASS	SIZE (IN)	RATED CAPACITY - LBS*		BASKET HITCH				EYE DIMENSIONS (APPROXIMATE)	
		VERTICAL	CHOKER**		30	60	90	WIDTH (IN)	LENGTH (IN)
6 × 19 IWRC	1/4	1,120	820	2,200	2,200	1,940	1,580	2	4
	5/16	1,740	1,280	3,400	3,400	3,000	2,400	2 1/2	5
	3/8	2,400	1,840	4,800	4,600	4,200	3,400	3	6
	7/16	3,400	2,400	6,800	6,600	5,800	4,800	3 1/2	7
	1/2	4,400	3,200	8,800	8,600	7,600	6,200	4	8
	9/16	5,600	4,000	11,200	10,800	9,600	8,000	4 1/2	9
	5/8	6,800	5,000	13,600	13,200	11,800	9,600	5	10
	3/4	9,800	(7,200)	19,600	19,000	17,000	13,800	6	12
	7/8	13,200	9,600	26,000	26,000	22,000	18,600	7	14
	1	17,000	12,600	34,000	32,000	30,000	24,000	8	16
	1 1/8	20,000	15,800	40,000	38,000	34,000	28,000	9	18
6 × 37 IWRC	1 1/4	26,000	19,400	52,000	50,000	46,000	36,000	10	20
	1 3/8	30,000	24,000	60,000	58,000	52,000	42,000	11	22
	1 1/2	36,000	28,000	72,000	70,000	62,000	50,000	12	24
	1 5/8	42,000	32,000	84,000	82,000	72,000	60,000	13	26
	1 3/4	50,000	38,000	100,000	96,000	86,000	70,000	14	28
	2	64,000	48,000	128,000	124,000	110,000	90,000	16	32
	2 1/4	78,000	60,000	156,000	150,000	136,000	110,000	18	36
	2 1/2	94,000	74,000	188,000	182,000	162,000	132,000	20	40

* Rated capacities for unprotected eyes apply only when attachment is made over An object narrower than the natural width of the eye and apply for basket hitches only when the d/d ratio is 20 or greater, where d=diameter of curvature around which the body of the sling is bent, and d=nominal diameter of the rope.

** See choker hitch rated capacity adjustment chart.

NCCER – *Basic Rigger*

an imaginary line that connects the two points of sling contact with the load. To determine the sling angle factor, the length (L) is divided by the height (H). The result is the factor to be applied to the weight of the load on each sling.

In the example shown in *Figure 25*, the height (H) is 74" (188 cm), and the length (L) is 86" (218 cm). (The length will always be longer than the height when the sling is at an angle.) The sling angle factor is calculated as follows:

US measure:

    Sling angle factor = L ÷ H
    Sling angle factor = 86" ÷ 74"
    Sling angle factor = 1.162

Metric:

    Sling angle factor = L ÷ H
    Sling angle factor = 218.4 cm ÷ 187.9 cm
    Sling angle factor = 1.162

The total weight of the load in this example is 2,000 lbs (907 kg). Since there are two slings, each sling must lift 1,000 lbs (454 kg). (2,000 lbs ÷ 2 slings = 1,000 lbs per sling.) Now apply the sling angle factor to determine the actual tension placed on each sling:

US measure:

    Sling tension = load per sling × sling factor
    Sling tension = 1,000 lbs × 1.162
    Sling tension = 1,162 lbs

Metric:

    Sling tension = load per sling × sling factor
    Sling tension = 453.5 kg × 1.162
    Sling tension = 527 kg

Figure 26 shows the effect of various sling angles on sling loading when 1,000 pounds (454 kg) of load are applied to each sling. Note that the tension on the slings is much higher when the legs are positioned at an angle of 30 degrees relative to the horizontal plane than when the legs are at an angle of 60 degrees. Optimum sling angles fall between 60 and 45 degrees to the horizontal plane. Angles of 30 degrees are occasionally required, but lesser angles are generally considered hazardous and unnecessary. At 30 degrees, the sling angle factor is 2.0, meaning that the tension on the sling is already double the actual load weight. If a lift plan leads to sling angles less than 30 degrees, changes in the rigging approach are likely needed.

These calculations for finding sling angle factor and tension help to determine whether slings are applied within their rated load capacity, and what that load actually is. However, they do not tell the rigger the sling angle. Tables have been developed that equate the sling angle factor to

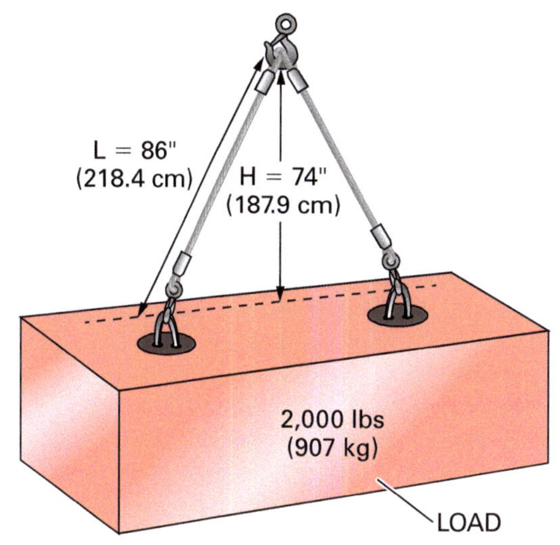

US MEASURE	METRIC
Sling angle factor = L ÷ H	Sling angle factor = L ÷ H
Sling angle factor = 86" ÷ 74"	Sling angle factor = 218.4 ÷ 187.9 cm
Sling angle factor = 1.162	Sling angle factor = 1.162

Figure 25 Determining the sling angle factor.

the angle. These tables can be used to determine the sling angle if the factor is known, or to determine the sling angle factor if the angle is known. *Table 2* is an example of such a table. Using the sling angle factor of 1.162 calculated in the previous example, it can be determined from the table that the sling angle is between 55 and 60 degrees. This is a very acceptable and safe sling angle for lifting.

2.4.3 Common Hitches

The way a sling is arranged to hold the load is referred to as a *hitch*. Hitches can be made using just the sling or by combining slings with connecting hardware. There are three basic types of hitches: vertical hitches, choker hitches, and basket hitches.

One of the most important parts of a rigger's job is making sure that the load is held securely. The type of hitch used depends on the nature of the load. For example, different hitches are used to secure a load of pipes, a concrete slab, or heavy machinery. Controlling the movement of the load once the lift is in progress is another extremely important part of the rigger's job. Therefore, the rigger must also consider the intended movement of the load when choosing a hitch. For example, some loads are lifted straight up and then lowered down to the same spot. Other loads may be lifted, turned 180 degrees in midair, and then set down in a completely different place. The

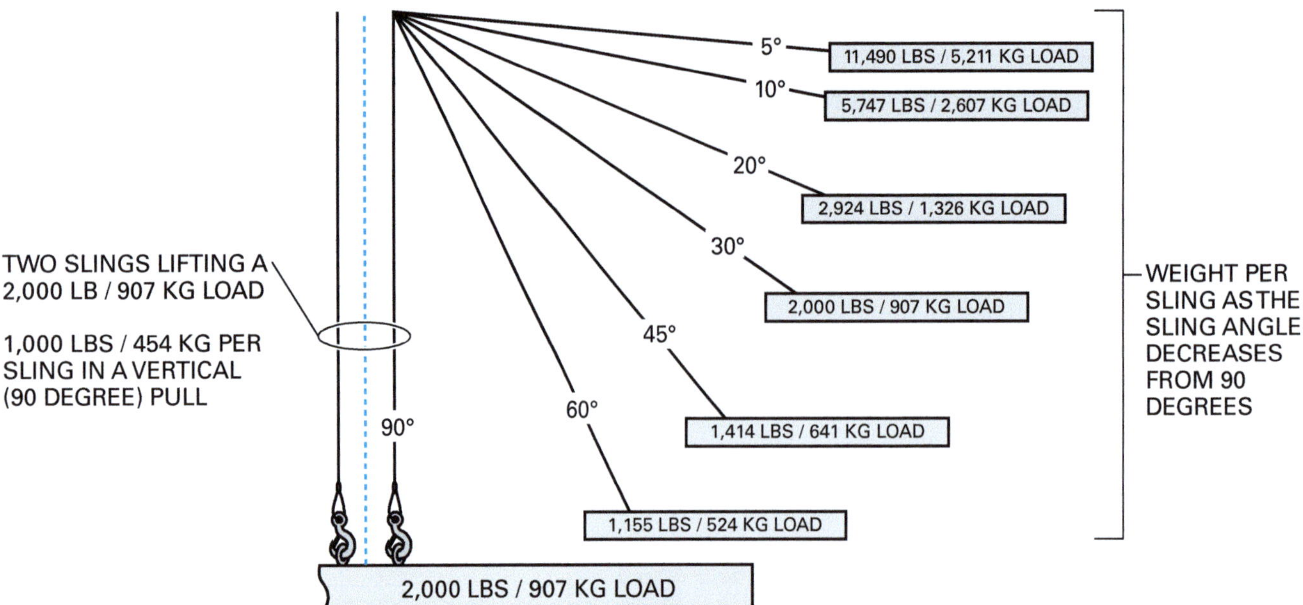

Figure 26 Effect of various sling angles.

Table 2 Sling Angle and Sling Angle Factor

Sling Angle	Sling Angle Factor
5	11.490
10	5.747
15	3.861
20	2.924
25	2.364
30	2.000
35	1.742
40	1.555
45	1.414
50	1.305
55	1.221
60	1.155
65	1.104
70	1.064
75	1.035
80	1.015
85	1.004
90	1.000

Figure 27 A vertical hitch using a plate clamp.

construction of the three basic types of hitches will be reviewed here.

The single vertical hitch (*Figure 27*) is used to lift a load straight up. With this hitch, some type of attachment hardware is needed to connect the sling to the load. The single vertical hitch allows the load to rotate freely. If you do not want the load to rotate freely, some method of load control must be used, such as a tagline.

Another version of the vertical hitch is the bridle hitch (*Figure 28*). The bridle hitch consists of two or more vertical hitches attached to the same hook, master link, or ring. This hitch allows the slings to be connected to the same load without the use of such devices such as spreader beams. Multiple-leg bridle hitches provide increased stability and balance for the load being lifted.

NCCER – *Basic Rigger*

The Ten-Inch Rule

A alternate way to determine the sling angle factor in the field is done using a tape measure. First, measure the distance up from a horizontal line between the connection points, and find the point where the sling is exactly 10" above the line (as shown in the image). Then measure the distance from that point down the angled sling to the connection point. That line will be longer, since it travels at an angle. In the example shown here, that distance measures 14". Now simply place a decimal point between the 1 and the 4 to arrive at a sling angle factor of 1.4. If the factor is 2.0 or greater, the sling angle is excessive and applies too much stress to the slings. Change the rigging to reduce the sling angle.

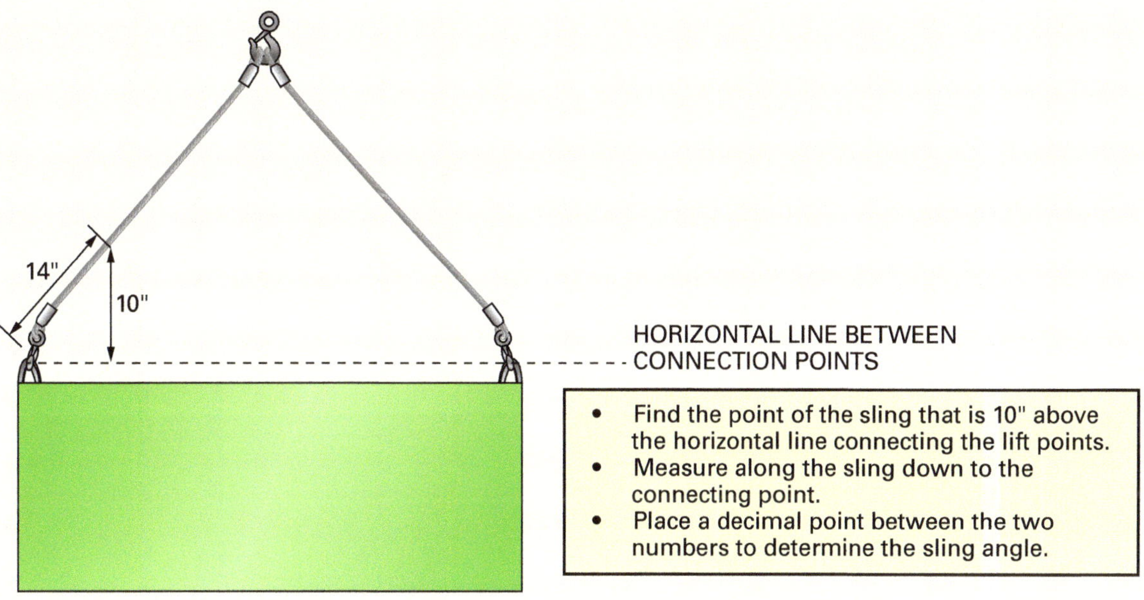

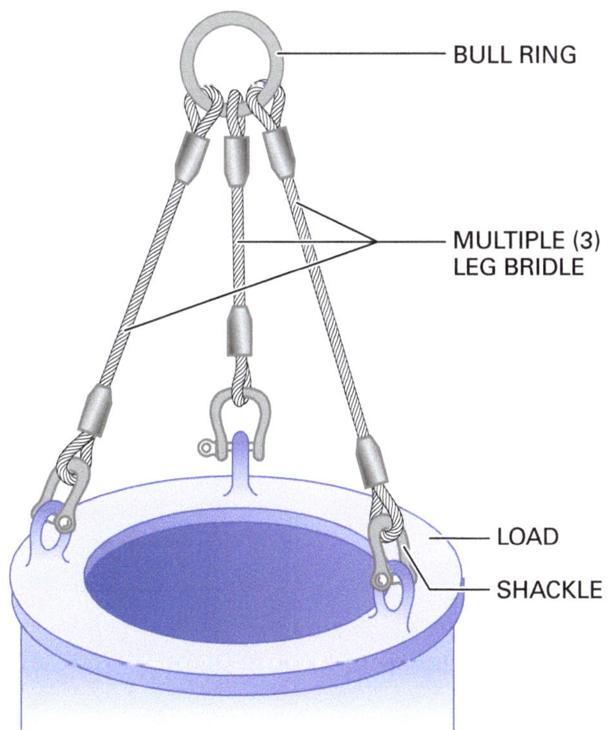

Figure 28 A multi-leg bridle hitch.

However, it is important to note that a bridle hitch results in slings that are at an angle other than 90 degrees to the horizontal plane. Therefore, the stress applied to the sling is increased and must be accounted for in sling selection.

A choker hitch is often used when a load has no attachment points or when the attachment points are not practical for lifting (*Figure 29*). The hitch is made by wrapping the sling around the

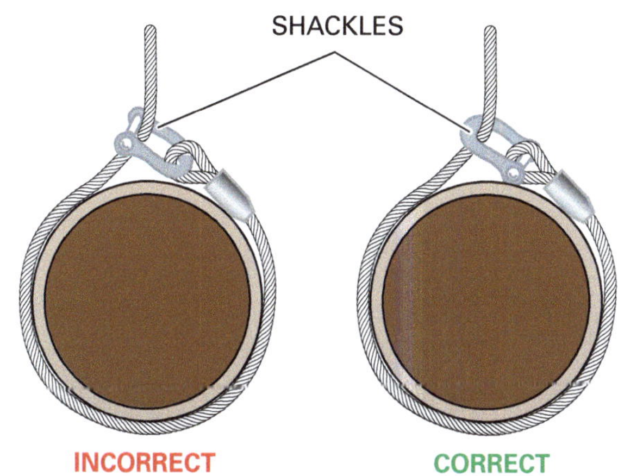

Figure 29 A choker hitch.

load and passing one eye of the sling through a shackle to form a constricting loop around the load. It is important that the shackle used in a choker hitch be oriented properly, as shown in *Figure 29*. It is also important to place a single choker hitch at the load's CG. Otherwise, the load will be unbalanced when lifted and will slip out of the hitch. The choker hitch affects the capacity of the sling, reducing it by a minimum of 25 percent. This reduction must be considered when choosing the proper sling.

A choker hitch does not grip the load as securely as the name implies. It is not recommended for loose bundles of materials because it tends to push loose items up and out of the choker. Many riggers use the choker hitch for loose bundles, mistakenly believing that forcing the choke down provides a tight grip. This actually increases the stress on the choked leg of the sling.

Instead, to gain gripping power, use a double-wrap choker hitch (*Figure 30*). The double-wrap choker uses the load weight to provide the constricting force, so there is no need to try and force the sling into a tighter choke. A double-wrap choker hitch is ideal for lifting bundles of items, such as pipes and structural steel. It will also keep the load in a certain position, which makes it ideal for equipment installation lifts.

When an item more than 12' (3.7 m) long is being rigged, the general rule is to use two choker hitches spaced far enough apart to provide the stability needed to transport the load. When two hitches are used, the hoist line should be positioned over the load's CG. To lift a bundle of loose items, or to maintain the load in a certain position during transport, remember to use the double-wrap choker hitch instead. Loads that are long enough to cause the sling angle to be too great should be rigged using a spreader beam.

Basket hitches (*Figure 31*) are very versatile and can be used to lift a variety of loads. A basket hitch is formed by passing the sling around the load and placing both eyes in the hook. Placing a sling into a basket hitch effectively doubles

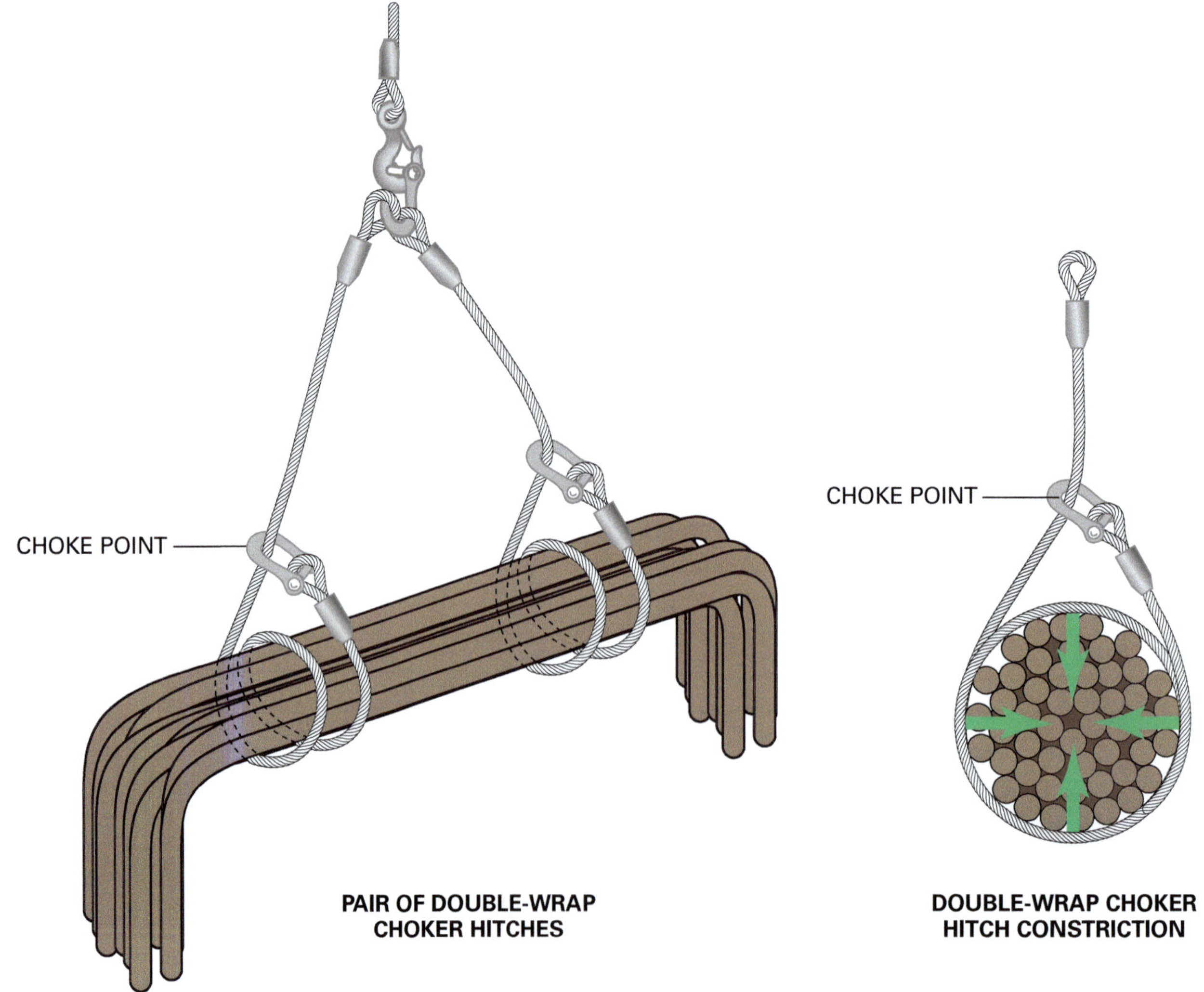

Figure 30 A double-wrap choker hitch.

NCCER – *Basic Rigger*

the capacity of the sling. This is because the basket hitch creates two sling legs from one sling. However, this does not provide secure control of the load.

The double-wrap basket hitch (*Figure 32*) combines the constricting power of the double-wrap choker hitch with the capacity advantages of a basket hitch. This means it is able to hold a larger load more tightly. The double-wrap basket hitch requires a considerably longer sling length than a double-wrap choker hitch, since both sling eyes must be connected to the crane hook. If it is necessary to join two or more slings together, the load must be in contact with the sling body only, not with the hardware used to join the slings. The double-wrap basket hitch provides support around the load. Just as with the double-wrap choker hitch, the load weight provides the constricting force for the hitch.

2.4.4 Finding the Load's Center of Gravity

The center of gravity (CG) of an object is the point around which the weight of the object is concentrated. The CG must be known when using any hitch in order to properly and safely position the load line over the load.

The CG of an object is the point where the weight times the length of the object on one side is equal to the weight times the length of the other side; the point in between these two sides is the CG. This is illustrated in *Figure 33*. When the two calculations are equal, the CG has been located. The symbol to identify the CG, also shown in *Figure 33*, will be seen on drawings for lifted components and equipment and sometimes on the load itself.

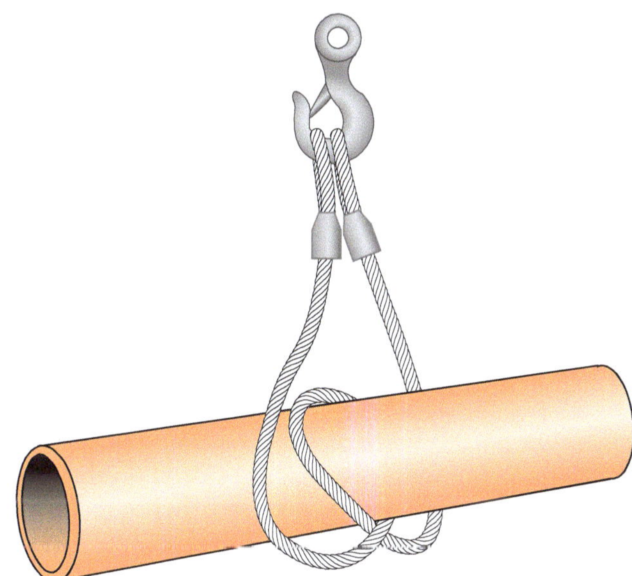

Figure 31 A simple basket hitch.

Figure 32 A double-wrap basket hitch.

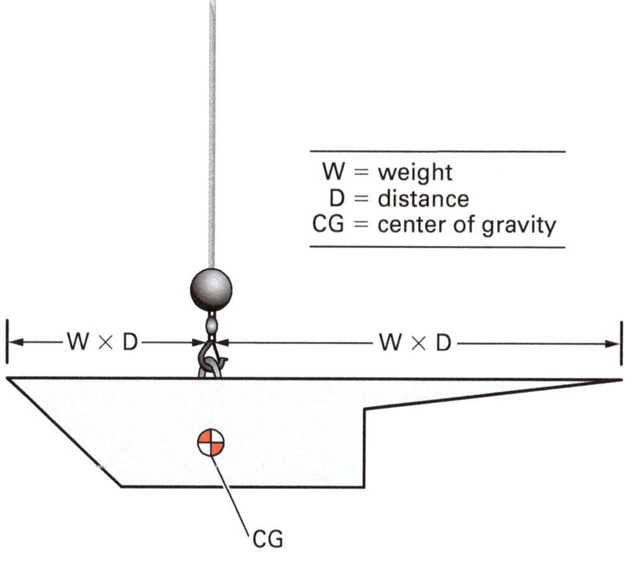

Figure 33 Calculating the center of gravity.

The simplest way of determining the CG of many objects is to experiment by lifting the object a few inches off the ground by a single point. If the object tilts, the single point by which the object is hanging is not directly over the CG. That means the lifting point needs to move in the same direction the load is tilting. When the object hangs level, the CG is vertically below the point from which the object is hung.

2.5.0 Rigging and Handling Piping Materials and Rebar

Reinforcing bars, piping, and similarly shaped construction materials are ordered from a supplier (usually the lowest bidder meeting the specification). Delivery of these materials is scheduled to coincide with the construction schedule as precisely as possible. Long-term storage of construction materials at the jobsite is not generally desirable.

These materials usually arrive at the job site on flatbed trucks or tractor trailers. If a spur track is available at the job site, shipments may be received on flatbed or gondola-style railroad cars. Where delivery of steel is scheduled to meet daily placement requirements, truckloads are delivered to the points of placement. Workers should be aware of safety factors that are necessary to achieve safe, efficient handling, storage, and hoisting of pipe and structural materials.

Trucks should be promptly unloaded to minimize jobsite traffic and potential safety hazards. Structural materials that are stored on the ground must be placed on timbers or other suitable blocking to keep them free from mud and to allow safe and easy handling.

Piping materials, rebar, and structural stock are normally stored by type and size first, then by length within each size if necessary. Materials such as pipe are usually of the same standard length unless sections or structural components have been precut or prefabricated before shipping. Tags are kept at the same end of each piece or bundle for easy identification. When a bundle is opened and part of the stock removed, the unit with the tag should remain with the bundle, and a new number reflecting the remaining quantity should be recorded. All mud and debris generally must be washed off before placement. The interior of piping materials should be kept clean and dry for most applications. For some applications, such as HVACR refrigerant piping, interior cleanliness is essential.

Use the following guidelines when rigging and handling a load of pipe, rebar, and similar products:

- Pipe of the same length should be lifted at the same time. If significantly different lengths of pipe must be lifted, lift each different length separately.
- Ensure that the open part of the hook is facing away from center when using slings with fixed or sliding hooks.
- Do not use carbon steel slings on stainless steel pipe to prevent cross-contamination of the metals. Carbon steel particles embedded in the stainless steel can rust and not only ruin the outer appearance, but also contaminate welds to the point of rejection. Use synthetic web slings on stainless steel and painted pipe.
- Lay the pipe on blocks so the slings can be worked on and off the load. Blocks should be made from hardwood and be thick enough to prevent the pipe from contacting uneven ground. Adequate ground clearance is also needed to slip a sling around the bundle. Pipe placed on blocks requires wedges to keep it from rolling off the blocking. Large diameter pipe requires larger chocks to keep it from rolling. In general, there should be 1" (2.5 cm) of chock per 1' (3 m) of diameter. For instance, a 42" (1,050 DN) pipe should have a chock that is about 3½" thick (9 cm) to prevent rolling. Note that a chock that is too thick can be pushed away by a rolling load.
- Use the hook and load line of the crane as the centerline with which to line up the load.
- As soon as the load clears the ground, check its orientation. If it is not level, the signal person should signal the operator to stop lifting the load, and guide the load back onto the blocks to adjust the position of the slings.
- Never stand underneath the load.
- Control all loads with one or more tag lines.
- Keep the load as low to the ground as possible.
- Store all pipe on level ground whenever possible.
- Stand to one side of the load, and guide it onto the blocks. Move to the end of the pipe before the rigging is released. Remember that a bundle of pipe can easily roll when the sling tension is released. Avoid pinch points when removing the sling, and do not stand in a position where the pipe may roll on you.
- Lay pipe side by side. Do not stack pipe if it can be avoided.

To select slings and properly rig a load, the weight of the load must be known. It is therefore beneficial to understand how to determine the weight of these materials.

2.5.1 Determining the Weight of Pipe

Every foot of pipe weighs a given amount, depending on the wall thickness and nominal size of the pipe. Therefore, if you know the weight per foot and the length, the total weight of the pipe can be determined.

Table 3 is an example of a weight chart for carbon steel pipe. Always be sure you consult a chart for the proper material as well as the correct size and weight. There are also separate charts for metric pipe sizes, which will report the weight as kilograms per meter rather than pounds per foot. Tables like the one shown here are available from a variety of sources online, including pipe distributors.

The numbers in the leftmost column represent the nominal pipe size in inches. The designations and numbers across the top represent wall thickness. The wall thickness of pipe is often designated by a schedule number; the numbers 10 through 160 across the top are schedule numbers. To use the table, find the point at which the nominal size and wall thickness, or schedule, of the pipe meet. The number in this block is the weight of one foot of pipe. To find the total weight, multiply the total number of feet by the weight per foot. Note that there is a significant difference in the weight of pipe made from other materials, such as PVC, copper, and stainless steel.

For example, to find the weight of five 10-foot lengths of 20" Schedule 40 carbon steel pipe, read right from the 20.0 in the Nominal Pipe Size column until reaching the Schedule 40 column. The number found there is 123.1. This is the weight, in pounds, of one foot of pipe. There is a total of 50' of pipe (5 lengths × 10' = 50'). Multiply the weight per foot from the table by the total length in feet to find the total weight of the pipe load (123.1 × 50' = 6,155 lbs).

Table 3 Carbon Steel Pipe Weights

Nominal Pipe Size (inches)	Wall Thickness								
	STD	XS	XXS	10	40	60	80	120	160
	Weight Per Foot in Pounds								
2.0	3.65	5.02	9.03	–	3.65	–	5.02	–	7.06
2.5	5.79	7.66	13.7	–	5.79	–	7.66	–	10.01
3.0	7.58	10.25	18.58	–	7.58	–	10.25	–	14.31
3.5	9.11	12.51	22.85	–	9.11	–	12.51	–	–
4.0	10.79	14.98	27.54	–	10.79	–	14.98	18.98	22.52
6.0	18.97	28.57	53.16	–	18.97	–	28.57	36.42	45.34
8.0	28.55	43.39	72.42	–	28.55	35.66	43.39	60.69	74.71
10.0	40.48	54.74	104.1	–	40.48	54.74	64.40	89.27	115.7
12.0	49.56	65.42	125.5	–	53.56	73.22	88.57	125.5	160.3
14.0	54.57	72.09	–	36.71	63.37	85.01	106.1	150.8	189.2
16.0	62.58	82.77	–	42.05	82.77	107.5	136.6	192.4	245.2
18.0	70.59	93.45	–	47.39	104.8	138.2	170.8	244.1	308.6
20.0	78.60	104.1	–	52.73	123.1	166.5	208.9	296.4	379.1
22.0	86.61	114.8	–	58.07	–	197.4	250.8	353.6	451.1
24.0	94.62	125.5	–	63.41	171.2	238.3	296.5	429.5	542.1
26.0	102.6	136.2	–	85.73	–	–	–	–	–
28.0	110.6	146.9	–	92.41	–	–	–	–	–
30.0	118.7	157.5	–	99.08	–	–	–	–	–
32.0	126.7	168.2	–	105.8	229.9	–	–	–	–
34.0	134.7	178.9	–	112.4	244.9	–	–	–	–
36.0	142.7	189.6	–	119.1	282.4	–	–	–	–
42.0	166.7	221.6	–	–	330.4	–	–	–	–

2.5.2 Determining the Weight of Rebar

The weight of rebar is determined the same way as the weight of pipe. *Table 4* shows a chart of rebar weight by bar size. This particular example shows the bar size and other characteristics in both imperial and metric units. Virtually all rebar is made from carbon steel, so it would be rare to encounter rebar made from another material. If that is the case however, the supplier would need to be contacted for the weight information.

Once the size and total length of rebar in a bundle is identified, the weight per foot (or per meter) is multiplied by the appropriate length to determine the weight. Always remember to work within the same system of units—kilograms per meter, or pounds per foot.

2.5.3 Rigging Valves

Rigging a valve correctly involves knowing where to place the sling, how to place the sling, what kind of sling to use, and what kind of valve is being lifted. Manufacturers of large valves often provide drawings to show how the part is best rigged for lifting. Very large valves and valve components may be equipped with factory-installed lifting lugs.

Synthetic slings are usually best for rigging all types of valves, since many are also painted and easily scarred by metal slings. Never use a carbon steel sling to rig a stainless steel valve to avoid surface contamination of the stainless steel.

To rig a valve like the one shown in *Figure 34*, place a synthetic sling on each side of the valve body between the bonnet and the flanges. If the handwheel remains attached, bring the slings up

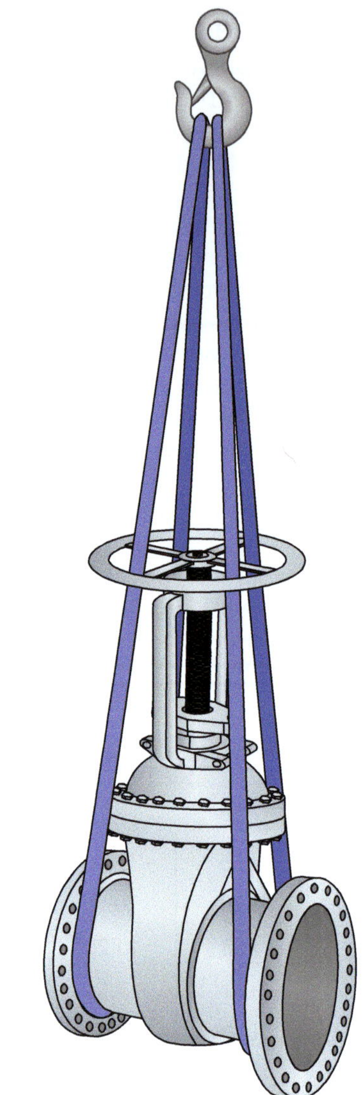

Figure 34 Rigging a valve.

Table 4 ASTM Standard Metric and US Measure Reinforcing Bar

| Bar Size | | Diameter | | Cross-Sectional Area | | Weight | |
Metric	[in-lb]	mm	[in]	mm	[in]	kg/m	[lbs/ft]
#10	[#3]	9.5	[0.375]	71	[0.11]	0.560	[0.376]
#13	[#4]	12.7	[0.500]	129	[0.20]	0.944	[0.668]
#16	[#5]	15.9	[0.625]	199	[0.31]	1.552	[1.043]
#19	[#6]	19.1	[0.750]	284	[0.44]	2.235	[1.502]
#22	[#7]	22.2	[0.875]	387	[0.60]	3.042	[2.044]
#25	[#8]	25.4	[1.000]	510	[0.79]	3.973	[2.670]
#29	[#9]	28.7	[1.128]	645	[1.00]	5.060	[3.400]
#32	[#10]	32.3	[1.270]	819	[1.27]	6.404	[4.303]
#36	[#11]	35.8	[1.410]	1006	[1.56]	7.907	[5.313]
#43	[#14]	43.0	[1.693]	1452	[2.25]	11.38	[7.65]
#57	[#18]	57.3	[2.257]	2581	[4.00]	20.24	[13.60]

*The equivalent nominal characteristics of inch-pound bars are the values enclosed within the brackets.

NCCER – *Basic Rigger*

through the handwheel spokes so that the valve cannot tilt from front to back. Do not place a sling around the handwheel or through the valve bore. The handwheel is not built to support the weight of the valve. A valve rigged around the handwheel is unsafe, even if the valve is going to be moved a short distance. Placing a sling through the bore of the valve can destroy the inner workings, even if soft synthetic slings are used.

2.6.0 Taglines and Knots

Taglines are natural fiber or synthetic ropes used to control the load (*Figure 35*). The absence or improper use of taglines can turn a simple hoisting operation into a hazardous situation. Taglines are used to maintain lateral control and prevent spinning of a suspended load as the crane and/or boom move. Loads of all shapes and sizes are subject to some level of dynamic and wind forces once in the air.

When selecting a tagline, several factors need to be considered. Natural fiber, or manila, is notably weaker than synthetic fibers such as nylon, polyester, polypropylene, or polyethylene. Although any rope that is wet becomes an electrical conductor, natural fiber rope absorbs water readily and may remain wet enough to conduct electricity for a long time. Most (but not all) synthetic ropes do not absorb moisture. A nonconductive tagline should be used when working in the vicinity of power lines.

Synthetic rope is lighter than natural fiber and has a high strength-to-weight ratio. Its resistance to water and reduced electrical conductivity do give it a distinct safety advantage. However, synthetic ropes are more easily damaged by heat. Significant contact with surfaces at 150°F (66°C) can result in a loss of strength, and many synthetic rope materials begin to melt at 300°F (149°C).

The diameter of a tagline should be large enough so that it can be gripped well when wearing gloves. Rope with a diameter of ½" (12 mm) is common, but ¾" (19 mm) and 1" (25 mm) diameter rope is sometimes used on heavy loads or where the tagline must be extremely long. Taglines should never have knots or loops tied in them. Terminating hardware, such as snaps and carabiners (*Figure 36*), may be added to make an easy connection to some loads. This hardware does not need to be designed and rated for lifting purposes, but must be substantial enough to handle the stress.

Always wear gloves when using a tagline to control a load. The momentum of a moving load can cause the rope to slide through the hands unexpectedly, causing severe rope burns. Never wrap a tagline around an arm or leg in an attempt to stop a load's swing. Never place yourself between a fixed object and a suspended load. Manning a tagline is a significant responsibility that requires careful thought and attention to execute the task safely and prevent serious injury to personnel and/or damage to property.

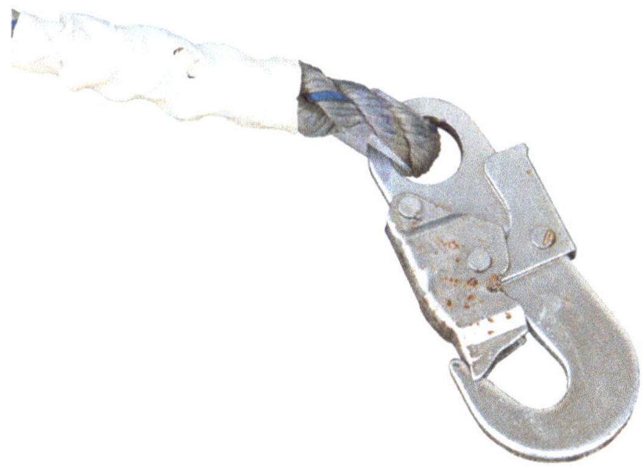

(A) HEAVY-DUTY SNAP

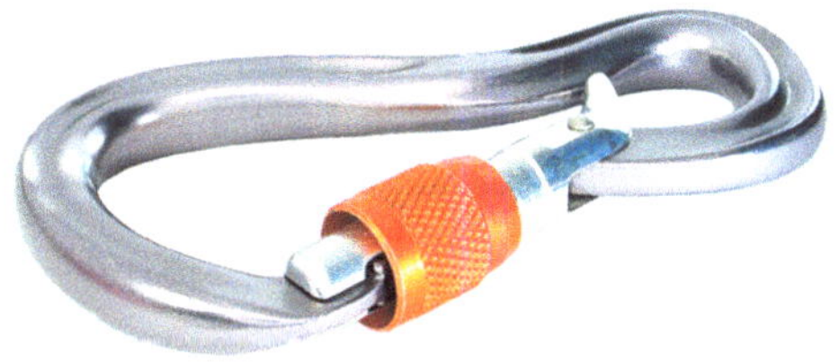

(B) CARABINER WITH LOCK

Figure 36 Rope hardware for quick connections.

Figure 35 Taglines in control of a load.

Taglines should be of sufficient length to allow control of the load from its original lift location until it is safely placed or until load control is transferred to other team members. Special consideration should be given to situations where a long tagline could interfere with the safe handling of loads, such as steel erection projects. Think through the lift and the material movement and consider where the tagline(s) will be and what obstructions might be encountered.

When working near power lines, there is always the hazard of electric shock or electrocution. An insulating link (*Figure 37*) can be added to taglines to help protect the user if the load or tagline contacts a power source.

Taglines should be attached to loads at a location that provides the best physical advantage in maintaining control. Long loads, for example, should have taglines attached as close to the ends as possible. The tagline should also be located in a place that allows personnel to access it for removal after the load is placed.

Figure 37 Insulating link.

To properly handle a tagline, you must determine the physical advantage intended. Consider the lift and which direction the load may be most likely to swing or rotate. With a long load, for example, try to maintain a position that is 90 degrees to the length of the object (*Figure 38*).

Whenever possible, keep yourself and the tagline in view of the crane operator. Stay alert. Do not become complacent during the lift. Be aware of the location of any excess rope and do not allow it to become fouled or entangled around your legs or on nearby objects.

Taglines should not to be used to pull or yank a load away from its natural vertical suspension. They also cannot be used in any way that results in them supporting or carrying any portion of the load. Large loads often require the use of multiple taglines. In these cases, tagline personnel must work as a team and coordinate their actions.

2.6.1 Knots

Knots used to attach taglines to loads should be tied properly to prevent slipping or accidental loosening, but they also must be easy to untie after the load is placed. Whenever possible, taglines should be of one continuous length and free of splices. If joining two taglines together becomes necessary, it is best done using the short-splice method. The short-splice method results in a knot diameter roughly twice that of the rope itself, and retains more rope strength than other methods. Larger knots tied in the middle of the taglines can sometimes create difficulties. However, the short splice can be a challenging knot to create, as it involves weaving of the individual rope strands. The short-splice method can be learned from numerous websites and internet videos if desired. In general, it is always best to avoid knots of any kind in a tagline and use a continuous length of rope instead.

Some recommended knots for rigging are the bowline and the clove hitch. The bowline (*Figure 39*) is used to form a secure loop in the end of a rope. It is sometimes called a *rescue knot* or the *king of knots* because it is reliable enough to be used for rescue work. It can be backed up with a second knot for extra security or tied twice, into a double bowline, for extra strength. A bowline does not slip or bind when under load, but it is easy to untie when there is no load.

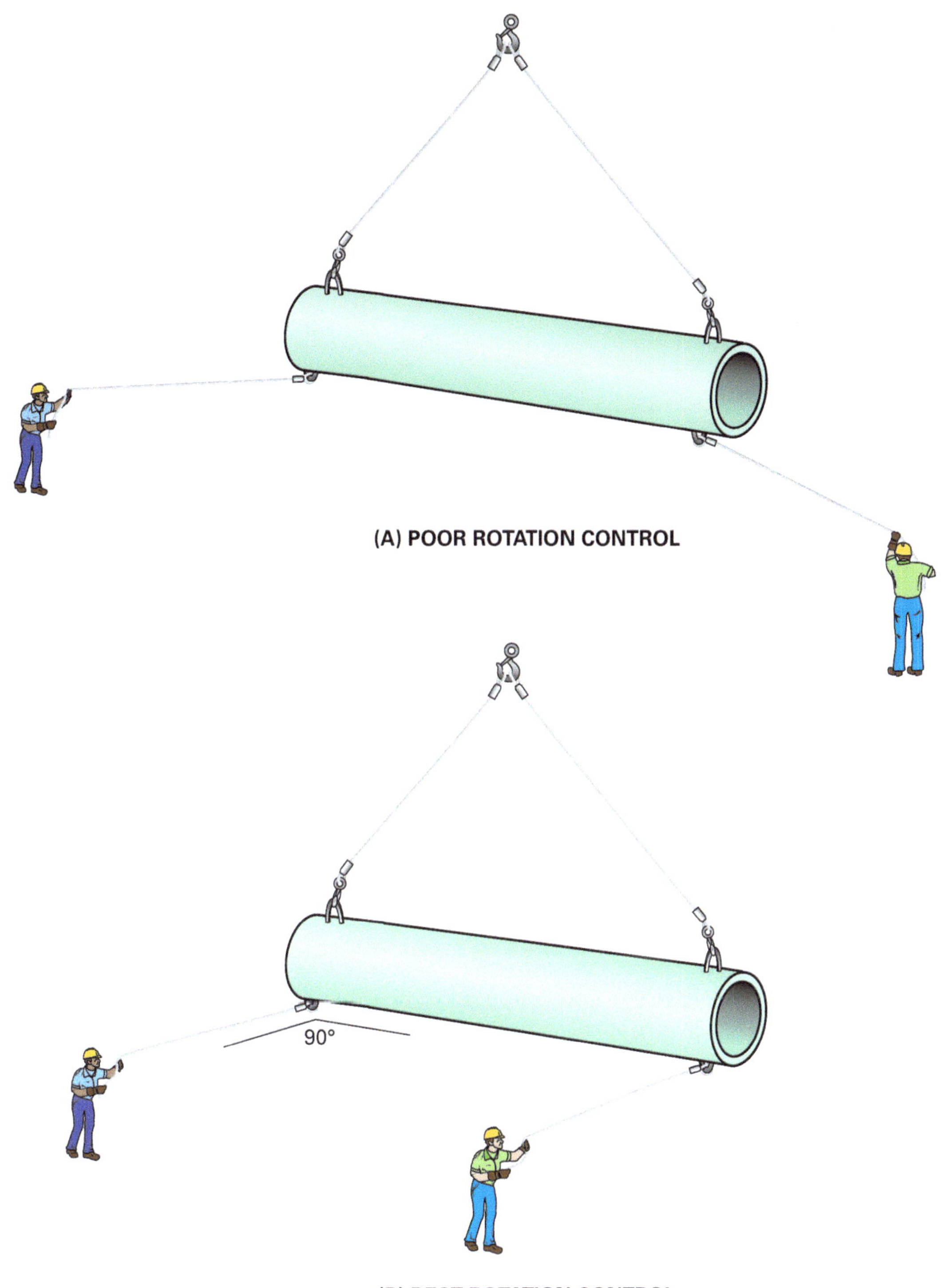

Figure 38 Best position for controlling load rotation.

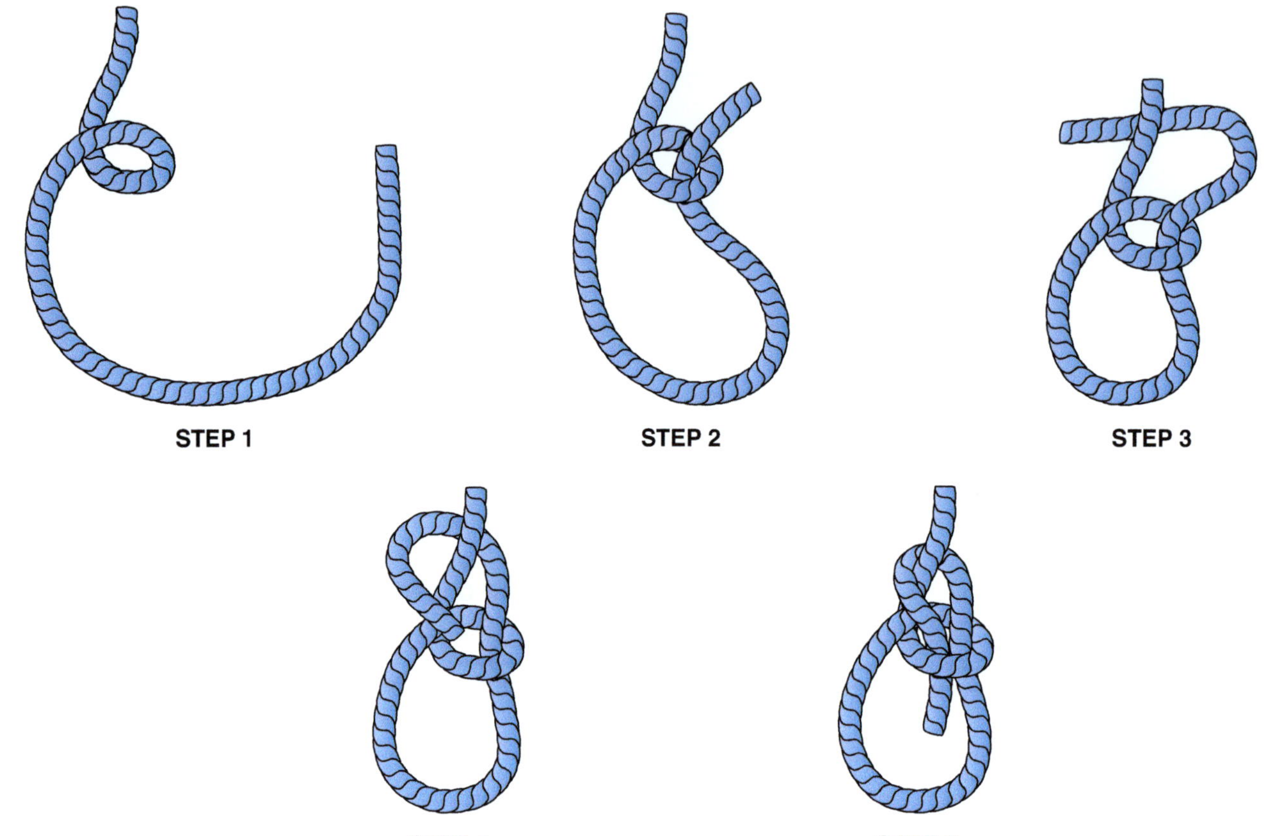

Figure 39 Tying a bowline.

The bowline reduces the strength of the rope that it is tied in by as much as 40 percent. Bowline knots also require a long tail for reliability.

The following steps here, shown in *Figure 39*, can be used to tie a bowline:

Step 1 Form a small loop in the rope, leaving a long enough tail to make the desired size of main loop.

Step 2 Pass the tail of the rope through the small loop.

Step 3 Pass the tail under the standing end.

Step 4 Pass the tail back down through the small loop.

Step 5 Adjust the main loop to the required size and tighten the knot, making sure to maintain a sufficiently long tail.

> **NOTE**
>
> Some people use this saying to help them remember how to tie a bowline: "The rabbit comes out of his hole, around a tree, and back into the hole."

A half hitch (*Figure 40*) can be used to tie a rope around an object such as a rail, bar, post, or ring. It is commonly used for tasks such as suspending items from overhead beams and carrying light loads that have to be removed easily. The half hitch is also widely used in making other knots, such as the clove hitch. Because the half hitch is not very stable by itself, it is not suitable for heavy loads or tasks in which safety is paramount. It is often used to back up and secure another knot that has already been tied.

The following steps, shown in *Figure 40*, can be used to tie a half hitch:

Step 1 Form a loop around the object.

Step 2 Pass the tail of the rope around the standing part and through the loop.

Step 3 Tighten the hitch by pulling on the working end and the standing part of the rope simultaneously.

A clove hitch (*Figure 41*) is one of the most widely used general hitches. It is typically used to make a quick and secure tension knot on a fixed object that serves as an anchor, such as a post, pole, or beam. A clove hitch can also be used as the first knot when lashing items together.

NCCER – *Basic Rigger*

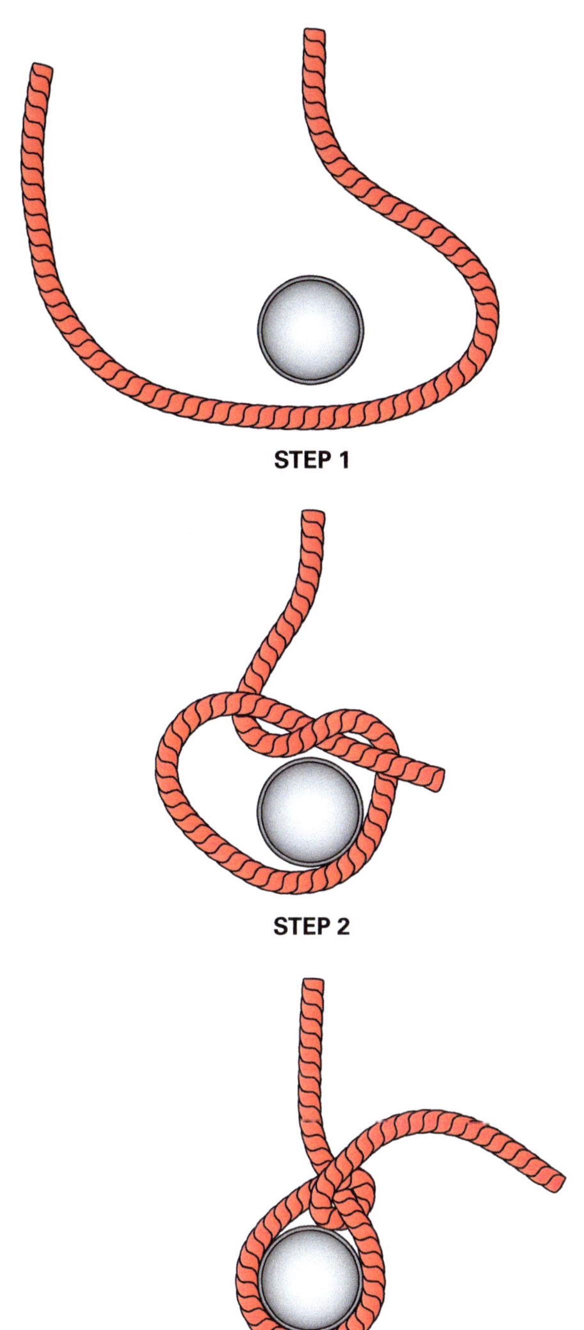

STEP 1

STEP 2

STEP 3

Figure 40 Tying a half hitch.

Because it is a tension knot, a clove hitch loosens when tension is removed from the rope. For these reasons, a clove hitch should not always be used as-is. Additional half hitches or an overhand safety knot should be added to make a clove hitch more secure, unless loosening when tension is released is intentional.

There are alternate techniques for tying a clove hitch. Regardless of the technique, the structure of the resulting knot consists of two half hitches made in opposite directions.

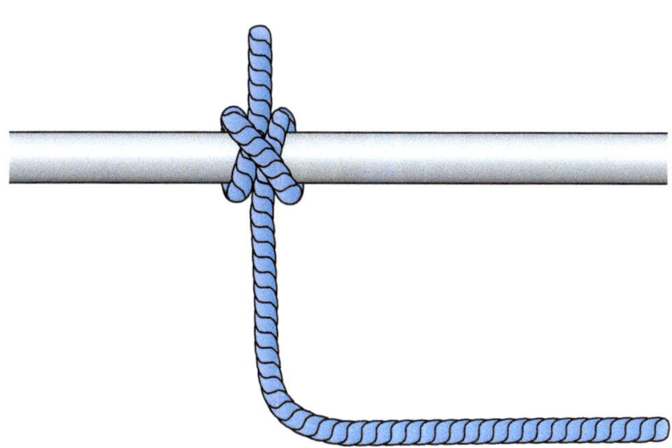

Figure 41 Clove hitch.

One common technique for tying a clove hitch that is used to attach a rope to a ring or upright structural component is to thread the end. This is called the *threading-the-end technique*. Use the following steps, shown in *Figure 42*, to tie a clove hitch using the threading-the-end technique:

Step 1 Pass the working end of the rope over the object.

Step 2 Pass the working end back over the standing part and then over the object. This forms a half hitch.

Step 3 Thread the working end back under itself and up through the loop to form a second half hitch.

Step 4 Pull both the working and standing ends evenly to tighten the hitch.

Another method for tying a clove hitch, the stacked-loops technique, allows the rope to be dropped quickly over a standing object, instead of tying the knot around the object. It also allows the hitch to be tied at any point in a rope, not just at an end. Use the following steps, shown in *Figure 43*, to tie a clove hitch the stacked-loops technique:

Step 1 Form two identical loops in a rope, one in the right hand and one in the left.

Step 2 Cross the loops one above the other to form a knot.

Step 3 Place the knot over the stake or post.

Step 4 Tighten the knot by pulling simultaneously on both ends of the rope. Note that the completed clove hitch consists of two half hitches stacked on each other.

Remember that an additional half hitch is often added for increased security.

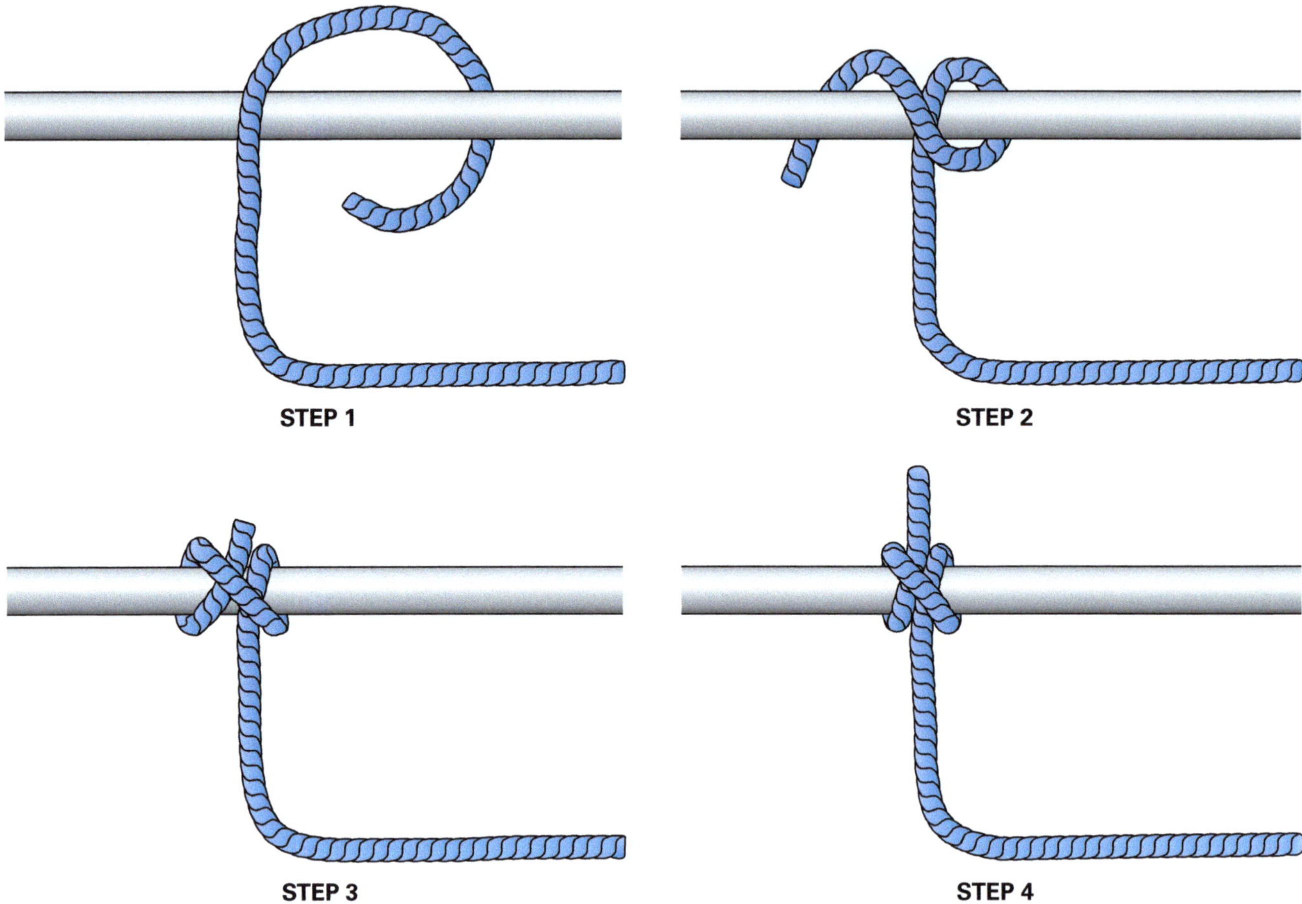

Figure 42 Threading-the-end technique for tying a clove hitch.

2.7.0 Rigging Safety Precautions

Workers on the job are responsible for their own safety and the safety of their fellow workers. Project and corporate management, in turn, has a responsibility to each worker. The responsibility of management and supervisors is to ensure that the workers who prepare and use the equipment, and those who work with or around it, are well trained in operating procedures and safety practices. Each worker is expected to put that training to good use. This section describes safety guidelines that are related to rigging.

> **NOTE**
>
> Safety guidelines pertaining to mobile cranes are covered in detail in NCCER Module 21106, "Crane Safety and Emergency Procedures" from *Mobile Crane Operations Level One*.

Did You Know?

The Ashley Book of Knots

Although *The Ashley Book of Knots* by Clifford W. Ashley is no longer in print, sailors, climbers, campers, macramé artists—anyone with more than a passing interest in knots—continue to find this book cited as the definitive work on the subject of knot tying. Originally published by Doubleday & Company in 1944, it contains the histories of and instructions for tying over 3,900 types of knots, accompanied by 7,000 pen-and-ink drawings.

Clifford Ashley was born in 1881, in New Bedford, Massachusetts. By trade he was an author and artist, but while sailing on many types of boats and researching knots he performed a wide variety of jobs. He spent 11 years writing and drawing the illustrations for his book. He died three years after the book's publication. Ashley continues to be regarded as a fine marine painter as well as one of the world's leading authorities on knot tying.

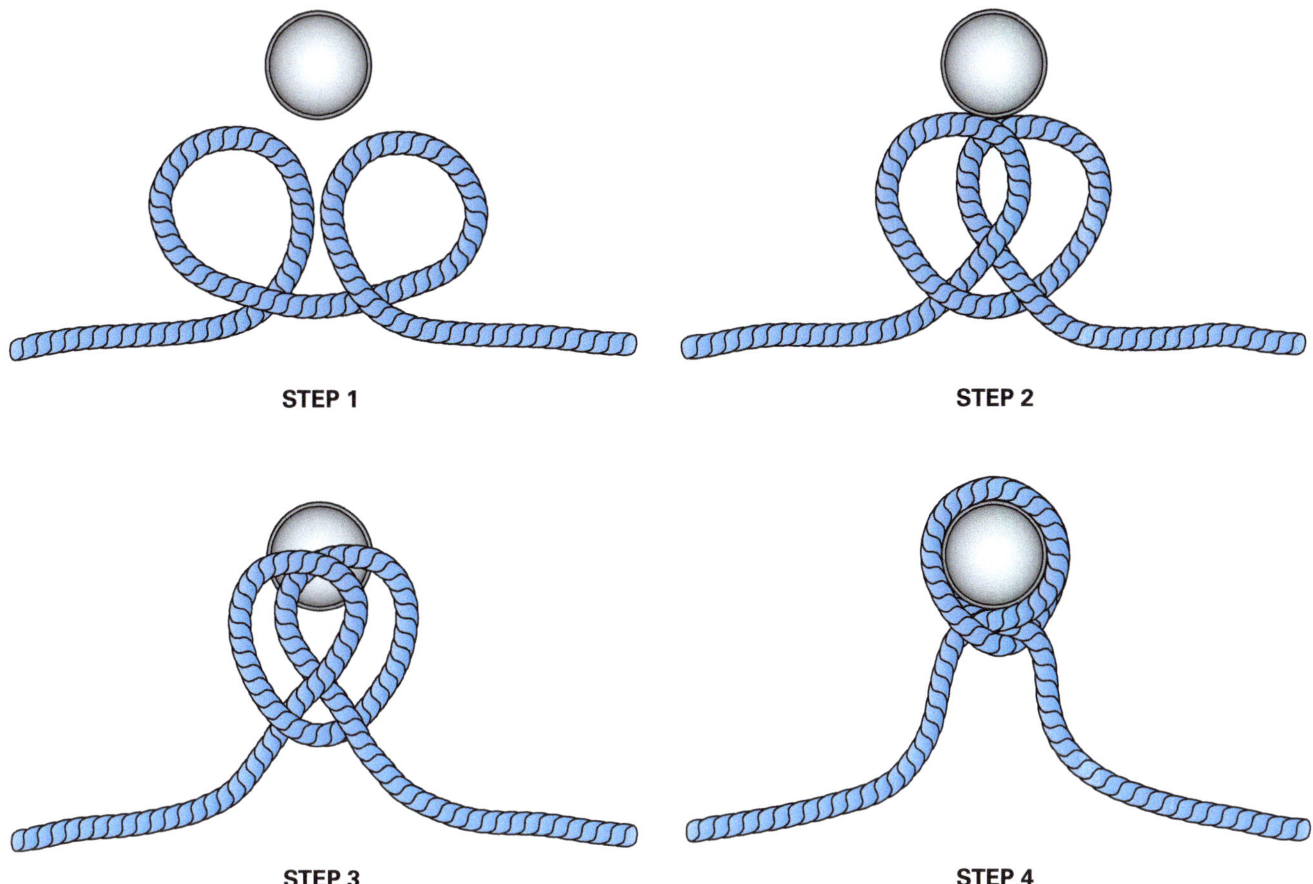

STEP 1   STEP 2

STEP 3   STEP 4

Figure 43 Stacked-loops technique for tying a clove hitch.

One very important rigging precaution is to determine the weight of every load before attempting to lift it. The weight of the load must obviously be within the capacity of the crane, but this is not the responsibility of the rigger. However, the load weight must be considered in the selection of the rigging components before the lift ever begins.

Unfortunately, accurately determining the weight is not always possible. This issue is addressed in 29 *CFR* 1926.1417, which requires that the crane operator verify the weight of the load in one of the following two ways:

- The weight can be determined from an industry-recognized source such as the manufacturer of the product(s), or by an industry-recognized calculation method. This latter method could be used, for example, to calculate the weight of an I-beam based on the weight per foot of such products.
- If the weight cannot be verified through the above methods, the operator may begin the lift using a load-weighing device, load-moment indicator, rated-capacity indicator, or rated-capacity limiter. This exercise is used to

determine if the lifted load is more than 75 percent of the crane's maximum rated capacity at the longest operating radius that the lift will require. If it is greater than 75 percent, the lift must be aborted until the actual weight can be verified through the first method.

It is equally important to rig the load so that it is stable and its center of gravity is below the hook. The personal safety of riggers and hoisting operators depends a lot on common sense. The following safety practices should be observed:

- Following the OSHA procedures from 29 *CFR* 1926.1417, determine the weight of loads before rigging whenever possible.
- Coordinate the rigging process as necessary with other personnel on the job site to minimize foot and vehicle traffic in the area.
- Know the rated load capacity of the equipment and tackle, and never exceed it. Remember that the rated load capacity of all hoisting and rigging equipment is based on ideal conditions. Normal working conditions are rarely ideal, so it is important to recognize factors that can reduce equipment capacity or increase the load imposed, such as sling angles.

- Examine all hardware, tackle, slings, and equipment before use. Remove defective components from service for evaluation and possible repair by others.
- Always wear gloves when handling slings, taglines, and similar rigging equipment.
- Never wrap hoist ropes around a load. Use only slings or other appropriate lifting devices.
- Ensure that all slings are made from the same material when using two or more slings on a single load.
- Never lift loads with one or two legs of a multi-leg sling until the unused slings are secured.
- Only use lifting beams for the purpose for which they were designed.
- All personnel must stand clear and not underneath loads being lifted and lowered, and when slings are being withdrawn from beneath the load.
- Never point-load a hook unless it has been specially designed for that task.
- Keep hands and feet away from pinch points as the slack is taken up by the crane.
- Use one or more taglines as necessary to keep the load under control.
- Immediately report defective equipment or rigging issues to the rigging supervisor or lift director, who must issue orders to proceed after safe conditions have been ensured.
- Prepare blocking at the load landing zone before the load arrives, rather than after it is suspended above. Allow the load to safely land before removing the slings.

Weather conditions can be a major factor in lifting operations. Stop hoisting and/or rigging operations when weather conditions such as the following present a hazard to property, workers, or bystanders:

- Winds exceeding crane manufacturer recommendations
- Lightning or thunder
- Poor visibility due to conditions such as darkness, dust, fog, rain, or snow, impairing view of rigger or hoist crew
- Temperature low enough to cause crane structures to fracture upon shock or impact

Crane Collapse in Manhattan

In early 2016, a lattice-boom crawler crane was attempting to lower its 565-foot (172-meter) boom and secure the crane due to high winds. Unfortunately, the wind gusts intensified and the boom came crashing into the streets of Manhattan. One bystander was killed and several others were injured. The crew had already begun clearing the streets of traffic to prepare to receive and secure the boom, minimizing the loss of life and injuries.

Figure Credit: © a katz/Shutterstock.com

Additional Resources

ASME Standard B30.5, Mobile and Locomotive Cranes. Current edition. New York, NY: American Society of Mechanical Engineers.

ASME Standard B30.9, Slings. Current edition. New York, NY: American Society of Mechanical Engineers.

ASME Standard B30.20, Below-The-Hook Lifting Devices. Current edition. New York, NY: American Society of Mechanical Engineers.

ASME Standard BTH-1, Design of Below-The-Hook Lifting Devices. Current edition, New York, NY: American Society of Mechanical Engineers.

29 *CFR* 1926, Subpart CC, **www.ecfr.gov**

29 *CFR* 1926.251, **www.ecfr.gov**

29 *CFR* 1926.753, **www.ecfr.gov**

Mobile Crane Operations Level One, NCCER. Third Edition. 2018. New York, NY: Pearson Education, Inc.

NCCER Module 00106, *Introduction to Basic Rigging*.

Mobile Crane Safety Manual (AEM MC-1407). 2014. Milwaukee, WI: Association of Equipment Manufacturers.

Willy's Signal Person and Master Rigger Handbook, Ted L. Blanton, Sr. Current edition. Altamonte Springs, FL: NorAm Productions, Inc.

Knots: The Complete Visual Guide, Des Pawson. First American Edition. 2012. New York, NY: DK Publishing.

North American Crane Bureau, Inc. website offers resources for products and training, **www.cranesafe.com**

2.0.0 Section Review

1. A sling with a wide, flat bearing surface that is especially well-suited for basket hitches where sharp bends are encountered is the _____.

    a. strand-core wire rope sling
    b. fiber-core wire rope sling
    c. braided-belt sling
    d. chain sling

2. If a sling is needed to lift a load with a polished surface, the best choice would be a _____.

    a. synthetic web sling
    b. chain sling
    c. metal-mesh sling
    d. wire-rope sling

3. Metal-mesh slings must have the approval of the manufacturer if they are to be used at temperatures above _____.

    a. 195°F (91°C)
    b. 375°F (191°C)
    c. 550°F (288°C)
    d. 750°F (399°C)

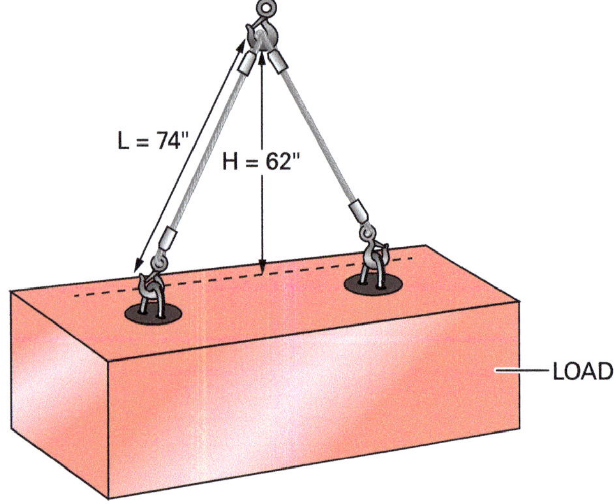

Figure SR01

4. Refer to *Figure SR01*. Based on the measurements shown, what is the sling angle factor that should be applied to the weight of the lifted load on each sling?

    a. 0.841
    b. 1.194
    c. 1.213
    d. 1.414

5. Refer to *Table 2* in Section 2.4.2. Based on the generally accepted safe range of sling angles, is the sling angle associated with a sling angle factor of 1.194 within the acceptable range?

    a. Yes
    b. No

6. To prevent 24" pipe material (600 DN) from rolling, the chock needs to be _____.

    a. 1" thick
    b. 2" thick
    c. 3½" thick
    d. 5" thick

7. One disadvantage of synthetic rope compared to natural fiber is that synthetic rope _____.

    a. is significantly heavier
    b. readily conducts electricity
    c. has a low strength-to-weight ratio
    d. is more easily damaged by heat

8. The procedures for determining the weight of a load to be lifted are covered in _____.

    a. *29 CFR 1910.140*
    b. *ASME Standard B30.9*
    c. *29 CFR 1926.1417*
    d. *ASME Standard BTH-1*

3.0.0 HOISTING EQUIPMENT AND JACKS

Objective

Identify and describe how to use various types of hoisting and jacking equipment.

a. Identify and describe how to use manual and powered hoisting equipment.
b. Identify and describe how to use jacks.

Performance Tasks

8. Select, inspect, and demonstrate the safe use of the following rigging equipment:
   - Block and tackle
   - Chain hoist
   - Ratchet-lever hoist
   - One or more types of jacks

Trade Terms

Gantry: A framed overhead structure supported by legs on each end, used to cross over obstructions. Gantries can be portable or permanent, providing support for hoisting equipment or raising and supporting lighting, cameras, and similar equipment.

Hauling line: The portion of a rope or chain on hoisting equipment that the operator uses to raise or lower the load. Also known as a hauling part.

Parts of line: The resulting number of lines that are supporting the load block when a line is reeved more than once.

Rigging isn't only about connecting loads to cranes. Riggers are also involved in equipment movement and placement inside of buildings and other locations where a crane is either unnecessary or impractical. This section presents various types of hoisting and jacking equipment.

3.1.0 Manual and Powered Hoisting Equipment

Both manual and powered hoisting equipment are used when it is necessary to lift components into position, or raise one component to insert something beneath it. Common hoisting equipment used includes block and tackle rigs, chain hoists, and ratchet-lever hoists.

3.1.1 Block and Tackle

The block and tackle is the most basic lifting device. It is used to lift or pull light loads. A block consists of one or more sheaves (pulleys) fitted into a wood or metal frame with a hook attached to the top. The tackle is the line or rope and end attachments connected to the block. Some block and tackle rigs have a brake that holds the load once it is lifted and others do not. The types that do not have a brake require continuous pull on the hauling line, or the hauling line must be tied off to hold the load.

There are two types of block and tackle rigs: simple and compound. A simple block and tackle consists of one sheave and a single line (*Figure 44*). It is used to lift or pull very light loads. The hook is attached to the load, and the load is lifted by pulling the line. The load capacity of this type of block and tackle is equal to the capacity of the load line. The block must be attached to a building structure or other support by a method that provides adequate load capacity to support the load and the tackle. Adequate capacity and stability of the supporting structure is crucial and it must be evaluated carefully by qualified personnel.

A compound block and tackle (*Figure 45*) uses more than one block. It has an upper, fixed block that is attached to the building structure or other support and a lower, traveling block that is attached to the load. Each block may have one or more sheaves. The more sheaves the blocks have, the more parts of line the block and tackle has, and the higher the lifting capacity. The compound block and tackle multiplies the power applied to the rope, so a worker can lift a much heavier load than is possible with a simple block and tackle.

3.1.2 Chain Hoists

A chain hoist, also called a *chain fall*, is a very useful and commonly used by a number of crafts. Chain hoists should be used for straight, vertical lifts only—just like a crane. They may be damaged if used for angled lifts or horizontal pulls. Although chain hoists are more popular, generally more durable, and capable of lifting heavier loads, there are electric and pneumatic cable hoists as well.

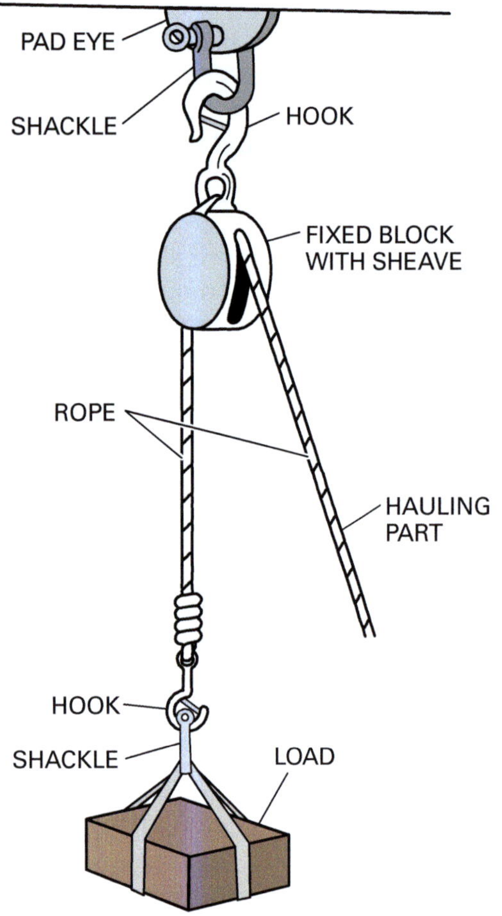

Figure 44 Simple block and tackle.

Figure 45 Compound block and tackle.

Chain hoists are standard equipment in many fabrication shops and rigging departments because they are dependable, portable, and easy to use. Some common types of chain hoists are the spur-geared (manual) chain and the electric (powered) chain.

The spur-geared chain hoist has two chains. An endless chain, the hand chain, drives a single pocketed sheave, which in turn drives a gear-reduction unit. The load chain is fitted to the gear-reduction unit and has a hook that attaches to the load. The gear-reduction unit provides the mechanical advantage, just as it does in many motor-driven conveyors and similar industrial systems. The spur-geared chain hoist is commonly used for significant loads and is convenient since it does not require a power source. Two common types are shown in *Figure 46*. The trolley-style chain hoist is designed to run on the rails of an I-beam, like the beam of a gantry.

A gantry (*Figure 47*) is often used indoors to hoist and move components. Gantries have maximum-load limitations just like chain hoists, but very large gantries with enormous capacities

Around the World

Block and Tackle History

Archimedes, a Greek mathematician, physicist, engineer, inventor, and astronomer, is credited with the development of the block and tackle around 250 BC. However, Archytas of Tarentum is thought to have developed the concept of a pulley—an essential component of the assembly to reduce friction. Block and tackle has been used in the construction of many structures that continue to fascinate us, including Stonehenge in the United Kingdom. Block and tackle systems are also considered to be the source of inspiration for modern cranes.

are used in shipping harbors and similar locations. These smaller versions are portable and very handy. A trolley-style chain hoist can be placed on the horizontal beam, or a separate trolley assembly can be purchased and a hook-type chain hoist can be attached to the trolley.

Electric chain hoists (*Figure 48*) work much like manual chain hoists, except that they have an electric motor instead of a pull chain to raise and lower the load. Electric chain hoists are faster and more efficient than manual chain hoists. Most are controlled by a pendant with pushbuttons, suspended from the hoist. There are a few battery-powered hoists on the market, but with limited capacity. As battery technology continues to advance, battery-operated hoists with more capacity may eventually be available. Most electric hoists operate on 120 and/or 240 VAC. Note that air-powered (pneumatic) chain hoists are also available.

Special care should be taken when raising a load with an electric chain hoist, since it is hard to tell how much force the electric motor is exerting. Motor sound may provide a clue. If the load gets caught on something immoveable, the chain hoist could be overloaded and damaged. A good electric chain hoist has reliable built-in overload protection that will open the motor circuit to prevent damage.

The load capacity of any chain hoist should be clearly marked on the data plate. The capacity should never be exceeded. Like a crane, it is necessary to know the weight of the load before beginning. Always use a sling to rig the load for a chain hoist if there is no single, dedicated lifting eye on the load. The chain of the hoist should never be wrapped around an object and used as a sling.

Like other lifting devices, chain hoists must be carefully inspected before each use. The chain must be checked to ensure that it has no significant defects. Before using a hoist, read the manufacturer's instructions and be aware of its limitations.

(A) HOOK-STYLE CHAIN HOIST

(B) TROLLEY-STYLE CHAIN HOIST

Figure 46 Manual chain hoists.

Figure 47 Portable gantry.

Figure 48 Electric chain hoist.

Follow these basic steps to select, inspect, use, and maintain a chain hoist:

Step 1 Select a chain hoist of adequate capacity to handle the load. Ensure that the overhead structure used to suspend the hoist also has adequate capacity.

Step 2 Inspect the load chain and hook to ensure that they are not excessively worn, bent, or deformed in any way.

Step 3 Inspect the sheaves to ensure that they are not bent or excessively worn.

Step 4 Ensure that the chain hoist has proper lubrication. Maintain per manufacturer's instructions.

Step 5 Hang the chain hoist from a suitable structure using the proper rigging. Provide an adequate power source for electric and pneumatic models, ensuring that it is routed to the hoist neatly and out of the way. The power cord or air hose should not be allowed to dangle straight down where it can interfere with the task.

Step 6 Position the hoist directly over the load. Lower the load hook, and connect it to the load using the proper rigging.

Step 7 Raise and place the load. To raise the load using a manual chain hoist, pull the hand chain. To raise the load using an electric chain hoist, press and hold the Up pushbutton on the handheld control.

Step 8 Disconnect the load hook from the load once it is in place.

Step 9 Remove the chain hoist from its support.

Step 10 Coil the chain so that it will not get tangled.

Step 11 Store the chain hoist in its proper place.

For detailed information about chain hoist inspection and operation, consult *ASME Standard B30.16, Overhead Hoists (Underhung)*.

3.1.3 *Ratchet-Lever Hoists and Come-Alongs*

Ratchet-lever hoists and come-alongs (*Figure 49*) are used for short pulls on heavy loads. You can see by the length of the chain in the figure that the ratchet-lever hoist cannot accommodate a long pull. The term come-along is widely used to identify both tools, but they are not the same thing. A come-along uses a cable, whereas a ratchet-lever hoist uses a chain. Linemen often use ratchet-lever hoists with a flat synthetic strap. Ratchet-lever hoists are often designed and rated for vertical lifts, but cable-type come-alongs can only be used to pull horizontally. Do not use a come-along or ratchet-lever hoist for lifting unless you are certain it is rated by the manufacturer to do so.

Never use a come-along for vertical lifts. They do not have the same safety-braking mechanisms as ratchet-lever hoists to prevent the load from slipping.

Come-alongs and ratchet-lever hoists are portable, easy to use, and available in varying capacities. Always read and follow the manufacturer's instructions. Follow these steps to select, inspect, use, and maintain a ratchet-lever hoist:

Step 1 Select a hoist of adequate capacity to handle the load. Ensure that it is rated for vertical lifting.

Step 2 Inspect the chain and hooks to ensure that they are not excessively worn, bent, or deformed in any way. Inspect the device overall for significant damage or signs of being overloaded.

Step 3 Hang the device from a suitable structure, and ensure that structure is adequate for the load.

Step 4 Turn the ratchet release to the mid position.

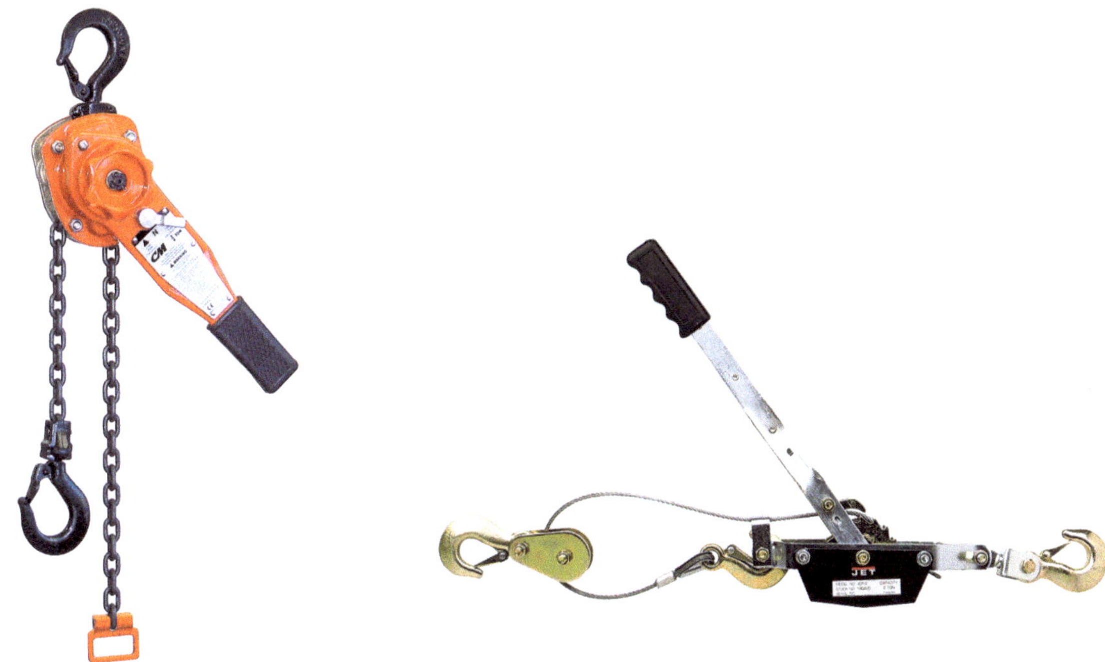

Figure 49 A ratchet-lever hoist and a come-along.

Step 5 Position the hoist directly over the load. Pull the chain out enough to attach it to the load.

Step 6 Attach the load hook to the load, using the proper rigging.

Step 7 Turn the fast-wind handle to take the slack out of the chain.

Step 8 Turn the ratchet control to the Up position.

Step 9 Pump the ratchet handle to raise the load.

Step 10 Turn the ratchet control to the Down position.

Step 11 Pump the ratchet handle to lower the load until there is slack in the chain.

Step 12 Disconnect the hook from the load.

Step 13 Dismount and store the hoist in its proper place.

3.2.0 Jacks

A jack is a device used to raise or lower equipment. Jacks are also used to move heavy loads a short distance, with good control over the movement. The following are the three basic types of jacks:

- Ratchet
- Screw
- Hydraulic

3.2.1 Ratchet Jacks

The ratchet jack (*Figure 50*), also called a *railroad jack*, is used to raise loads under 25 tons. It uses the lever-and-fulcrum principle. The downward stroke of the lever raises the rack bar one notch at a time. A latching mechanism, called a pawl, automatically springs into position, holding the load and releasing the lever for the next lifting stroke. They can lift full jack capacity on the toe or on the cap. Ratchet jacks are rated by lifting capacity and the length of their stroke, or lifting distance.

3.2.2 Screw Jacks

Screw jacks (*Figure 51*) are used to lift heavier loads than ratchet jacks, but more slowly. A simple screw jack uses the screw-and-nut principle to lift the stem. A simple lever is placed into the hole and turned to raise or lower the threaded stem. For heavier loads and jacks, a gear-reduction unit is used to reduce the amount of power required to turn the nut.

3.2.3 Hydraulic Jacks

Hydraulic jacks (*Figure 52*) are operated by the pressure of pumped fluid. These jacks can lift a surprising amount of weight. Simple bottle jacks have a hand-operated lever to pump the jack and raise it. A bypass valve is opened to allow fluid to flow out of the cylinder and lower the jack.

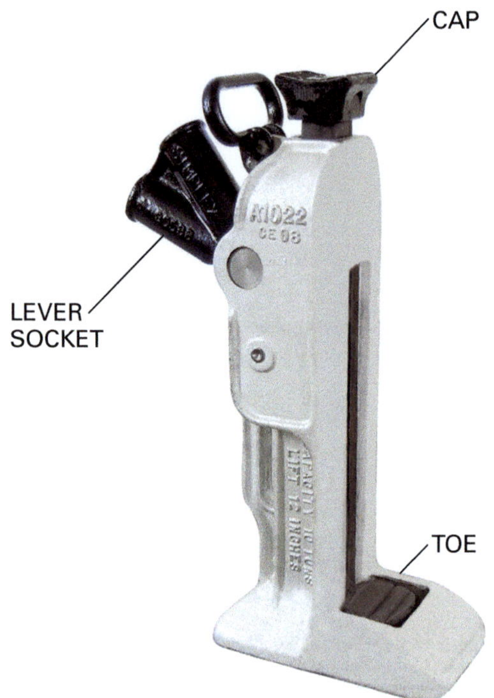

Figure 50 Ratchet jack.

Figure 51 Screw jack.

Toe jacks are a different version of bottle jacks that provide a lifting toe at the base. The toe can be pushed under a load that is much closer to the floor, while a bottle jack requires a lot of clearance to fit under the load.

Jacks with higher lifting capacities use an external pump. The pump is connected to the jack by a hose and a quick-disconnect coupling. The pump can be hand-operated, foot-operated, electric, air-operated, or even gasoline-engine driven. It is not uncommon to see portable hydraulic jacks that are rated for 1,000 tons.

3.2.4 Jack Use and Maintenance

In time, a jack can become damaged or worn and fail under a load. To avoid such failures, all jacks should be carefully inspected before each use. Apply the following general guidelines when using jacks:

- Inspect jacks before using them to ensure that they are not damaged in any way. For jacks that use an external pump, also inspect the pump, hoses, and couplings carefully. The connecting hose is the most fragile part of the system and is subject to cuts and deep scarring during normal use.
- Thoroughly clean all hydraulic hose connectors before connecting them.
- Never exceed the load capacity of the jack.
- Use wood softeners when jacking against metal.
- Never place jacks directly on earth when lifting; provide a solid footing.
- Position jacks so the direction of force is perpendicular to the base and the surface of the load. Then raise the load evenly to prevent the load from shifting or falling.
- Use the proper jack handle, and remove it from the jack when it is not in use. Do not use extensions to the jack handles. If the added power of an extension is necessary, the wrong jack is being used.
- Never step on a jack handle to create additional force; use a foot-operated pump and an appropriate jack.
- Always apply blocking or cribbing under a raised load when jacking. Never leave a jack under a load without having the load blocked up so that it will not fall if the jack fails suddenly.
- Brace loads to prevent the jacks from tipping.
- Lash or block jacks when using them in a horizontal position to move an object.
- Never jack against any kind of roller or wheel.
- Match multiple jacks for uniform lifting of a single object.

Figure 52 Hydraulic jacks.

Additional Resources

29 *CFR* 1926.251, **www.ecfr.gov**

Willy's Signal Person and Master Rigger Handbook. Current edition. Ted L. Blanton, Sr. Altamonte Springs, FL: NorAm Productions, Inc.

North American Crane Bureau, Inc. website offers resources for products and training, **www.cranesafe.com**

3.0.0   Section Review

1. The mechanical advantage gained by using a spur-geared chain hoist is provided by _____.

   a. multiple parts of chain
   b. a trolley
   c. an operating lever
   d. the gear-reduction unit

2. Which of the following is a true statement about the use of jacks?

   a. If a load is difficult to lift, you can use your foot to operate the jack handle.
   b. Always apply blocking or cribbing under a raised load.
   c. The capacity of a hydraulic jack is only limited by the hydraulic pressure applied.
   d. A toe jack is just a different version of a screw jack.

SUMMARY

Selecting and setting up hoisting equipment, hooking cables to the load to be lifted or moved, and helping guide the load into position are all part of the rigging process. Performing this process safely and efficiently requires the rigger to properly select equipment, use it in the correct way, understand safety hazards, and know how to prevent accidents while working efficiently.

Riggers must have great respect for the hardware and equipment of their trade because their lives may depend on them functioning correctly.

Although they are not required to be rigging professionals, crane operators must also understand rigging practices and be able to distinguish between proper and improper methods for the safety of all jobsite personnel.

1. Threaded blind holes used with eyebolts should have a minimum depth of _____.

    a. 1¼ times the bolt diameter
    b. 1½ times the bolt diameter
    c. 1¾ times the bolt diameter
    d. 2 times the bolt diameter

2. A device used as a connection point for a load that has no reduction in rated load when an angular pull is applied is a _____.

    a. shouldered eyebolt
    b. swivel hoist ring
    c. rigging hook
    d. shoulderless eyebolt

3. Custom-fabricated spreader and equalizer beams are tested at _____.

    a. 125 percent of their rated load
    b. 200 percent of their rated load
    c. 250 percent of their rated load
    d. 350 percent of their rated load

4. To keep them in place while the sling is positioned, manufactured protective materials for slings may be equipped with _____.

    a. adhesive strips
    b. magnets
    c. hook-and-loop fasteners
    d. wooden edges

5. A wire rope sling should be removed from service if localized abrasion and scraping has reduced the diameter more than _____.

    a. 25 percent of the original rope diameter
    b. 15 percent of the original rope diameter
    c. 10 percent of the original rope diameter
    d. 5 percent of the original rope diameter

6. When the eyes of a synthetic web sling are sewn at right angles to the plane of the sling body, it is called a(n) _____.

    a. round sling
    b. endless sling
    c. standard eye-and-eye sling
    d. twisted eye-and-eye sling

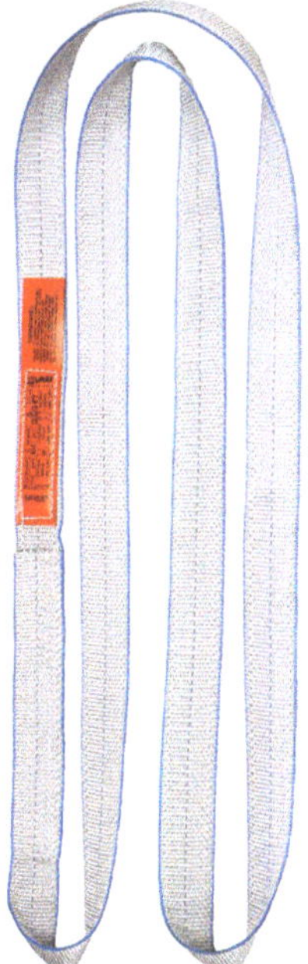

Figure RQ01

7. The type of sling shown in *Figure RQ01* is a(n) _____.

    a. standard eye-and-eye
    b. twisted eye-and-eye
    c. endless
    d. round

8. Which of the following is a disadvantage of chain slings when compared to wire-rope slings?

    a. Chain slings do not handle abrasion and corrosion as well as wire rope.
    b. Chain slings cannot be used in high-heat applications.
    c. Chain slings have less reserve strength and are more likely to fail without warning.
    d. Chain slings are more easily damaged by sharp corners than wire rope.

9. Which of the following slings is best suited for situations where the load is abrasive, hot, and/or tends to cut?

    a. Wire-rope sling
    b. Metal-mesh sling
    c. Synthetic web sling
    d. Endless grommet sling

10. Which of the following types of damage to a metal-mesh sling is allowed to be more extensive than the other before it is removed from service?

    a. Abrasion
    b. Corrosion

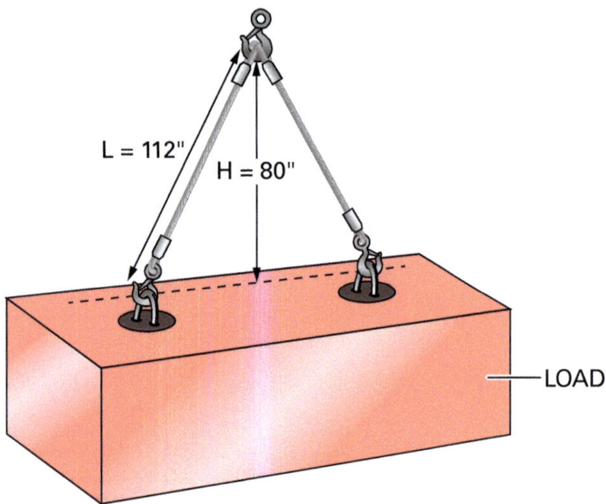

Figure RQ02

11. Refer to *Figure RQ02*. Based on the measurements shown, what is the sling angle factor that should be applied to the weight of the lifted load on each sling?

    a. 0.902
    b. 1.003
    c. 1.213
    d. 1.400

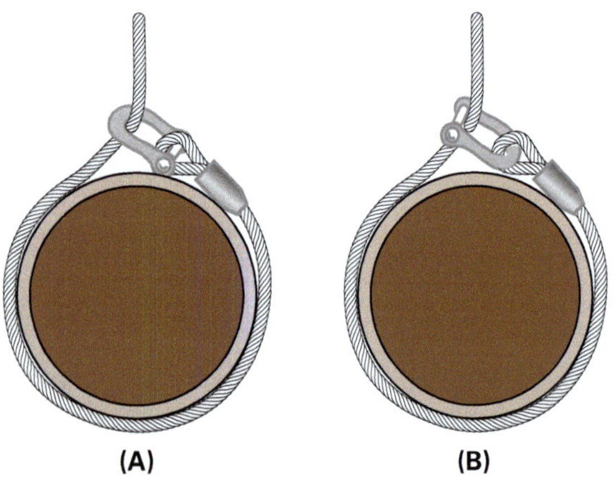

Figure RQ03

12. Refer to *Figure RQ03*. Which of the two drawings, A or B, represent the proper way to orient the shackle?

    a. Drawing A
    b. Drawing B

13. Regardless of how a clove hitch is actually tied, the result is the same as ______.

    a. two bowlines tied one after another
    b. the combination of a bowline and a half hitch
    c. two half hitches made in the same direction
    d. two half hitches made in opposite directions

14. When the weight of a load cannot be determined in other ways, it can be lifted slightly while monitoring the weight using instruments in the crane. The lift can proceed if the load is less than ______.

    a. 40 percent of the crane's capacity at the longest required operating radius
    b. 75 percent of the crane's capacity at the longest required operating radius
    c. 85 percent of the crane's capacity at the longest required operating radius
    d. 100 percent of the crane's capacity at the longest required operating radius

15. Which of the following is a correct statement?

    a. All come-alongs are rated and best used for vertical lifts.
    b. A ratchet-lever hoist can accommodate any length of chain.
    c. A come-along uses a cable while a ratchet-lever hoist uses a chain.
    d. Come-alongs and ratchet-lever hoists use the same type of braking mechanisms.

NCCER – *Basic Rigger*

Trade Terms Quiz

Fill in the blank with the correct term that you learned from your study of this module.

1. Damage that occurs to wire rope identified by the separation of strands that then balloon outward like vertical bars is called __________.

2. The type of wire rope construction considered to be the most durable for rigging applications is the __________ type.

3. A(n) __________ is attached to a lifted load for the purpose of controlling load spinning and swinging.

4. Rigging devices used to distribute the weight of a load on multi-crane (tandem) lifts are called __________.

5. If a hole does not fully penetrate the material, resulting in a hole with a bottom, it is referred to as a(n) __________.

6. The portion of a hook directly below the center of the lifting eye is called the __________.

7. A common hitch made by passing a sling through a load or connection and attaching both sling eyes to the hoist line is called a(n) __________.

8. The __________ identifies the amount of stress required to bring a rigging component to its breaking point.

9. The terms *safe working load*, *working load limit*, and *rated capacity* are all synonyms for __________.

10. A simple hitch that uses one end of a sling to connect to a point on the load and the opposite end to connect to the hoist line is the __________.

11. The angle formed by the legs of a sling with respect to the horizontal plane when tension is placed on the rigging is referred to as the __________.

12. A(n) __________ is made by passing a sling around the load and then passing one eye of the sling through the other.

13. A(n) __________ is comprised of 2 or more single-leg hitches that is used for lifting objects equipped with lifting lugs or similar points of connection.

14. Rigging devices that are often used when the object being lifted is too long or large to be lifted from a single point, or when the use of slings around the load may crush the sides, are __________.

15. __________ have three or more holes in them and are used to level loads when sling lengths are not equal.

16. Plates with two holes, referred to as __________, are used as termination hardware to connect rigging to specific lift points.

17. A hook should always be positioned over a load's __________.

18. A major load of pipe or similar materials might be delivered to a site on a __________.

19. A chain hoist or ratchet lever hoist used for lifting can be suspended under a __________.

20. The __________ have a lot to do with the lifting assistance provided by a block and tackle.

21. To raise a load using a block and tackle, the user must pull on the __________.

Trade Terms

Basket hitch
Bird caging
Blind hole
Bridle hitch
Center of gravity (CG)
Choker hitch

Equalizer beams
Equalizer plates
Gantry
Hauling line
Independent wire rope core (IWRC)

Minimum breaking strength (MBS)
Parts of line
Rated load
Rigging links
Saddle

Sling angle
Spreader beams
Spur track
Tagline
Vertical hitch

Harold "Ed" Burke

Rigging Training Specialist,
Mammoet USA South

Please give a brief synopsis of your construction career and your current position.

I began working as a rigger with Mammoet in 1997. I had never been around cranes before, so I was green, for sure. I was fortunate, though, to learn and grow under the guidance of some of the best riggers in the industry—people like Dean and Mark Bell, Jeff Bland, and Jason Marks, just to name a few of those I respect and owe a debt of gratitude. I've been a part of some great projects across the United States and abroad, such as The Netherlands, Belgium, Italy, Taiwan, and Chile. In October 2013, I was approached by our safety director and asked if I would consider serving as a trainer for Mammoet USA. I accepted and I'm proud to still be a part of it.

How did you get started in the construction industry?

I had been doing general industry work since high school, with some time working in residential construction. I was fortunate to get an opportunity with Mammoet through a connection there. When I was hired, the same person told me, "I got you the job—you'll have to keep it."

Who or what inspired you to enter the construction industry?

Honestly, it was exactly what I needed at that point in my life. I was looking to better myself. When I started with Mammoet, I made more money for the remaining eight months of the year than in any full year before that. The earning potential was a key factor, but I really needed some positive changes in my life, and the construction industry surely provided that.

How has training in construction impacted your life and career? What types of training have you completed?

When I got into crane and rigging in the late 1990s, there was not so much emphasis on formal training. Most training happened through mentoring on the job. The industry has changed dramatically in the 20 years since I became involved. I really believe quality training contributes to a safer and more efficient work force. Knowledge truly is power; I say that all the time.

Why do you think credentials are important in the construction industry?

I believe valid credentials provide benchmarks, showing where an individual's skill is in relation to an accepted standard of competency. Credentials provide a glimpse of their knowledge foundation on a given topic or craft.

What do you enjoy most about your career?

The people, without a doubt. I have had the pleasure of working with some of the best people across the world from my days in the field. I continue to work with some of the best in my training environment today. I wouldn't be an effective educator without the people who have been a part of my career and growth as a craft professional. If you surround yourself with great professionals, you have a much better chance of becoming one yourself.

Would you recommend construction as a career to others? Why?

Absolutely! Why, you ask? Not everyone is suited for the college experience or for a career in an office. The earning potential of young men and women graduating from high school who are willing to work hard, learn, and be an asset to their chosen craft is really fantastic. They can start to build real wealth immediately instead of taking on crippling college-loan debt.

What advice would you give to someone who is new to the construction industry?

I see new hires all the time, and those that are also new to the industry. What I tell them is this: Show up early, stay late, do more than is expected, be the first to volunteer, learn all you can by keeping your ears open and your mouth closed, and do a good enough job to convince your supervisor he can't do the next job without you. You must build a reputation for yourself, and it must be a good one.

How do you define craftsmanship?

Many would say that it is skill in a particular craft, and I agree with that. But there is a noticeable level of pride and passion evident in the work and attitude of a true craftsman. You really have to love what you do to be happy and productive, and you pour yourself into it to be a craftsman. That's my take on it.

Basket hitch: A common hitch made by passing a sling around a load or through a connection and attaching both sling eyes to the hoist line.

Bird caging: A deformation of wire rope that causes the strands or lays to separate and balloon outward like the vertical bars of a bird cage.

Blind hole: A hole that does not penetrate the material completely, leaving a hole with a bottom.

Bridle hitch: A type of hitch comprised of 2 or more single-leg hitches, used for lifting objects equipped with lifting lugs or other points of connection.

Center of gravity (CG): The point at which the entire weight of an object is considered to be concentrated, such that supporting the object at this specific point would result in its remaining balanced in position.

Choker hitch: A hitch made by passing a sling around the load, and then passing one eye of the sling through the other. The one eye is then connected to the hoist line, creating a choke-hold on the load.

Equalizer beams: Beams used to distribute the load weight on multi-crane lifts. The beam attaches to the load below, with two or more cranes attached to lifting eyes on the top.

Equalizer plates: A type of rigging plate that has three or more holes, used to level loads when sling lengths are unequal.

Gantry: A framed overhead structure supported by legs on each end, used to cross over obstructions. Gantries can be portable or permanent, providing support for hoisting equipment or raising and supporting lighting, cameras, and similar equipment.

Independent wire rope core (IWRC): Wire rope with a core consisting of wire rope, as opposed to a fiber or single-stranded core; considered to be the most durable for rigging applications.

Hauling line: The portion of a rope or chain on hoisting equipment that the operator uses to raise or lower the load. Also known as a *hauling part*.

Minimum breaking strength (MBS): The amount of stress required to bring a rigging component to its breaking point. The MBS is a factor in determining a components' rated load capacity.

Parts of line: The resulting number of lines that are supporting the load block when a line is reeved more than once.

Rated load: The maximum working load permitted by a component manufacturer under a specific set of conditions. Alternate names for rated load include *working load limit* (WLL), *rated capacity*, and *safe working load* (SWL).

Rigging links: Links or plates with two holes used as termination hardware to appropriate lifting points.

Saddle: The portion of a hook directly below the center of the lifting eye.

Sling angle: The angle formed by the legs of a sling with respect to the horizontal plane when tension is placed on the rigging.

Spreader beams: Beams or bars used to distribute the load of a lift across more than one point to increase stability. Spreader beams are often used when the object being lifted is too long or large to be lifted from a single point, or when the use of slings around the load may crush the sides.

Spur track: A relatively short branch leading from a primary railroad track to a destination for loading or unloading. A spur is typically connected to the main at its origin only (a dead end).

Tagline: A rope attached to a lifted load for the purpose of controlling load spinning and swinging, or used to stabilize and control suspended attachments.

Vertical hitch: A simple hitch that uses one end of a sling to connect to a point on the load and the opposite end to connect to the hoist line. Also known as a *straight-line hitch*.

Additional Resources

This module presents thorough resources for task training. The following reference material is recommended for further study.

ASME Standard B30.5, Mobile and Locomotive Cranes. Current edition. New York, NY: American Society of Mechanical Engineers.

ASME Standard B30.9, Slings. Current edition. New York, NY: American Society of Mechanical Engineers.

ASME Standard B30.10, Hooks. Current edition. New York, NY: American Society of Mechanical Engineers.

ASME Standard B30.16, Overhead Hoists (Underhung). Current edition. New York, NY: American Society of Mechanical Engineers.

ASME Standard B30.20, Below-The-Hook Lifting Devices. Current edition. New York, NY: American Society of Mechanical Engineers.

ASME Standard BTH-1, Design of Below-The-Hook Lifting Devices. Current edition, New York, NY: American Society of Mechanical Engineers.

29 *CFR* 1926, Subpart CC, **www.ecfr.gov**

29 *CFR* 1926.251, **www.ecfr.gov**

29 *CFR* 1926.753, **www.ecfr.gov**

Mobile Crane Operations Level One, NCCER. Third Edition. 2018. New York, NY: Pearson Education, Inc.

NCCER Module 00106-15, *Introduction to Basic Rigging*.

Mobile Crane Safety Manual (AEM MC-1407). 2014. Milwaukee, WI: Association of Equipment Manufacturers.

Willy's Signal Person and Master Rigger Handbook, Ted L. Blanton, Sr. Current edition. Altamonte Springs, FL: NorAm Productions, Inc.

Knots: The Complete Visual Guide, Des Pawson. First American Edition. 2012. New York, NY: DK Publishing.

The following websites offer resources for products and training:

Occupational Safety and Health Administration (OSHA), **www.osha.gov**

Electronic Code of Federal Regulations, **www.ecfr.gov**

North American Crane Bureau, Inc. website offers resources for products and training, **www.cranesafe.com**

Link-Belt Construction Equipment Company, Module opener
Columbus McKinnon Corporation, Figures 1, 4A, 4C, 46, 48, 49A
Konecranes Americas, Inc., SA01
Courtesy of The Crosby Group LLC, Figures 4B, 27
J. C. Renfroe & Sons, Figure 9
Vaculift™, Inc. d.b.a. Vacuworx®, SA02
Lift-All Company, Inc., Figures 14, SA03, 19, 20, 23, Review Question Figure 1, Exam Figure 1
Lift-It Manufacturing Co., Inc., Figure 17A
Linton Rigging Gear Supplies LLC - www.lrgsupplies.com, Figure 17B
Mazzella Companies, Figure 18
Ed Gloninger, Figure 21
© Donvictorio/Dreamstime.com, Figure 35
© iStock.com/krungchingpixs, Figure 36A
© iStock.com/Cajarima01, Figure 36B
Insulatus Company, Inc., Figure 37
© a katz/Shutterstock.com, SA05
© Steve Norman/Shutterstock.com, Figure 45
Vestil Manufacturing, Figure 47
Walter Meier Manufacturing Americas, Figures 49B, 51, 52A, 52B, Exam Figure 3
Photos courtesy of Enerpac, Figures 50, 52C, 52D

Answer	Section Reference	Objective
Section One		
1. a	1.1.2	1a
2. a	1.2.4	1b
Section Two		
1. c	2.1.1	2a
2. a	2.2.1	2b
3. c	2.3.2	2c
4. b	2.4.2	2d
5. a	2.4.2	2d
6. b	2.5.0	2e
7. d	2.6.0	2f
8. c	2.7.0	2g
Section Three		
1. d	3.1.2	3a
2. b	3.2.4	3b

Section Review Calculations

2.0.0 SECTION REVIEW

Question 4

Divide the length (L) by the height (H) to determine the sling angle factor:

Sling angle factor = L ÷ H
Sling angle factor = 74" ÷ 62"
Sling angle factor = 1.194

The sling angle factor is **1.194**.

This page is intentionally left blank.

NCCER CURRICULA — USER UPDATE

NCCER makes every effort to keep its textbooks up-to-date and free of technical errors. We appreciate your help in this process. If you find an error, a typographical mistake, or an inaccuracy in NCCER's curricula, please fill out this form (or a photocopy), or complete the online form at **www.nccer.org/olf**. Be sure to include the exact module ID number, page number, a detailed description, and your recommended correction. Your input will be brought to the attention of the Authoring Team. Thank you for your assistance.

Instructors – If you have an idea for improving this textbook, or have found that additional materials were necessary to teach this module effectively, please let us know so that we may present your suggestions to the Authoring Team.

NCCER Product Development and Revision

13614 Progress Blvd., Alachua, FL 32615

Email: curriculum@nccer.org
Online: www.nccer.org/olf

❏ Trainee Guide   ❏ Lesson Plans   ❏ Exam   ❏ PowerPoints   Other _______________________

Craft / Level: Copyright Date:

Module ID Number / Title:

Section Number(s):

Description:

Recommended Correction:

Your Name:

Address:

Email: Phone:

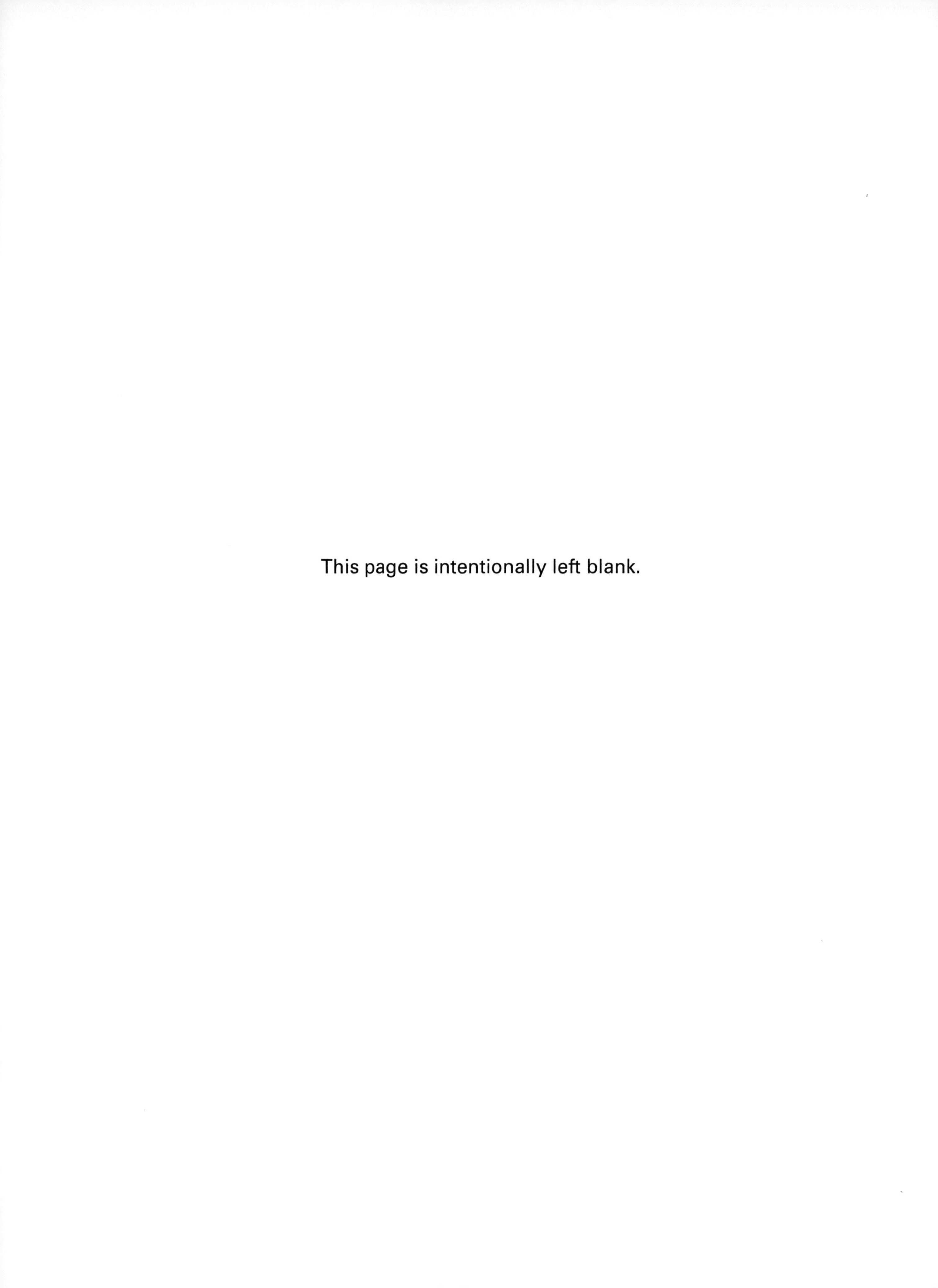

Glossary

Abate: To reduce or minimize.

Accident: As defined by OSHA, an unplanned event that results in personal injury or property damage.

Aerial lifts: Mobile work platforms designed to transport and raise personnel, tools, and materials to overhead work areas.

As-built drawings: Drawings of a completed installation that reflect the structure as it was constructed and show any unexpected changes during construction.

Asphalt shingles: Shingles manufactured by coating a reinforcing material (paper felt or fiberglass mat) with an asphalt-based coating and having mineral granules on the side exposed to the weather.

Base flashing: Plies or strips of roof membrane material used to close off and/or seal a roof at the horizontal-to-vertical intersections, such as at a roof-to-wall juncture. Base flashing covers the edge of the field membrane and extends up the vertical surface.

Basket hitch: A common hitch made by passing a sling around a load or through a connection and attaching both sling eyes to the hoist line.

Battens: Horizontal strips of solid material, typically wood, that provide a fixing point for connecting a roof system.

Bird caging: A deformation of wire rope that causes the strands or lays to separate and balloon outward like the vertical bars of a bird cage.

Bitumen: A dark, sticky, cement-like substance found in asphalts, tars, pitches, and asphaltites. May also refer to any material composed mainly of bitumen, such as asphalt or coal tar.

Blind hole: A hole that does not penetrate the material completely, leaving a hole with a bottom.

Body harness: Straps that may be secured about the worker in a manner that will distribute the fall-arrest forces over at least the thighs, pelvis, waist, chest, and shoulders, with means for attaching it to other components of a personal fall-arrest system.

Bridle hitch: A type of hitch comprised of 2 or more single-leg hitches, used for lifting objects equipped with lifting lugs or other points of connection.

Building envelope: The systems of a building that separate the interior of the building from the outside, including the roof, walls, windows, and doors.

Building information modeling (BIM): An intelligent process that uses computer databases and computer-aided drafting (CAD) modeling to create a dynamic, interactive model of an entire structure and track its lifecycle.

Built-up roof (BUR) systems: Roof systems in which multiple layers of bitumen, reinforcement, and surfacing material are applied in place on the deck.

Capacity: The total amount of weight capable of being lifted. This includes personnel, tools and materials, and/or equipment.

Carpenter: A craftworker who constructs, assembles, installs, and repairs structures made of wood or other materials.

Center of gravity (CG): The point at which the entire weight of an object is considered to be concentrated, such that supporting the object at this specific point would result in its remaining balanced in position.

Choker hitch: A hitch made by passing a sling around the load, and then passing one eye of the sling through the other. The one eye is then connected to the hoist line, creating a choke-hold on the load.

Clay tiles: A roof covering made by molding clay into a tile shape and baking it.

Cleats: Continuous metal strips, or angled pieces, used to secure metal components.

Closed-cell: The structure of cellular insulation in which the tiny cellular structures are packed closely together but are not connected to each other. The cells of the material themselves may be solid or hollow.

Closed-cut valleys: Valleys with shingles that extend across the valley from one side while the shingles on the other side are trimmed back from the centerline.

Combustible: Capable of burning.

Competent person: As defined by OSHA, an individual who is capable of identifying existing and predictable hazards in the surroundings or working conditions which are unsanitary, hazardous, or dangerous to employees, and who has the authorization to take prompt corrective measures to eliminate such hazards.

Competent person: As defined by OSHA, one who is capable of identifying existing and predictable hazards in the surroundings or working conditions which are unsanitary, hazardous, or dangerous to employees, and who has authorization to take prompt corrective measures to eliminate them.

Computer-aided drafting (CAD): The making of a set of construction drawings with the aid of a computer.

Concrete tiles: A roof covering made from shaping a mixture of cement, sand, and water into tiles.

Condensation: The conversion of water vapor or other gas to liquid phase as the temperature drops or atmospheric pressure rises.

Controlled access zone (CAZ): A designated work area in which certain types of masonry work may take place without the use of conventional fall protection systems.

Controlled decking zone (CDZ): An area in which certain work (for example, initial installation and placement of metal decking) may take place without the use of guardrail systems, personal fall arrest systems, fall restraint systems, or safety net systems and where access to the zone is controlled.

Copings: Coverings on top of a wall exposed to the weather, usually made of metal, brick, or stone. Copings are usually sloped to shed water.

Corrosion: The breakdown or destruction of a material, especially metal, caused by chemical reactions. The most common form of corrosion is rust, which occurs when iron combines with oxygen to create iron oxide.

Counter-battens: Grids of vertical wood or metal strips installed under horizontal battens that help with ventilation and system installation.

Counterflashing: Secondary flashing installed along the top edge of primary flashing to protect the surface and its fasteners; A roof system component, usually composed of metal, used to cover or shield the upper edges of the membrane base flashing or wall flashing.

Crickets: Components used to divert water away from a chimney, wall, expansion joint, or other interruption in the field of a roof.

Cut: A common term for a scaffold level.

Deck: A structural component of the roof of a building, capable of safely supporting the weight of the roof or waterproofing system, as well as the additional live loads required by the governing building codes. The deck provides the substrate to which the roof or waterproofing system is applied.

Detail drawings: Drawings shown at a larger scale in order to show specific features or connections.

Dew point: The temperature at which air becomes saturated with water vapor and has a relative humidity of 100 percent.

Diameter: An invisible, straight line segment starting on one end of a circle, passing through the center point, and reaching the other end of the circle.

Downspouts: Pipes that carry water from the gutter to the ground or a drain.

Drawing set: The set of detailed drawings or plans drawn to scale by an architect and/or engineer, showing all information and dimensions necessary to build or remodel a structure.

Eaves: The horizontal edges at a roof's perimeter.

Elbows: Help direct the flow of a downspout.

Equalizer beams: Beams used to distribute the load weight on multi-crane lifts. The beam attaches to the load below, with two or more cranes attached to lifting eyes on the top.

Equalizer plates: A type of rigging plate that has three or more holes, used to level loads when sling lengths are unequal.

Evaporation: The natural change of liquid to vapor at a temperature below its boiling point.

Excavations: Man-made cuts, cavities, trenches, or depressions in the earth's surface, formed by removing earth. Excavations can be made for the purpose of building anything from basements to highways.

Felt: A flexible sheet made by the interlocking of fibers with a binder or through a combination of mechanical work, moisture, and heat.

Field: The main, uninterrupted surface area of a roof.

Flash off: When an adhesive cures or dries just enough for two pieces to stick together. It is tacky to the touch but will not leave any residue on your glove.

Flashings: Components used to weatherproof or seal edges of a roof system at perimeters, penetrations, walls, expansion joints, valleys, drains, and other places where the roof covering is interrupted or terminated.

Free fall: The act of falling before a personal fall-arrest system begins to apply force to arrest the fall.

Gable roof: A roof style named after its prominent gables, or the triangular shapes formed between edges on the opposing sides of an intersecting slope.

Galvanic corrosion: A form of corrosion (rusting) resulting from the creation of an electric current flowing between dissimilar metals in contact with each other.

Gambrel roof: Also called a *Dutch roof*; this roof style is usually symmetrical with two slopes on each side—its lower slope steeper than its upper slope.

Gantry: A framed overhead structure supported by legs on each end, used to cross over obstructions. Gantries can be portable or permanent, providing support for hoisting equipment or raising and supporting lighting, cameras, and similar equipment.

Guarded: Enclosed, fenced, covered, or otherwise protected by barriers, rails, covers, or platforms to prevent dangerous contact.

Guardrails: Railings installed around the perimeter of a worksite to prevent falls from a roof's edge.

Gutters: Channeled components installed along the perimeter of a roof to carry water to drains or downspouts; are usually attached with straps and support brackets.

Hand line: A line attached to a tool or object so a worker can pull the tool up after climbing a ladder or scaffold.

Hauling line: The portion of a rope or chain on hoisting equipment that the operator uses to raise or lower the load. Also known as a *hauling part*.

Hazard Communication Standard (HAZCOM): The standard that requires contractors to educate employees about hazardous chemicals on the job site and how to work with them safely.

Hip roof: A roof style formed by slopes on all four sides of a building coming together at the top of the roof to form a ridge.

Hydrokinetic: Shedding liquid or snow off of a roof.

Incident: As defined by OSHA, an unplanned event that does not result in personal injury but may result in property damage or is worthy of recording.

Independent wire rope core (IWRC): Wire rope with a core consisting of wire rope, as opposed to a fiber or single-stranded core; considered to be the most durable for rigging applications.

Infrared: A type of invisible light energy that is emitted by all objects. This type of energy is most effective at producing radiant heating when absorbed.

Joinery: The process of joining two or more pieces of sheet metal.

Lanyard: A short section of rope or strap, one end of which is attached to a worker's safety harness and the other to a strong anchor point above the work area.

Legend: In maps, plans, and diagrams, an explanatory table defining all symbolic information contained in the document.

Lifeline: A component consisting of a flexible line that is either connected vertically to an anchorage at one end (vertical lifeline) or connected horizontally to an anchorage at both ends (horizontal lifeline). A lifeline serves as a means for connecting other components of a personal fall arrest system to the anchorage.

Limited access zones: Restricted areas alongside a masonry wall that is under construction.

Low-slope: A category of roofs that generally includes weatherproof membrane types of roof systems installed on slopes of 3:12 or less.

Management system: The organization of a company's management, including reporting procedures, supervisory responsibility, and administration.

Mansard roof: Also called a *French roof*; a decorative steep-slope roof on the perimeter of a building.

Maximum intended load: The total weight of all people, equipment, tools, materials, and loads that a ladder can hold at one time.

Membrane roof systems: Roof systems containing a membrane material (weatherproof covering). Membrane roof systems are usually installed as low-slope systems.

Membrane: A flexible or semiflexible roof covering or waterproofing whose primary function is to exclude water.

Midrail: A mid-level, horizontal board required on all open sides of scaffolds and platforms that are more than 14 in. (35 cm) from the face of the structure and more than 10 ft. (3.05 m) above the ground. It is placed halfway between the toeboard and the top rail.

Mil: A unit of measurement equal to one-thousandth of an inch, or 0.001" (0.0254 mm).

Minimum breaking strength (MBS): The amount of stress required to bring a rigging component to its breaking point. The MBS is a factor in determining a components' rated load capacity.

Nominal size: Approximate or rough size by which materials are known and sold. Nominal size is usually slightly larger than the actual size.

Occupational Safety and Health Administration (OSHA): An agency of the US Department of Labor whose mission is to set occupational safety and health standards for all places of employment, enforce these standards, ensure that employers provide and maintain a safe workplace for all employees, and provide research and educational programs to support safe working practices.

Oil canning: Physical distortions in the flatness of metal. This condition only effects the appearance of the metal and does not have negative effects on structural integrity.

On-the-job learning (OJL): Job-related learning an apprentice acquires while working under the supervision of journey-level workers. Also called *on-the-job training (OJT)*.

Open valleys: Valleys that trim the roof-covering material on both sides to expose metal valley flashing that helps shed the water down a roof.

Oxidize: To combine with oxygen, such as in burning (rapid oxidation) or rusting (slow oxidation).

Parts of line: The resulting number of lines that are supporting the load block when a line is reeved more than once.

Patina: A green or brown oxide film that develops on the surface of copper and copper alloys.

Pavers: Precast slabs, usually made of concrete, used to weigh down (or ballast) a roof system and provide a functional finish and walking surface.

Penetration: Any construction, such as pipes, conduits, or HVAC supports, passing through a roof or waterproofing system.

Personal fall arrest system (PFAS): A system used to stop an employee in a fall from a working level. It consists of an anchorage, connectors, and a body harness, and may include a lanyard, deceleration device, lifeline, or suitable combinations of these.

Personal protective equipment (PPE): Equipment or clothing designed to prevent or reduce injuries.

Planked: Having pieces of material 2 in. (5 cm) thick or greater and 6 in. (15 cm) wide or greater used as flooring, decking, or scaffold decks.

Ply: A layer of felt or thin sheet of wood in a built-up membrane roof or waterproofing system.

Positive drainage: A condition in which a roof is designed and shaped to ensure proper drainage of the roof area within 48 hours after rainfall.

Project manual: Documents prepared by the architect/engineer for the owner in order to execute a project. Documents in a project manual include bidding documents, contracts, project details, and a list of construction drawings. Also called a *specification book* or *spec book*.

Purlins: Horizontal secondary structural members that transfer loads to the primary structural framing.

Putlogs: Horizontal scaffold members on which the scaffold platform rests.

Qualified person: As defined by OSHA, one who, by possession of a recognized degree, certificate, or professional standing, or who by extensive knowledge, training, and experience, has successfully demonstrated his ability to solve or resolve problems relating to the subject matter, the work, or the project.

R-value: A measurement of how well a material resists the transfer of heat through itself. Also called the *thermal resistance.*

Rated load: The maximum working load permitted by a component manufacturer under a specific set of conditions. Alternate names for rated load include *working load limit (WLL), rated capacity,* and *safe working load (SWL).*

Record drawings: Drawings that show a record of any unexpected changes made during the construction of a structure.

Reroofing: The process of re-covering or tearing off and replacing an existing roof system.

Rigging links: Links or plates with two holes used as termination hardware to appropriate lifting points.

Riser diagram: A schematic drawing that depicts the layout, components, and connections of a piping system.

Saddle: The portion of a hook directly below the center of the lifting eye; small, sloped structures that help to channel surface water to drains, frequently located in a valley. Saddles are often constructed like a small hip roof or pyramid with a diamond-shaped base. Also called *diamond crickets.*

Safety data sheet (SDS): A form that lists the hazards, safe handling practices, and emergency control measures for a specific substance. Also called a *material safety data sheet (MSDS).*

Safety monitoring: A safety system that assigns monitoring duties to a competent person who has the knowledge and authority to make safety decisions.

Safety nets: Nets installed around and underneath a working area that can catch falling objects or workers.

Scaffolding: A temporary built-up framework or suspended platform or work area designed to support workers, materials, and equipment at elevated or otherwise inaccessible job sites.

Schedules: Tables that describe and specify the various types and sizes of construction materials used in a building. Door schedules, window schedules, and finish schedules are the most common types. Other types include equipment schedules, beam schedules, column schedules, and footing schedules.

Scupper: A drainage outlet through a wall, parapet wall, or raised roof edge, typically lined with a sheet-metal sleeve.

Self-retracting lanyard (SRL): A deceleration device containing a drum-wound line that can be slowly extracted from, or retracted onto, the drum under slight tension during normal employee movement, and which, after the onset of a fall, automatically locks the drum and arrests the fall.

Shed roof: A roof style that slopes only once in one direction; frequently featured on small buildings or integrated with other roof styles.

Single-ply roof systems: Roof systems in which the primary roof covering is a single layer of flexible membrane material (i.e., EPDM or thermoplastic).

Site-specific safety program: A safety program developed for a jobsite that identifies and takes into account any specific potential hazards that may be encountered.

Six-foot rule: A rule stating that platforms or work surfaces with unprotected sides or edges that are 6 ft. (≈2 m) or higher than the ground or level below it require fall protection.

Slate: A naturally occurring roofing material made from rocks.

Sling angle: The angle formed by the legs of a sling with respect to the horizontal plane when tension is placed on the rigging.

Slope: The angle of a roof surface, usually expressed as a ratio of vertical rise to horizontal length (sometimes referred to as *run*). When dimensions are given in inches, slope may be expressed as a ratio of rise over a distance of 12" (for example, 4:12) or as an angle in degrees.

Spreader beams: Beams or bars used to distribute the load of a lift across more than one point to increase stability. Spreader beams are often used when the object being lifted is too long or large to be lifted from a single point, or when the use of slings around the load may crush the sides.

Spur track: A relatively short branch leading from a primary railroad track to a destination for loading or unloading. A spur is typically connected to the main at its origin only (a dead end).

Steep-slope: A category of roofing that generally includes water-shedding types of roof coverings installed on slopes greater than 3:12.

Step flashing: Weatherproofing installed from under roof shingles or tiles and up to sidewall.

Substrate: The surface upon which a roofing or waterproofing membrane is applied (such as the structural deck or rigid board insulation).

Suspension trauma: A serious medical condition that can occur when a worker is suspended for a short time in a body harness.

Tagline: A rope attached to a lifted load for the purpose of controlling load spinning and swinging, or used to stabilize and control suspended attachments.

Takeoff: The process of surveying, measuring, itemizing, and accounting for all materials and equipment needed for a construction project.

Tensile strength: The resistance of a material to a force tending to tear it apart.

Thermal expansion: The increase in the dimension or volume of a body because of temperature variations.

Top rail: A top-level, horizontal board required on all open sides of scaffolds and platforms that are more than 14 in. (36 cm) from the face of the structure and more than 10 ft. (3 m) above the ground.

Underlayment: An asphalt-saturated felt or other composite or synthetic sheet material (sometimes self-adhering) installed between a roof deck and roof covering, usually used in a steep-slope roof construction. Underlayment is primarily used to separate the roof covering from the roof deck, shed water, and provide fire protection and secondary weather protection.

Vertical hitch: A simple hitch that uses one end of a sling to connect to a point on the load and the opposite end to connect to the hoist line. Also known as a *straight-line hitch*.

Water vapor: Water in a vapor (gas) form, especially when below the boiling point and diffused in the atmosphere.

Water-shedding: Able to depend on gravity for quick drainage to prevent water from entering into the roof system. Steep-slope roof systems are generally water-shedding.

Wood shakes: A roof covering split from lumber logs; they typically have rough and textured surfaces.

Wood shingles: A roof covering cut from wood; they typically have flat and smooth surfaces.

This page is intentionally left blank.

This page is intentionally left blank.